Medicinal Chemistry:
Today and Tomorrow

International Union of Pure and Applied Chemistry
in association with The Pharmaceutical Society of Japan
and the Asian Federation for Medicinal Chemistry

International Union of Pure and Applied Chemistry

Medicinal Chemistry: Today and Tomorrow

Proceedings of the
AFMC International Medicinal Chemistry Symposium
held in Tokyo, Japan, 3–8 September 1995

EDITED BY

MIKIO YAMAZAKI

Faculty of Pharmaceutical Sciences
Chiba University
Chiba, Japan

**Blackwell
Science**

© 1997 International Union of
Pure and Applied Chemistry
and published for them by
Blackwell Science Ltd
Editorial Offices:
Osney Mead, Oxford OX2 0EL
25 John Street, London WC1N 2BL
23 Ainslie Place, Edinburgh EH3 6AJ
238 Main Street, Cambridge
    Massachusetts 02142, USA
54 University Street, Carlton
    Victoria 3053, Australia

Other Editorial Offices:
Arnette Blackwell SA
    224, Boulevard Saint Germain
    75007 Paris, France

Blackwell Wissenschafts-Verlag GmbH
    Kurfürstendamm 57
    10707 Berlin, Germany

    Zehetnergasse 6
    A-1140 Wien
    Austria

First published 1997

Printed and bound in Great Britain
at the University Press, Cambridge

The Blackwell Science logo is a
trade mark of Blackwell Science Ltd,
registered at the United Kingdom
Trade Marks Registry

DISTRIBUTORS

Marston Book Services Ltd
PO Box 269
Abingdon
Oxon OX14 4YN
(*Orders*:   Tel: 01235 465500
             Fax: 01235 465555)

USA
Blackwell Science, Inc.
238 Main Street
Cambridge, MA 02142
(*Orders*:   Tel: 800 215-1000
                   617 876-7000
             Fax: 617 492-5263)

Canada
Copp Clark Professional
200 Adelaide Street West, 3rd Floor
Toronto, Ontario M5H 1W7
(*Orders*:   Tel: 416 597-1616
                   800 815-9417
             Fax: 416 597-1617)

Australia
Blackwell Science Pty Ltd
54 University Street
Carlton, Victoria 3053
(*Orders*:   Tel: 3 9347 0300
             Fax: 3 9347 5001)

A catalogue record for this title
is available from the British Library

ISBN 0-632-04272-9

Library of Congress
Cataloging-in-Publication Data

AFMC International Medicinal Chemistry Symposium
(1995: Tokyo, Japan)
    Medicinal chemistry: today and tomorrow: proceed-
ings of the AFMC International Medicinal Chemistry
Symposium held in Tokyo, Japan, 3–8 September
1995/edited by Mikio Yamazaki
                p.   cm.
    At head of title: International Union of Pure and
Applied Chemistry.
    Includes bibliographical references.
    ISBN 0-632-04272-9
    1 Pharmaceutical chemistry–Congresses.
    I. Yamazaki, Mikio, 1931–   .
    II. Asian Federation for Medicinal Chemistry.
    III. International Union of Pure and Applied Chem-
istry.
    IV. Title.
    [DNLM: 1. Chemistry, Pharmaceutical–congresses.
    2. Chemistry, Clinical–congresses. QV 744 A257
    1997]
    RS401. A35 1995
    615'. 19–dc21
    DNLM/DLC
    for Library of Congress   96-38942
                                                    CIP

Contents

Preface

The AFMC International Medicinal Chemistry Symposium (AIMECS 95) was held at Keio Plaza Hotel, Tokyo from September 3rd to 8th, 1995 with great success. As chairman, I am honoured to write the preface to the proceedings of the lectures which made the symposium a significant occasion.

The Pharmaceutical Society of Japan organized this symposium on behalf of the Asian Federation for Medicinal Chemistry (AFMC). The European Federation for Medicinal Chemistry (EFMC) and the American Chemical Society, Division of Medicinal Chemistry (USMC) kindly joined us and organized the two sessions: the session on Enzyme Inhibitors and that on Molecular Diversity. I would like to express my gratitude to both organizations. Also I would like to give my hearty thanks to the Advisory Board, the lecturers and chairmen, and all the members of the Organizing Committee and the Scientific Committee.

The symposium consisted of eight plenary lectures, 40 invited lectures in 13 sessions, and 185 discussions stimulated by posters exhibited on the occasion. Although research on medicinal chemistry is broad and various, the plenary lecturers are all leading scientists of today with outstanding achievements in their own fields. I believe they made a great impact on the floor. Invited lecturers explained current issues on important subjects of medicinal chemistry. I hope their lectures gave good bearings to the audience for their future work.

With the recognition that the mission of chemistry in society at present is to contribute to the promotion of health and the protection of the environment, the International Union of Pure and Applied Chemistry (IUPAC) has been carrying out its work. Its subdivision, the Medicinal Chemistry Section, is requested to support and pursue the work: to promote health and welfare.

Indeed, medicinal chemistry is an interdisciplinary science covering a huge range, with chemistry as its core and with the purpose of producing excellent new pharmaceuticals. Accompanied by the recent rapid progress in related sciences, medicinal chemistry has made remarkable achievements. I think, therefore, that today is a good time to try to get an enlarged, synthetic view and to produce innovative drugs making the best use of recent developments in pharmaceutical research and skills. The most important thing for us to do is to establish international cooperative systems with close relationships among scientists in industry and in academic and governmental institutions.

As to the future system of the medicinal chemistry federation, broad international cooperation should be formed, for the federation consists of societies and institutions in various countries. For the future of medicinal chemistry, research and development to make original, new pharmaceuticals is considered essential. In addition, the enrichment of basic pharmaceutical education, the training of leaders and proper guidance of industrial chemistry to mass-produce pharmaceuticals will be an important problem for developing countries.

I would like to thank those who collaborated with me in the symposium. All its success is due to all the participants. I would like to express my gratitude again to the Pharmaceutical Society of Japan who held the symposium, especially to the members of the Organizing Committee who spared their busy time and helped me, and also to the Ministry of Health and

Welfare of Japan and Japan Health Sciences Foundation, and to the Research Foundation for Pharmaceutical Sciences. I am grateful to the pharmaceutical and other related companies and various organizations both home and abroad who kindly supported the symposium. And finally I would like to express my hearty thanks to the IUPAC for its sponsorship.

NAOFUMI KOGA

Antitumor taxoids

Françoise Guéritte-Voegelein, Daniel Guénard and Pierre Potier

Centre National de la Recherche Scientifique, Institut de Chimie des Substances Naturelles, F-91198 Gif-sur-Yvette, France

Abstract: The isolation of 10-deacetylbaccatin III $\underline{2}$ from a renewable natural source: the leaves of the European yew, *Taxus baccata* L., solved the major supply problem of the antitumor drug paclitaxel (Taxol™) $\underline{1}$. The availablity of 10-deacetylbaccatin III $\underline{2}$ also facilitated the synthesis of a number of taxoids and led to the discovery of docetaxel (Taxotere™) $\underline{3}$ which showed significant anticancer properties. Details concerning the chemistry and structure-activity relationships of taxoids will be reported.

Chemotherapy represents one of the principle strategies for the treatment of cancer which is used either alone, or in combination with other therapies such as surgery or radiotherapy. Unfortunately, many cancers are, or become, incurable. One of the main reasons behind this later phenomen is the resistance of certain cancer cells to the drugs used. This problem is of major importance, and renders imperative the discovery of new antitumor agents. Our laboratory has been engaged for almost a quarter of a century in the search for anticancer drugs, and in this regard particular emphasis has been placed on *spindle poisons*. Spindle poisons are considered to be the only compounds which are truly antimitotic as they interact directly with tubulin the major constituent of the mitotic spindle. This specificity distinguishes them from alkylating agents or antimetabolites for example. Although *spindle poisons* are toxic, they are not mutagenic and their secondary effects are, in general, reversible. Tubulin is a heterodimeric protein which polymerizes into microtubules. These filamentous structures performs a variety of functions in living cells. Besides their role in mitosis, they form part of the cytoskeleton and are involved in intracellular transport, motility and in determining the cell shape. *In vitro*, the interaction of drugs with tubulin can be easily evaluated using a highly sensitive and specific *in vitro* assay, the " tubulin assay".

This assay was first used to the search for new antitumor alkaloids of the vinblastine group. This led to the discovery of vinorelbine (Navelbine®) (1) which is currently used, worldwide in the treatment of non-small lung cancer and breast cancer. The discovery, more than twenty years ago in the US, of the antitubulin properties (2) of paclitaxel (Taxol™) $\underline{1}$ (Fig. 1), a diterpene isolated in very low yield from the trunk bark of the American yew tree, *Taxus brevifolia* Nutt. (3), convinced us to investigate the European yew tree, *Taxus baccata* L., in order to find a suitable precursor to this molecule, as well as new antitumor compounds belonging to the taxoid family. These studies led first to the isolation in fairly good yields of 10-deacetylbaccatin III $\underline{2}$ (Fig. 1) (4, 5) which proved to be a convenient precursor for the synthesis of paclitaxel $\underline{1}$ (5, 6, 7). Moreover, the availability of 10-deacetylbaccatin III $\underline{2}$ facilitated the synthesis of new taxoids. Among these synthetic taxoids, docetaxel (Taxotere®) $\underline{3}$ (Fig. 1) (6-7) showed impressive biological activity (8). This paper will focus mainly on the chemistry and structure-activity relationships of taxoids, with particular emphasis on results obtained from our own laboratory.

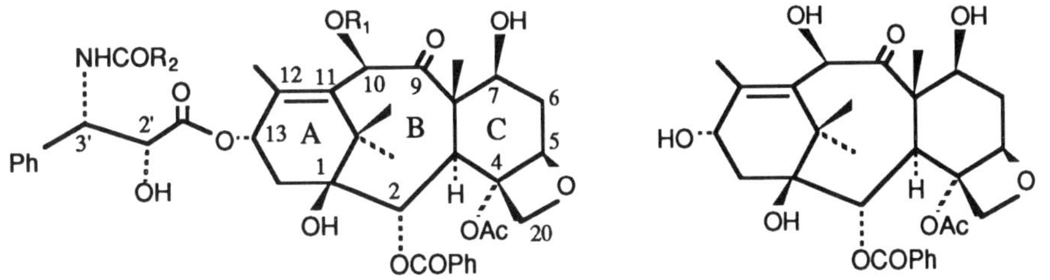

$\underline{1}$: R_1= Ac, R_2= Ph $\underline{3}$: R_1= H, R_2= OtBu $\underline{2}$

Fig. 1 Structures of paclitaxel $\underline{1}$, 10-deacetylbaccatin III $\underline{2}$ and docetaxel $\underline{3}$

1- 10-deacetylbaccatin III: Isolation and Chemistry

1.1 Bioassay-guided isolation of taxoids from *Taxus baccata*

T. baccata L. is the most common species of the genus *Taxus* found in Europe. This slow-growing tree has been known since antiquity for its deleterious effects. The first significant chemical studies on *Taxus* were reported in 1856 by Lucas (9) who isolated a mixture of alkaloids called taxine. These compounds were considered as being responsible for the high toxicity of the leaves of the tree (10). It has since been shown that the poisonous properties of taxine are due to their calcium antagonizing properties (11). A century after the isolation of taxine, the discovery of the antitumor activity of the yew tree led to the isolation of paclitaxel <u>1</u> (3) by Wall and its coworkers and to the discovery of its unique mechanism of action on microtubules by Horwitz et al. (2). Paclitaxel <u>1</u> favours microtubules assembly and inhibits their disassembly into tubulin. Our initial study started in 1979 with the antitubulin activity-guided isolation of the constituents of the European yew tree, *Taxus baccata* L (4, 12). The biological activity of the extracts, various fractions and pure compounds was evaluated on the disassembly process of microtubules into tubulin. The most active substances were isolated from the trunk bark and identified as paclitaxel-like compounds bearing a side chain at carbon 13.

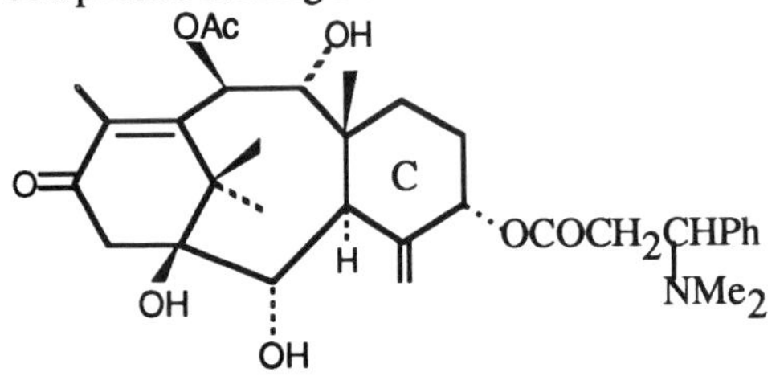

Fig. 2 Structure of Taxine B <u>4</u>

The least active compound was found to be 10-deacetylbaccatin III <u>2</u> which was isolated from the leaves with a yield of 0.02% (12) to 0.1% (5). The taxine type compounds such as taxine B <u>4</u> (Fig. 2) which are the main components of the leaves of *T.baccata* (13), were not isolated during these studies since they do not interact with microtubules. Thus, 10-deacetylbaccatin III <u>2</u> appeared as an easily and constantly accessible paclitaxel precursor. Taxine B <u>4</u> can also be considered as an interesting precursor and has been used for the preparation of derivatives bearing the oxetan ring on the C ring (14).

1.2 Chemical studies of 10-deacetylbaccatin III

The three-dimensional structure of 10-deacetylbaccatin III <u>2</u> (Fig. 3) shows the locked cup-like conformation of the taxane skeleton. This unique feature together with the presence of the various oxygenated functions on the skeleton lead to particular chemical reactivity. Figure 4 shows some chemical modifications which have been performed on 10-deacetylbaccatin III <u>2</u>.

- Reactivity of the hydroxyl groups and acylation at C-13.

10-deacetylbaccatin III <u>2</u> possesses three secondary hydroxyl groups at carbons 7, 10 and 13 and one tertiary hydroxyl group at carbon 1. Because of its hindered position, the C-13 hydroxyl group is the least reactive group of the secondary alcohols (15).

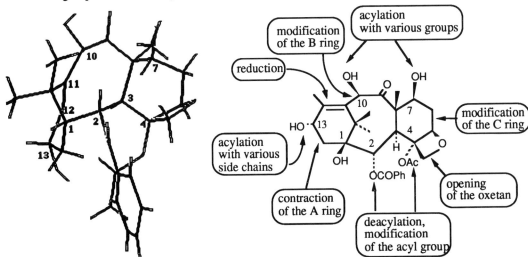

Fig. 3 3D Structure of 10-deacetylbaccatin III   Fig. 4 Chemical Modifications of 10-deacetylbaccatin III

After protection of the hydroxyl groups at C-7 and C-10, chemical conversion of 10-deacetylbaccatin III <u>2</u> into paclitaxel <u>1</u> was achieved *via* two routes. The first one involves the esterification of the C-13 hydroxyl group with cinnamoyl chloride or cinnamic acid and subsequent hydroxyamination at

2' and 3' with osmium tetroxide and *t*-butyl-N-chloro-N-sodio-carbamate (6). Direct deprotection at C-7 and C-10 led to docetaxel **3** (6). On the other hand, paclitaxel **1** was obtained after replacement of the *t*-butyloxycarbonyl by a benzoyl group at C-3' and acetylation at C-10. Improved methods which have been reviewed recently (16-19), were then developed for a large scale production of these drugs. Direct esterification at C-13 with the corresponding acid side chains afforded paclitaxel **1** and docetaxel **2** in excellent yields.

-10-deacetylbaccatin III under basic or reductive conditions.

It is well known that 10-deacetylbaccatin III **2** epimerizes at C-7 after treatment with weakly basic reagents (19). This easy epimerization occurs *via* a retro-aldol process involving the free hydroxyl group at C-7 and the ketone at C-9. In order to avoid this reaction, the hydrolysis of the ester groups at C-2 and C-4 was studied on the 7-protected derivative of 10-deacetylbaccatin III. Alkaline hydrolysis, lithium aluminium hydride reduction, and methanolysis led first to deacylation at C-2 and C-4 without selectivity (20-22). However recent studies (23-26) have provided different procedures for the regioselective deacylation at C-2 and C-4 (Scheme 1). These procedures take advantage of the close proximity of the C-1 hydroxyl to the C-2 benzoyl group and of the C-13 hydroxyl to the C-4 acetyl group (see Fig. 3). Thus, super hydride reduction (23) or deprotonation with potassium *tert*-butoxide (26) of the C-7 protected derivative of 10-deacetylbaccatin III or baccatin III led only to 4-deacetyl derivatives such as **5**. On the other hand, the 2-debenzoyl baccatin III derivative **6** was obtained when the 13-protected derivative of baccatin III was treated by potassium *tert*-butoxide (26) or by Red-Al (24). These results shows that the free hydroxyl groups at C-1 or C-13 play a major role in the intramolecular deacylation of the molecule. Moreover, selective C-4 deacetylation can also be obtained from baccatin III protected at C-1, C-7 and C-13 after Red-Al reduction (25). In that case, it has been suggested that coordination of Red-Al occured with the oxetan oxygen and led to selective deacetylation when the C-1 hydroxyl group is protected (25).

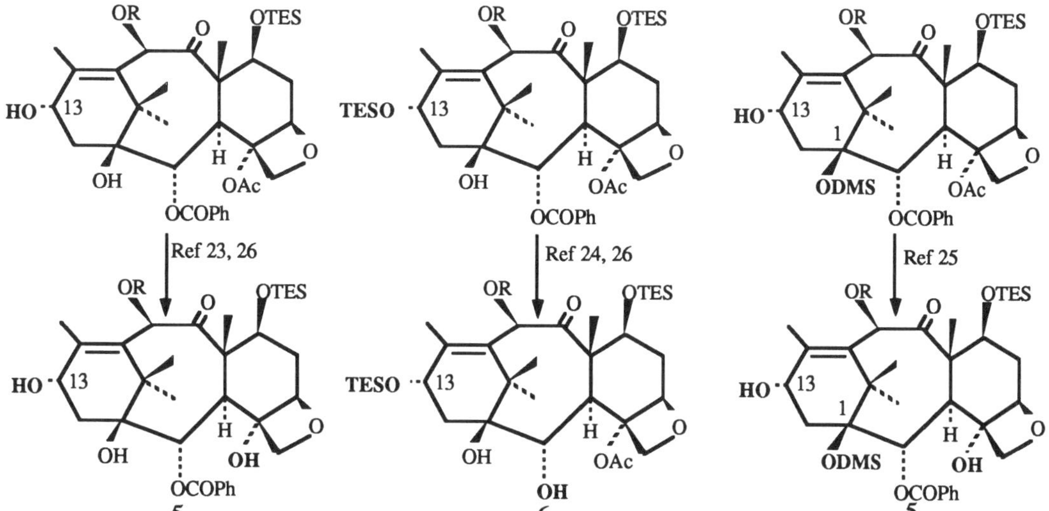

Scheme 1 Selective hydrolysis of 10-deacetylbaccatin III (R=H) or baccatin III (R=Ac) derivatives

-Modification of the A-, B-, C- and D (oxetan) rings.

Under acidic or electrophilic conditions, chemoselective rearrangements occured on the A ring and/or on the oxetan of 10-deacetylbaccatin III (20, 23, 27-28).

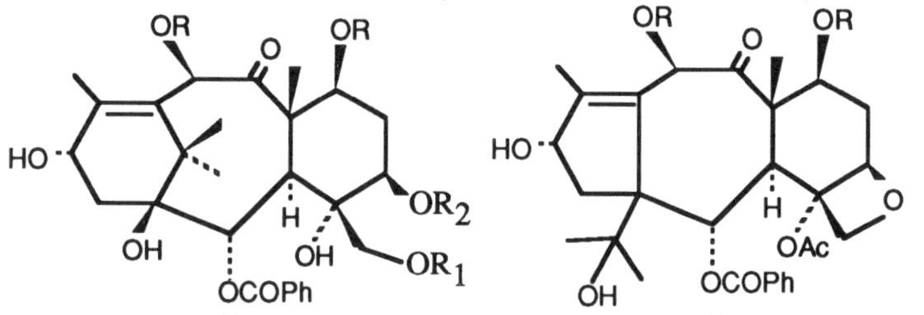

Fig. 5 Structures of analogues modified on the A and D rings.

Depending upon the reactions conditions, 10-deacetylbaccatin III was converted to oxetane ring opened derivatives **7** and / or to A-ring contracted derivatives **8** (Fig. 5). Opened oxetane analogues can also be obtained from the oxidation of the C-7 hydroxyl group (23, 29). For example, oxidation of 13-oxo-baccatin III with PCC led to the unsaturated ketone **9** (Fig. 5). The structure of this product was confirmed by X-ray analysis (unpublished result). Another modification of the A-ring concerns the reduction of the C_{11}-C_{12} double bond. 10-Deacetylbaccatin III **2** as well as paclitaxel **1** are resistant to hydrogenation. The double bond could be reduced only after zinc-promoted reduction of 13-oxo baccatin III derivatives. This gave 11,12-dihydro analogues such as **10** (30). Upon certain reaction conditions the reduced products are unstable, and a skeleton rearrangement occurs leading to the cleavage of the C-10, C-11 double bond (Fig 6, compound **11**) (30).

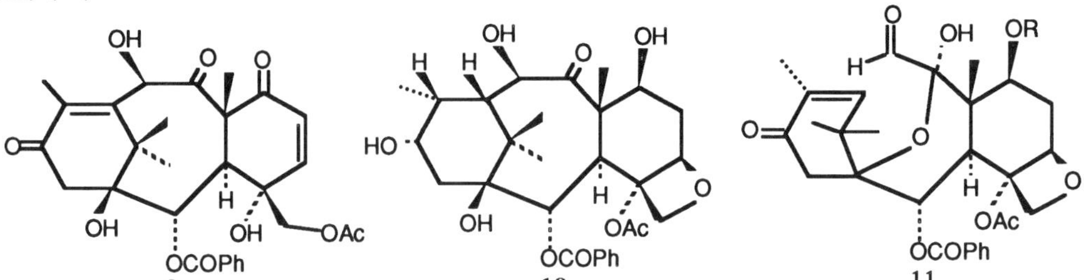

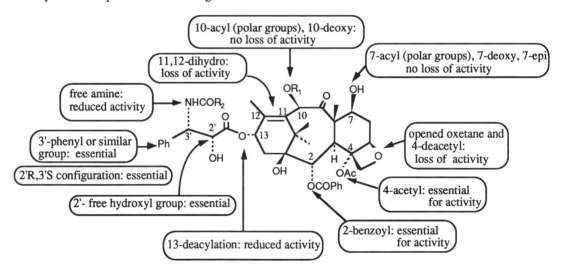

Fig 6 10-deacetylbaccatin III analogues with modifications on the A, B and C-rings.

A number of other derivatives have been prepared from 10deacetylbaccatin II **2**, by us and others. These new taxoids allowed the study of structure-activity correlations.

2- Structure-activity relationships and conformational studies of taxoids.

In this chapter, we will focus only on some significant SAR results concerning the antitubulin activity. Using the two synthetic approaches mentioned above (oxyamination or direct coupling with various side chains), a number of taxoids with a modified side chain and a modified taxane skeleton have been prepared (For reviews see Ref. 17-19). The important features concerning the structure-antitubulin activity relationships are outlined in Fig.7.

Fig. 7 Structure-antitubulin activity relationships in the taxoid series

Concerning the modifications on the taxane skeleton, it was shown that modifications on the North part of the molecule are generally compatible with antitubulin activity. Epimerisation or reduction of the hydroxyl group at C-7 led to active compounds on tubulin. Replacement of the two hydroxyl groups at C-7 and C-10 by polar groups provided active analogs, whereas the presence of apolar substituents at these two positions reduces the activity (unpublished results from our laboratory). Taxoids possessing a suitable side chain at C-13 and a contracted A-ring such as in **8** possess a conformation similar to that of paclitaxel and therefore interact with microtubules. The 11,12-dihydro derivative is no longer active. This inactivity has been related to the conformational change of the A-ring which, after reduction at C-11 and-12, possess a chair conformation instead of a boat conformation (30). The acyl groups on the Southern sector are also essential for activity. Indeed, deacylation at C-2 and C-4 led to inactive compounds. Noticeable is the strong antitubulin activity of compounds bearing a *meta*-substituted benzoyl group at C-2 (31) or a 4-

cyclopropyl ester group (32) instead of the acetyl function. Opening of the oxetan in electrophilic medium led also to deacylation at C-4 (see Fig 5, compound 7). Taxol derivatives bearing these two structural modifications do not interact with microtubules.

The presence of the side chain at C-13 is essential for a strong interaction with tubulin, although 10-deacetylbaccatin III does interact with this protein. Removal of the 2' and / or 3' substituents results in compounds which exhibit significantly less activity than paclitaxel. The 2'R, 3'S configuration is also essential for a strong binding with microtubules. A free hydroxyl group at C-2' is also needed for interaction with microtubules but the acyl derivatives at this position are easily hydrolyzed *in vivo* and can be used as prodrugs. On the other hand, active taxoids can be obtained by replacing the phenyl group or the N-acyl moiety with similar substituents without significant loss of activity.

From these SAR studies it was found that the substituents on the South part of the molecule at C-2 and C-4, and the substituents of the side chain play an important role in the interaction with tubulin. Moreover, conformational studies (33-37) show that these substituents are in close proximity. Two structural types have been proposed from these studies to correspond to active taxoids. One conformation possesses the 3'-amido group close to the 2-benzoyl group. This structure also was obtained from X-ray analysis (38) as well as NMR experiments in apolar solvents and molecular modeling. The other structure is found when paclitaxel or docetaxel are in a polar environment. In that case, the 3'-phenyl is close to the 2'-benzoyl groups. Moreover, depending on the environment, different intermolecular interactions in docetaxel (33) and paclitaxel (39) were found to occur between the oxygenated functions of the North part of the molecule and the side chain or the ester group at C-2. Further studies are thus needed to determine unambiguously the active conformation of antitumor taxoids.

3. Conclusion

Paclitaxel 1 has been approved in the US for its use in the treatment of ovarian cancer (40). In clinical studies, docetaxel 3 appears to be a strong active drug in the treatment of breast cancer and gave also very good results in ovarian and small-cell lung cancers (41). Today, these two products appear to be the first of a new series of potent anticancer drugs. Moreover, taxoids are very efficient tools to understand some aspects of cell division as they interact with microtubules which are involved in the major process of cell division. For these reasons, we continue to search for new taxoids which could have improved biological activities.

Acknowledgments

It is a great pleasure to acknowledge our colleagues who have made important contributions to this work. Their names are cited in the references. We also express our thanks to Rhône-Poulenc Rorer S.A. for the financial support and to Drs. M-C Bissery, A. Commerçon, J-L. Fabre, F.Lavelle, and their coworkers for their contributions and helpful discussions. We are also grateful to Dr. D. Grierson for critical reading of this manuscript.

References

1. P.Potier. *Sem Oncol.* **16** (2) (Suppl 4), 2 (1989).
2. Schiff, P.B.; Fant, J. and Horwitz, S.B. *Nature* **277,** 665 (1979).
3. Wani, M.C.; Taylor, H.L.; Wall, M.E.; Coggon, P. and McPhail, A.T. *J.Am.Chem.Soc.* **93,** 2325 (1971).
4. Chauvière, G.; Guénard, D.; Picot, F.; Sénilh, V. and Potier, P. *C.R.Acad.Sci.Paris.* **293,** 501 (1981).
5. Denis, J-N.; Greene, A.; Guénard, D.; Guéritte-Voegelein, F.; Mangatal, L. and Potier, P. *J.Am.Chem.Soc.***110,** 5917 (1988).
6. Mangatal, L; Adeline, M-T.; Guénard, D.; Guéritte-Voegelein, F. and Potier, P. *Tetrahedron,* **45,** 4177 (1989).
7. Guéritte-Voegelein, F.; Guénard, D.; Lavelle, F.; Le Goff, M-T.; Mangatal, L. and Potier, P. *J.Med.Chem.***34,** 992 (1991).
8. Lavelle, F. *Curr Opin Invest Drugs* **2,** 627 (1993).
9. Lucas, H. *Arch. Pharm.* **95,** 145 (1856).
10. Bryan-Brown, T. *J.Pharm.Pharmacol.* **5,** 205 (1932).
11. Tekol, Y. *Planta Medica* 357 (1985).
12. Sénilh, V.; Blechert, S.; Colin, M.; Guénard, D.; Picot, F.; Potier, P. and Varenne, P. *J.Nat.Prod.,* **47,** 131 (1984).
13. Ettouati, L.; Ahond, A.; Poupat, C. and Potier, C. *J.Nat.Prod.* **54,** 1455 (1991).
14. Ettouati, L.; Ahond, A.; Poupat, C. and Potier, C. *Tetrahedron,* **47,** 9823 (1991).
15. Guéritte-Voegelein, F.; Sénilh, V.; David, B.; Guénard, D. and Potier, P. *Tetrahedron* **42,** 4451 (1986)
16. Guénard, D.; Guéritte-Voegelein, F. and Potier, P. *Acc.Chem.Res.* **26,** 160 (1993).
17. Georg, G.I.; Ali, S.M.; Zygmut, J. and Jayasinghe, L.R. *Exp.Opin.Ther.Pat.* **4,** 109 (1994).

18. Nicolaou, K.C.; Dai, W-M. and Guy, R.K. *Angew.Chem.Int.Ed.Engl.* **33**, 15 (1994).
19. Kingston, D.G.I.; Molinero, A.A. and Rimoldi, J.M. *Progress in the Chemistry of Organic Natural Products*, **61**, p. 1, W. Herz; G.W. Kirby; R.E. Moore; W. Steglish and C. Tamm, Ed.; Springer-Verlag, New York (1993).
20. Wahl, A.; Guéritte-Voegelein, F.; Guénard, D.; Le Gogg, M-T. and Potier, P. *Tetrahedron* **48**, 6965 (1992).
21. Chen, S-H.; Wei, J-M. and Farina, V. *Tetrahedron Lett.* **34**, 3205 (1993).
22. Samaranayake, G.; Neidigh, K.A. and Kingston, D.G.I. *J.Nat.Prod.* **56**, 884 (1993).
23. Guéritte-Voegelein, F.; Guénard, D.; Dubois, J.; Wahl, A.; Marder, R.; Muller, R.; Lund, M.; Bricard, L. and Potier, P. *Taxane Anticancer Agents, Basic Science and Current Status*, p.189, G.I. Georg; T.T. Chen; I. Ojima; D.M. Vyas, Ed.; ACS Symposium Series 583, Washington, DC (1995)
24. Chen, S.H.; Farina, V.; Wei, J-M.; Long, B.; Fairchild, C.R.; Mamber, S-W.; Kadow, J.F.; Vyas, D. and Doyle, T.W. *Bioor.Med.Chem.Lett.* **4**, 479 (1994).
25. Chen, S-H.; Kadow, J.F.; Farina, V.; Fairchild, C.R. and Johnston, K.A. *J.Org.Chem.* **59**, 6156 (1994).
26. Datta, A.; Jayasinghe, L.R. and Georg, G.I. *J.Org.Chem.* **59**, 4689 (1994).
27. Samaranayake, G.; Magri, N.F.; Jitrangsri, C. and Kingston, D.G.I. *J.Org.Chem.* **56**, 5114 (1991).
28. Chen, S.H.; Huang, S.; Wei, J.M. and Farina, V. *Tetrahedron* **49**, 2805 (1993).
29. Magri, N.F. and Kingston, D.G.I. *J.Org.Chem* **51**, 797 (1986).
30. Marder, R.; Dubois, J.; Guénard, D.; Guéritte-Voegelein, F. and Potier, P. *Tetrahedron* **51**, 1985 (1995).
31. Chaudhary, A.G.; Gharpure, M.M.; Rimoldi, J.M.; Chordia, M.D.; Gunatilaka, A.A.L. and Kingston, D.G.I. *J.Am.Chem.Soc.* **116**, 4097 (1994).
32. Chen, S-H.; Fairchild, C. and Long, B.H. *J.Med.Chem.* **38**, 2263 (1995).
33. Dubois, J.; Guénard, D.; Guéritte-Voegelein, Guedira, N.; Potier, P.; Gillet, B.and Beloeil, J-C. *Tetrahedron* **49**, 6533 (1993).
34. Williams, H.J.; Scott, A.I.; Dieden, R.A.; Swindell, C.S.; Chirlian, L.E.; Francl, M.M.; Heerding, J.M.and Krauss, N.E. *Tetrahedron* **49**, 6545 (1993).
35. Vander Velde, D.G.; Georg, G.I.; Grunewald, G.L.; Gunn, K. and Mitscher, L.A. *J.Am.Chem.Soc.* **115**, 11650 (1993).
36. Baker, J.K. *Spectroscopy Lett.* **25**, 31 (1992).
37. Paloma, L.G.; Guy, R.K.; Wrasidlo, W. and Nicolaou, K.C. *Chem. Biol.* **1**, 107 (1994).
38. Guéritte-Voegelein, F.; Mangatal, L.; Guénard, D.; Potier, P.; Guilhem, J.; Cesario, M. and Pascard, C. *Acta Crystallogr.* **C46**, 781 (1990).
39. Balasubramanian, S.V.; Alderfer, J.L. and Straubinger, R.M. *-J.Pharm.Sci* **83**, 1470 (1994).
40. Pazdur, R.; Kudelka, A.P.; Kavanagh, J.J.; Cohen, P.R. and Raber, M.N. *Cancer Treatment Reviews* **19**, 1 (1993).
41. Fabre, J.L.; Locci-Tonelle, D. and Spiridonidis, C.H. *Drugs of the Future* **20**, 464 (1995).

Prospects for pharmaceuticals from marine organisms

Kenneth L. Rinehart

Roger Adams Laboratory, University of Illinois, Urbana, IL 61801, U.S.A.

Abstract. Marine species provide enticing sources of potential new pharmaceutical agents. An additional incentive is the lure of novelty in active compounds from marine sources. In spite of their promise, however, relatively few marine natural products have reached clinical trials. Promising drug candidates have been discovered in the areas of antibacterial, antifungal, antiviral, immunomodulatory, anti-inflammatory and anti-ischemic agents, but the greatest effort has been directed toward developing antitumor agents. Examples of promising antitumor agents at various stages of development include bryostatin 1, didemnin B, dolastatin 10, ecteinascidin 743, halichondrin B, and halomon. Providing an adequate supply of a successful marine-derived pharmaceutical agent has been solved by re-collection of large quantitites of invertebrates thus far, but aquaculture and chemical synthesis are attractive alternatives. Although progress has apparently been slow, "drugs from the sea" are alive and well and should soon provide successful pharmaceuticals or at the very least new pharmacophores for development.

Relative to terrestrial plant-derived and fermentation-derived compounds pharmaceutically active compounds of marine origin are a relatively recent addition to man's pharmacopoeia. Mainly this is a result of the difficulty of obtaining marine species, which only became easy with the advent of SCUBA gear for collecting marine organisms routinely to 50 meters and scientific submersibles able to collect to the depths of the ocean. These capabilities became realities beginning in the 1950's and by the 1960's regular symposia on marine-derived pharmaceuticals ("Drugs from the Sea") were held.[1] By 1980 it was clear that marine-derived compounds were every bit as likely to be active (at least in in vivo antitumor assays) as terrestrially-derived plant products.[2] A recent review identified a number of promising marine-derived compounds from macroorganisms (sponges, etc.) but also noted that marine-derived microorganisms provide a new source of potential drugs from the sea.[3] A high percentage of marine-derived compounds have novel structures and consequently identify new pharmacophores for structure-activity relationship studies.[4]

Activities Other than Antitumor. Most of the present discussion will deal with marine-derived antitumor compounds, since they constitute the most extensively studied category, but other activities have also been identified. These include antibacterial compounds, such as laurinterol, active against gram-positive organisms,[5] and ptilocaulin,[6] the most potent broad-spectrum antimicrobial agent identified from a survey of nearly 500 Caribbean species. Plakortic acid is a strongly antifungal compound[7] and holothurin A[8] is marketed in Japan as part of a mixture of related antifungal compounds. Antiviral agents didemnin B[9] (to be discussed more fully later) and eudistomin C[10] are active in vivo against both RNA and DNA viruses, while avarone and avarol have been described as having anti-HIV activity.[11] Among anti-inflammatory agents pseudopterosin A[12] analogs and manoalide[13] have been indicated to have topical activity and similar anti-inflammatory activity has been recently demonstrated for topsentin.[14] Among immunomodulators, didemnin B has been shown to be far more active than cyclosporin A in some assays for immunosuppression.[9]

7

It should be noted here that in addition to being a rich source of pharmaceutical agents the sea has also provided a number of potent toxins, several of which have proved to be useful biochemical tools, especially in mode of action studies. These include cyclic guanidines--tetrodotoxin[15] and saxitoxin,[16] polypeptides like ω-conotoxin,[17] which is in clinical trial for ischemia, a host of highly oxygenated toxins--palytoxin,[18] ciguatoxin,[19] brevetoxin,[20] and okadaic acid,[21] as well as the cyclic peptides microcystin and nodularin.[22] Interestingly, the mode of action of microcystin and nodularin is inhibition of protein phosphatases 1 and 2A, the same mechanism demonstrated for okadaic acid.

Compounds with Antitumor Activity. The remainder of the present discussion will be devoted to antitumor agents, whose activity has been most widely studied, thanks in large measure to the extensive program carried out by the National Cancer Institute in the U.S. over a period of 25 years.[2] In discussing antitumor activity one must distinguish among the several stages of testing of a potential antitumor agent. Very many compounds demonstrate activity in _in vitro_ assays for cytotoxicity, the ability to kill cancer cells, or in inhibition of specific enzymes, and many of these initial in vitro assays have been carried out on shipboard in remote locations on numerous expeditions. Crambescidins[23a] and myriaporones[23b] are recent examples of compounds with $IC_{50} < 1$ µg vs. L1210 leukemia cells. Far fewer compounds have demonstrated activity in _in vivo_, which is defined by the NCI as the ability to extend the lifetime of mice infected, e.g., with leukemia, by at least 25%, or the ability to reduce the size of solid tumors by at least 60%. An example of a compound with especially good therapeutic activity in mice is the recently discovered aplyronine A (Fig. 1).[24] The best compounds in _in vivo_ assays are then recommended for

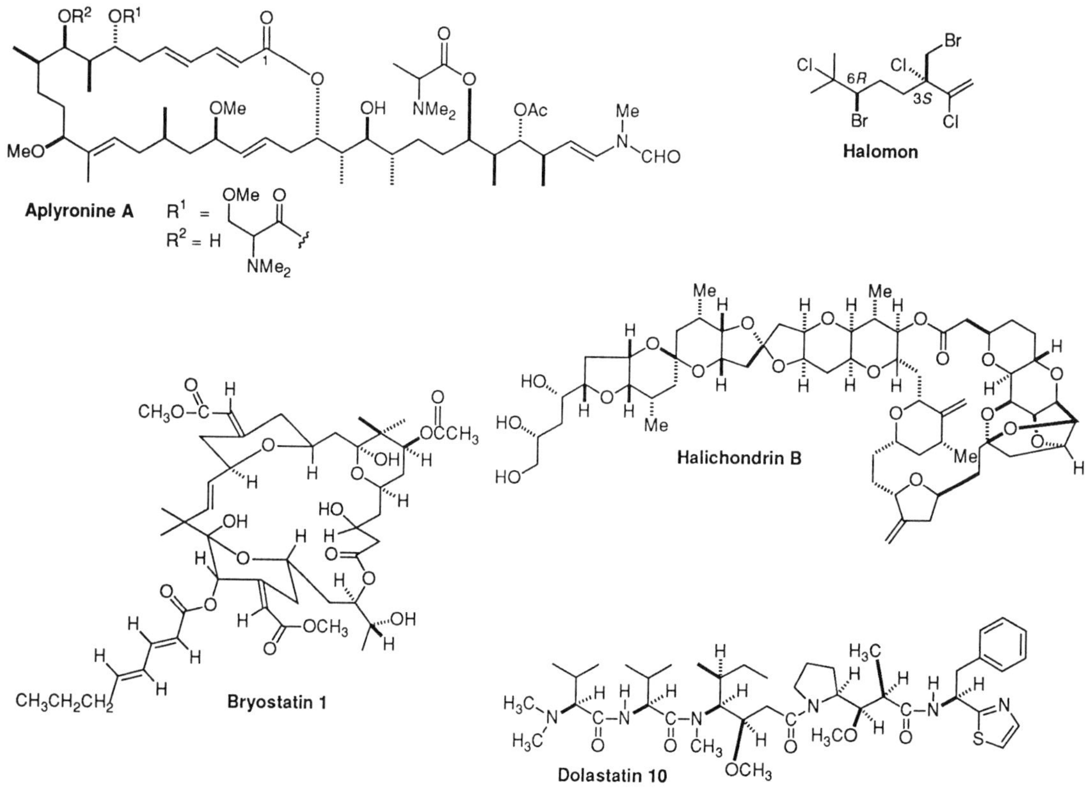

Fig. 1. Antitumor compounds in preclinical or clinical testing

further preclinical studies--for toxicity testing in a variety of animal species and for formulation and quantitation. At the present time three compounds are in some stages of advanced preclinical development at the NCI. These include halomon[25] and halichondrin B,[26] as well as ecteinascidin 743 (Et 743), which will be discussed later. Of these, Et 743 is by far the most advanced. The next stage, a Phase 1 clinical study, involves testing for human toxicity and determining a maximum tolerated dose in humans at one or two cancer centers, and dolastatin 10 is in that stage.[27] A Phase 2 clinical study involves testing for efficacy with some 10 to 20 different tumors in a like number of cancer centers. Bryostatin 1[28] has just entered Phase 2 and didemnin B (DB)[29] has been in Phase 2 trials now for approximately eight years. A Phase 3 clinical study involves a larger number of hospitals (perhaps 100); no marine-derived compound is presently at that stage. Only one marine-related compound, Ara C, has reached the marketplace and we shall return to it at the end.

It is of some interest to note that antitumor activity is not restricted to one phylum: crambescidins and halichondrin B come from sponges, ecteinascidins and didemnins from tunicates, bryostatins and myriaporones from bryozoans, dolastatins and aplyronines from sea hares, halomon from a red alga.

Didemnins. The remainder of the present discussion will be directed toward two groups of compounds from our laboratory, didemnins and ecteinascidins. The cytotoxicity and antiviral activity of Trididemnum solidum were found during an expedition in 1978 on board the R/V Alpha Helix sponsored by the U.S. National Science Foundation.[30] The producing organism is a compound tunicate, pancake-like, which grows on and ultimately kills corals, sponges, algae, etc., as well as on rocks at depths from 10-40 meters in numerous sites in the Western Caribbean. A single didemnin, dehydrodidemnin B, is produced by quite a different tunicate, Aplidium albicans, which was obtained in the Balearic islands at a depth of 50 meters.

Some of the biologically active compounds found in Trididemnum solidum are shown in Fig. 2. The didemnins are cyclic depsipeptides containing six amino or hydroxy acids in the ring and at least one in the side chain. In addition to the normal amino acids present--threonine, proline and leucine--a number of modified amino acids, including N-methyl-D-leucine and N,O-dimethyl-L-tyrosine, and the highly unusual isostatin and Hip (hydroxyisovalerylpropionic acid, the latter a hydroxy keto acid, rather than an amino acid), are present. Structures of didemnins were assigned by hydrolysis of the intact peptides to individual amino acids with identification of standard derivatives and stereochemical assignment of the amino acids being based on gas chromatography/mass spectrometry and identification of the unusual components by NMR experiments.

All in all more than 10 didemnins have now been found in the tunicate and many others have been prepared in either total or partial syntheses.[31] Stereochemical modifications have been made in the isostatin, the Hip, the proline, the N-methylleucine, and the leucine units, and structural modifications in the Hip and dimethyltyrosine portions of the molecule. In addition, compounds with various substituents attached to the methylleucine have been obtained chemically or from the tunicate. Particularly significant are the oligoglutaminyl residues capped by a pyroglutamyl unit (didemnins D, E, and M) or a 3-hydroxydecanoyl unit (didemnins X and Y). Total syntheses of didemnins A and B have been achieved in

3S, 4R, 5S–Ist
2S, 4S–Hip
D–MeLeu
L–Thr
L–Leu
L–Pro
L–Me2Tyr
Didemnin A: R = H
Didemnin B: R = CH3CHOHC—N—C —
L–Lac
L–Pro
Didemnin C: R = L–Lac
Didemnin D: R = L–pGlu—(L–Gln)3—L–Lac—L–Pro—
Didemnin E: R = L–pGlu—(L–Gln)2—L–Lac—L–Pro—
Didemnin M: R = L–pGlu—L–Gln—L–Lac—L–Pro—
Didemnin G: R = CHO
Didemnin X: R = Hydec—(L–Gln)3—L–Lac—L–Pro—
Didemnin Y: R = Hydec—(L–Gln)4—L–Lac—L–Pro—
Hydec = n–C7H15
Dehydrodidemnin B: R = CH3C-C—N—C —
Pyruv
L–Pro

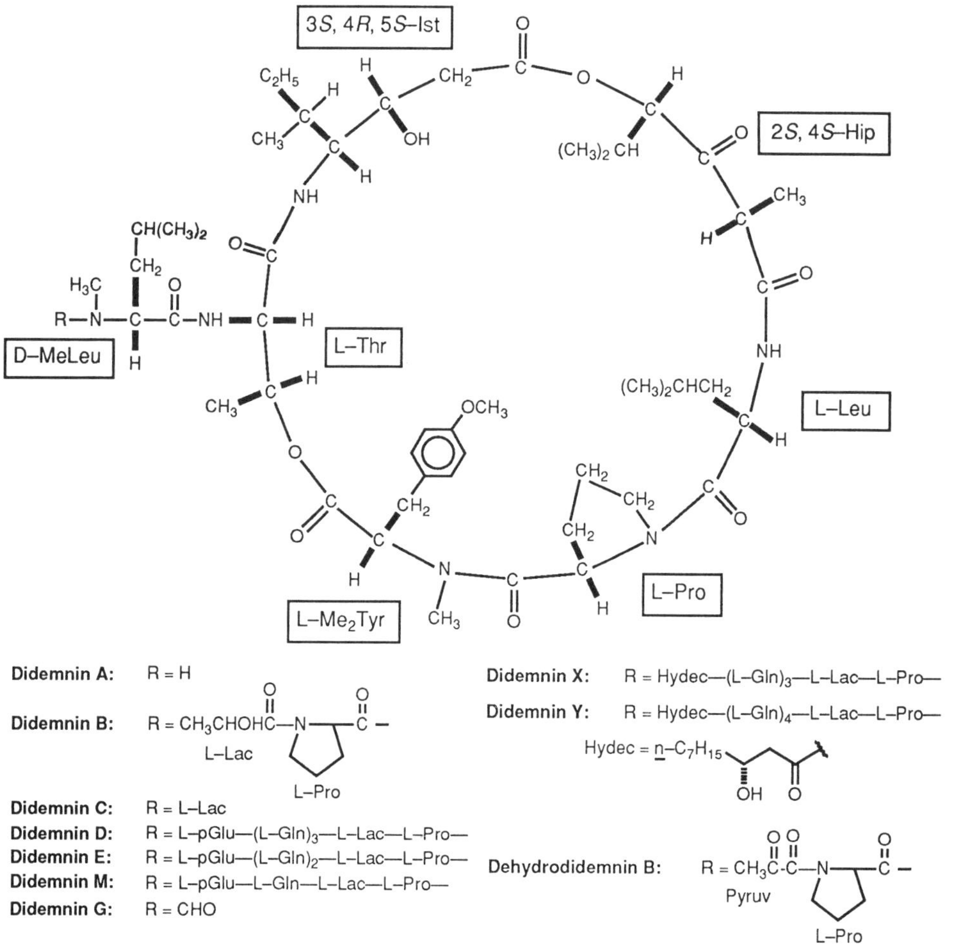

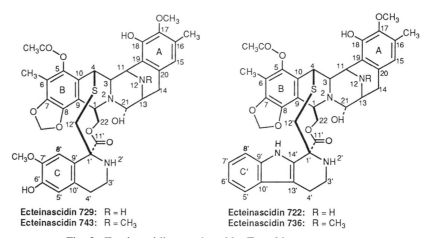

Fig. 2. Didemnins produced by T. cyanophorum and A. albicans (dehydrodidemnin B)

Ecteinascidin 729: R = H
Ecteinascidin 743: R = CH_3

Ecteinascidin 722: R = H
Ecteinascidin 736: R = CH_3

Fig. 3. Ecteinascidins produced by E. turbinata

our laboratory and others in yields suitable for large-scale commercial syntheses of the compounds. Didemnin B and dehydrodidemnin B, for example, have been prepared in good yields from didemnin A.[32]

Some 42 didemnins have thus far been studied for structure/activity relationships.[31] Cytotoxicity of the various didemnins, which can be taken as a measure of the potential antitumor activity of the individual congeners, varies widely. Especially active are dehydrodidemnin B, didemnin B analogues in which the terminal lactyl group has been replaced by a less polar acyl group, and derivatives of didemnin A in which the secondary amino group has been acylated by a short-chain acid. That DDB is a highly active compound in vitro can be demonstrated by its activity against a variety of cell lines, which compares well with that of a standard, tamoxifen. Didemnin A registers intermediate activity, while compounds of relatively low activity (but still active at 0.5 to 2.0 µg/mL) are long-chain acyl derivatives of didemnin A and analogues in which the stereochemistry of the isostatin or Hip groups has been modified. That cytotoxicity is a useful predictor of activity can be shown by the relative in vivo activities of dehydrodidemnin B and didemnin B, where dehydrodidemnin B shows greater life extension against P388 leukemia and B16 melanoma and also has activity against Lewis lung carcinoma, a solid tumor.

We noted above that didemnin B has been in Phase 2 clinical trials for several years. In the Phase 2 trials, most tumors did not respond to didemnin B, but complete or partial response was obtained in studies with non-Hodgkins lymphoma and a glioblastoma.[29]

It was indicated above that didemnins have antiviral activity,[9] and this, too, varies with the substituents and modifications of the didemnins.[31] The most active compounds are N-acetyldidemnin A and didemnin B as well as analogues of didemnin B in which the lactyl group is replaced by a short-chain nonpolar acyl group. There is some difference between anti-VSV (an RNA virus) and anti-HSV (a DNA virus) activities. Didemnins containing a pyroglutamyl or a glutaminyl group appear to be especially active against HSV. At the bottom of the antiviral activity scale are acyclo-didemnin A, with the ring opened, a variety of didemnins with modified stereochemistry in the Hip unit, and didemnin N, in which the N,O-dimethyltyrosine unit has been modified. Most of these compounds have not been tested in vivo but didemnin B was active against both RNA and DNA viruses including Rift Valley fever and vaginal herpes.[9]

The final characteristic of didemnins extensively studied is their immunosuppressive activity, with enormous variation in activity.[31] Especially active is didemnin M, in which the hydroxyl group of lactic acid has been substituted by a pyroglutamyl-glutaminyl unit. Other compounds with a pyroglutamyl substituent are also among the most immunosuppressive--O-pGlu DB and didemnin E. All three of these compounds are much more active in suppressing the mixed lymphocyte reaction (MLR) than didemnin B, which was previously reported for its activity.[9] For example, didemnin M is 20,000 times as active as didemnin B in this assay. Moreover, the selectivity comparing cytotoxicity against lymphocytes (LcR) vs. activity in the MLR is approximately 500 times as great as for didemnin B. The selectivity remains high in comparing the MLR vs. toxicity to lymphoblasts.

Didemnins B and M and O-pGlu-didemnin B have all been tested in the graft vs. host reaction and have shown significant immunosuppression. The optimum results were manifested for didemnin M, at two orders of magnitude lower concentration than for the two others. Obviously, further studies for didemnin M are in order. Earlier studies, as indicated, have involved only didemnin B, but one of those studies indicated activity in rat heart transplantation.[9]

The other tunicate-derived anticancer agents, ecteinascidins, are produced by Ecteinascidia turbinata, a colonial tunicate which grows preferentially on mangrove roots and looks much like a bunch of colorless grapes with pink or orange tips.[33] The tunicate is relatively abundant in mangrove swamps throughout the Caribbean at depths of 1-3 meters. Extracts of Ecteinascidia turbinata were recognized by the NCI as early as 1968 as one of their most potent antitumor marine organisms in in vivo assays but the compounds responsible resisted isolation for over 15 years. In order to isolate the compounds and solve the structure, development of new techniques was necessary--countercurrent chromatography, bioautography of tissue cultures, fast atom bombardment mass spectrometry, and long-range heteronuclear correlation techniques in NMR spectroscopy.[33] The utility of these techniques can be illustrated by the identification of the ion cascade in a tandem mass spectrometry study and by an NMR study identifying heteronuclear long-range correlations. The MS/MS fragmentation cascade correlated particularly well with fragmentation of the fermentation-derived antibiotic safracin B for the ions at m/z 204 and 218 and this was confirmed by long-range correlation spectroscopy for the A ring and its adjacent tetrahydropyridine unit.[34] Ecteinascidins differ from the smaller saframycins and safracins, however, by containing an additional C ring unit (another tetrahydroisoquinoline in Et's 729 and 743) linked via a methylene sulfide bridge to the B ring system (Fig. 3). Later, a second series of ecteinascidins was discovered (Et 722, 736) containing a tetrahydro-β-carboline instead of a tetrahydroisoquinoline as a C ring (Fig. 3).[35] We have designated the ecteinascidins by numbers indicating the highest mass of the ions originally observed for the compounds; loss of the C-21 OH group gives rise to these ions. Following the initial structure assignment, X-ray crystallography on two derivatives of ecteinascidins 729 and 743 confirmed the overall structures and assigned the relative stereochemistry.[35]

Still later, another series of ecteinascidins lacking the C ring system but retaining the methylene sulfide bridge was found and cleavage of that bridge yielded L-cysteine, identified by gas chromatography.[36] The biosynthesis of ecteinascidins is presumed to resemble that demonstrated earlier for saframycin A,[37] with minor modifications, as shown in Fig. 4. Recent studies elsewhere[38] confirm some points earlier proposed,[33-35] but the Scheme shown there should be modified to take account of the newly isolated compounds[36] lacking the C ring. The mode of action of the ecteinascidins is presumed to involve the loss of the C-21 hydroxyl (a carbinolamine hydroxyl) to form an iminium ion, which binds to DNA reversibly. Molecular modeling experiments have shown a good fit of ecteinascidins in the minor grove of DNA.[35]

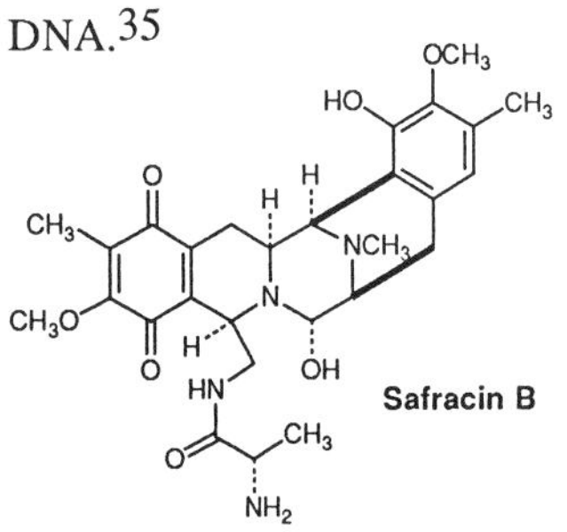

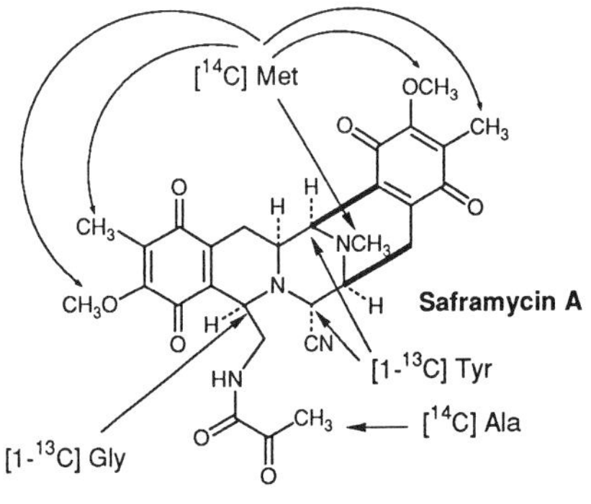

Fig. 4. Safracin B and Saframycin A

Ecteinascidins are exceedingly potent cytotoxic agents, approximately an order of magnitude more cytotoxic than the related saframycins and safracins.[39] Of the ecteinascidins, Et 729 is generally the most cytotoxic, followed by Et 743, Et 722, and Et 736.[40] It was initially planned that Et 729 should be advanced to clinical trials, but the choice was switched to Et 743 due to its greater abundance in the tunicate. There is some differential activity among the ecteinascidins in vivo as well, Et 722 being more active against P388 leukemia, but Et 729 being more active against B16 melanoma, M5076 ovarian sarcoma, Lewis lung tumors, and human xenografts on nude mice (mammary cancer, MX1, and lung cancer, LX1.[35] The compounds do not appear to be active against colon cancer (CX1 xenograft, C38 colon tumor). Especially impressive results were obtained in xenografts with advanced stage MX1, which gave 9/10 mice tumor free on day 23 and four of ten animals still tumor free at day 58.[41]

We noted at the beginning of the discussion that only one marine-related compound has advanced thus far to the marketplace--Ara C (cytosine arabinoside, cytarabine). In fact that compound was not originally isolated from marine species. Rather it was modeled after nucleosides found in the sponge Cryptotethya crypta in which arabinose replaced ribose--spongouridine and spongothymidine.[40] We can note here the time elapsed (nearly 20 years) between the discovery of the sponge's nucleosides and the introduction of Ara C into the marketplace.[41] An even longer time was required (nearly 30 years) for the development and introduction of Ara A (adenosine arabinoside) as an antiviral agent.[42]

We can also note here that approximately 30 years elapsed between the identification of the cytotoxicity of an extract of Taxus brevifolia[43] and the FDA approval of the use of taxol in ovarian cancer.

A final question frequently asked regarding marine-derived compounds has to do with supplies of the compounds. Thus far all those compounds which have entered clinical trials or are about to do so-- didemnin B, bryostatin 1, dolastatin 10, ecteinascidin 743, halomon, halichondrin B, manoalide, pseudopterosin A, holothurins--have been obtained from natural sources, even though the amount of natural material required can be quite large--40,000 liters of Bugula neratina, 500 kg of Trididemnum solidum, several tons of Ecteinascidia turbinata. However, other sources will presumably be employed in the future. Culturing of microorganisms, of course, is relatively routine and methods are being developed for the cultivation of tunicates and sponges. Chemical synthesis is also an accomplished fact; didemnins, dolastatin 10, halichondrin B, aplyronine A, eudistomins, ptilocaulin, manoalide have all been synthesized, so the desired drugs should be available from a variety of sources.

In conclusion, we can summarize that the development of marine natural products as sources of pharmaceutical agents has been slow, but not unusually so, and that an array of potent and potentially therapeutically useful compounds stand on the brink of introduction as pharmaceutical agents.

Acknowledgment. I would like to acknowledge here the chemical efforts of my colleagues on didemnins (Drs. J. B. Gloer, V. Kishore , R. Sakai) and ecteinascidins (Drs. T. G. Holt, R. Sakai, N. L. Fregeau), and assistance with the mass spectrometry (F. Sun) and X-ray crystallography (Dr. A. H. J. Wang). We also acknowledge the assistance of biologists with the assays (Drs. R. G. Hughes, L. H. Li, G. R. Wilson, and G. Faircloth), with taxonomy (Drs. F. Monniot and R. C. Brusca), and the support of the National Institutes of Allergy and Infectious Diseases and of General Medical Sciences, the National Cancer Institute, PharmaMar, S.A., and The Upjohn Company.

References

1. <u>Drugs from the Sea Symposium</u>, Aug. 27-29, 1967; H. G. Freudenthal, Ed.; Marine Technology Society, Washington, D.C., 1968; and subsequent symposia with the same or similar titles.

2. M. Suffness and J. Douros, <u>J. Nat. Prod.</u> **1982**, <u>45</u>, 1-14.

3. J. L. Fernandez Puentes and K. L. Rinehart, <u>Pharmaceutical Manufacturing International</u> 1995, 17-19.

4. (a) W. Bradner. In <u>Cancer & Chemotherapy</u>; S. Crooke and A. Prestayko, Eds.; Academic Press: New York, 1980; Vol. 1. (b) J. Douros and M. Suffness. In <u>New Anticancer Drugs</u> (Recent Results in Cancer Research, Vol. 70); S. Carter and T. Sakuri, Eds.; Springer-Verlag, 1980.

5. (a) T. Irie, et al., <u>Tetrahedron Lett.</u> **1966**, 1837-1840. (b) S. M. Waraszkiewicz and K. L. Erickson, <u>Tetrahedron Lett.</u> **1974**, 2003-2006.

6. G. C. Harbour, et al., <u>J. Am. Chem. Soc.</u> **1981**, <u>103</u>, 5604-5606.

7. D. W. Phillipson and K. L. Rinehart, Jr., <u>J. Am. Chem. Soc.</u> **1983**, <u>105</u>, 7735-7736.

8. I. Kitagawa, et al., <u>Tetrahedron Lett.</u> **1979**, 1419-1422.

9. K. L. Rinehart. In <u>Peptides, Chemistry and Biology</u>, Proceedings of the Tenth American Peptide Symposium; G. R. Marshall, Ed.; ESCOM: Leiden, 1988; pp 626-631.

10. K. L. Rinehart, Jr., et al., <u>J. Am. Chem. Soc.</u> **1984**, <u>106</u>, 1524-1526.

11. P. S. Sarin, et al., <u>J. Natl. Cancer Inst.</u> **1987**, <u>78</u>, 663-666.

12. S. A. Look, et al, <u>Proc. Natl. Acad. Sci. USA</u> **1986**, <u>83</u>, 6238-6240.

13. (a) E. D. De Silva and P. J. Scheuer, <u>Tetrahedron Lett.</u> **1980**, <u>21</u>, 1611-1614. (b) K. B. Glaser and R. S. Jacobs, <u>Biochem. Pharmacol.</u> **1986**, <u>35</u>, 449-453.

14. (a) S. Tsuji, et al, <u>J. Org. Chem.</u> **1988**, <u>53</u>, 5446-5452. (b) R. S. Jacobs, in Ref. 23b.

15. (a) T. Yasumoto, et al., <u>Agric. Biol. Chem.</u> **1986**, <u>50</u>, 793-795. (b) T. Noguchi, et al., <u>J. Biochem.</u> **1986**, <u>99</u>, 311-314.

16. (a) E. J. Schantz, et al., <u>J. Am. Chem. Soc.</u> **1975**, <u>97</u>, 1238-1239. (b) J. Bordener, et al., <u>J. Am. Chem. Soc.</u> **1975**, <u>97</u>, 6008-6012.

17. (b) R. A. Meyers, et al. <u>Chem. Rev.</u> **1993**, <u>93</u>, 1923-1936. (b) S. S. Bowersox and R. R. Luther, <u>Drugs Fut.</u> **1994**, <u>19</u>, 128-130.

18. (a) R. E. Moore and G. Bartolini, <u>J. Am. Chem. Soc.</u> **1981**, <u>103</u>, 2491-2494. (b) D. Uemura, et al., <u>Tetrahedron Lett.</u> **1981**, <u>22</u>, 2781-2784.

19. M. Murata, et al., <u>J. Am. Chem. Soc.</u> **1989**, <u>111</u>, 8929-8931.

20. (a) Y. Y. Lin, et al., <u>J. Am. Chem. Soc.</u> **1981**, <u>103</u>, 6773-6775. (b) Y. Shimizu, et al., <u>J. Am. Chem. Soc.</u> **1986**, <u>108</u>, 8929-8931.

21. K. Tachibana, et al., <u>J. Am. Chem. Soc.</u> **1981**, <u>103</u>, 2469-2471.

22. K. L. Rinehart, et al., <u>J. Appl. Phycol.</u> **1994**, <u>6</u>, 159-176.

23. (a) E. A. Jares-Erijman, et al., <u>J. Org. Chem.</u> **1991**, <u>56</u>, 5712-5715. (b) K. L. Rinehart and K. Tachibana, <u>J. Nat. Prod.</u> **1995**, <u>58</u>, 344-358.

24. K. Yamada, et al., <u>J. Am. Chem. Soc.</u> **1993**, <u>115</u>, 11020-11021.

25. R. W. Fuller, et al., <u>J. Med. Chem.</u> **1992**, <u>35</u>, 3007-3011.

26. Y. Hirata and D. Uemura, <u>Pure Appl. Chem.</u> **1985**, <u>58</u>, 701-710.

27. G. R. Pettit, et al, <u>J. Am. Chem. Soc.</u> **1987**, <u>109</u>, 6883-6885.

28. G. R. Pettit, <u>Prog. Org. Chem. Nat. Prod.</u> **1991**, <u>57</u>, 153-195.

29. <u>Ann Rept. to the FDA: Didemnin B, NSC 325319, IND 24505</u>, Division of Cancer Treatment, NCI, Bethesda, MD, Aug. 1994.

30. K. L. Rinehart, et al., <u>Pure Appl Chem.</u> **1981**, <u>53</u>, 795-817.

31. R. Sakai, et al., <u>J. Med. Chem.</u>, submitted.

32. (a) K. L. Rinehart, et al., <u>J. Am. Chem. Soc.</u> **1987**, <u>109</u>, 6846-6848. (b) B. Kundu, University of Illinois, unpublished results.

33. (a) T. G. Holt, Ph.D. Thesis, University of Illinois, Urbana, 1986. (b) K. L. Rinehart, et al., <u>J. Nat. Prod.</u> **1990**, <u>53</u>, 771-792.

34. (a) K. L. Rinehart, et al., <u>J. Org. Chem.</u> **1990**, <u>55</u>, 4512-4515. (b) A. E. Wright, et al., <u>J. Org. Chem.</u> **1990**, <u>55</u>, 4508-4512.

35. R. Sakai, et al., <u>Proc. Natl. Acad. Sci. USA</u> **1992**, <u>89</u>, 11456-11460.

36. R. Sakai, et al., <u>J. Am. Chem. Soc.</u> **1995**, <u>117</u>, in press.

37. Y. Mikami, et al., <u>Biol. Chem.</u> **1985**, <u>260</u>, 344-348.

38. R. G. Kerr, <u>J. Nat. Prod.</u> **1995**, <u>58</u>, in press.

39. (a) T. Arai, et al., <u>Gann</u> **1980**, <u>71</u>, 790-796. (b) T. Okumoto, et al., <u>J. Antibiot.</u> **1985**, <u>38</u>, 767-771.

40. R. Sakai, Ph.D. Thesis, University of Illinois, Urbana, 1991.

41. G. Faircloth, et al., Symposium on Advances in the Treatment of Breast Cancer, Madrid, June 5-6, 1995.

42. W. Bergman and R. J. Feeney, <u>J. Org. Chem.</u> **1951**, <u>16</u>, 981-987.

43. G. P. Bodey, et al., <u>Cancer Chemother.</u> **1969**, <u>53</u>, 59-66.

44. R. A. Buchanan and F. Hess, <u>Pharmacol. Ther.</u> **1980**, <u>8</u>, 143-171.

45. M. C. Wani, et al., <u>J. Am. Chem. Soc.</u> **1971**, <u>93</u>, 2325-2327.

New lead structures in pharmacophore-based drug design

N. Claude Cohen

CIBA-GEIGY Limited, Pharmaceutical Division, Research Department, Klybeckstrasse 141, CH-4002 Basel, Switzerland.

Abstract: Designing molecules which assemble, in 3-D, a set of well-defined but disconnected functional groups is not a trivial task and is one important aspect of pharmacophore-based drug design. In this perspective, impressive advances were recently made in providing automated methods for lead-finding in drug research. One of the strengths of such approaches is that the new structures which are generated can be chemically unrelated to the reference molecules from which the initial 3-D query was derived. This provides an unbiased starting point for the design of new lead structures in drug research. In practice lead structures are either the results of tailor-made constructions (the construction approach) or extracted from existing databases (the database approach). Several computerized systems were developed in the last few years, based on both approaches. Current methods, including the ones we have conceived, will be discussed to show how automated methods have opened new perspectives to molecular mimicry, an activity of central importance in drug design.

INTRODUCTION

Molecular mimicry is a concept of central importance in drug design. Very often the molecules created are conceived as mimics of substances known to interact with biological macromolecules. The strategies used in rational drug design are simple and can be clearly understood even by the layman: new molecules are conceived either on the basis of similarities with known reference structures (pharmacophore-based drug design), or on the basis of their complementarity with the 3-D structure of known active sites (receptor structure-based drug design). In the first case, optimal working conditions require the knowledge of a set of structurally different molecules binding at the same receptor site, and in the second case, the knowledge of an x-ray structure of the macromolecular target or, of at least one complex between the protein and a ligand.

Pharmacophores consist of the spatial arrangement of chemical moieties which are responsible for a given biological activity by their interaction with specific biological receptors. In current work a pharmacophore can also be defined operationally as a rule for superimposing molecules in 3-D. Along these lines pharmacophore-based drug design provides useful means for the design of new lead structures in pharmaceutical research. Designing molecules which assemble, in 3-D, a set of well-defined but disconnected functional groups (see fig. 1) is not a trivial task and remains very subjective.

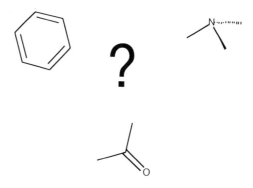

Figure 1: How to link the recognition fragments of a pharmacophore ?
Reproduced with permission from reference 23 (N.C. Cohen and V. Tschinke, Progress in Drug Research, vol. 45, in press).

In this perspective impressive advances were recently made in providing automated methods for lead-finding in drug research. New lead structures are now obtained using either tailor-made constructions (the construction approach) or from existing databases (the database approach).

CONCEPTS

The visualization in 3-D of molecular properties (steric aspects, electrostatic potentials, hydrophobicity, etc.) of active and inactive molecules is one step towards the identification of similarities and differences between these compounds. The easiest way to reveal 3-D structural features common to a series of compounds is to use superimposition procedures. This technique has been applied extensively since the early seventies. The superimpositions are evident when topological analogues are considered; but they are not obvious when chemically unrelated structures are compared. Extensive analyses made in a variety of therapeutic areas have shown how different molecules can elicit similar biological activities when they share common stereochemical features in 3-D. This has led to the development of 3-D pharmacophoric models, which are necessary for the rational design of new lead compounds. The structures are usually aligned on the basis of their atomic positions, but the alignment can be very subjective. The best way to reduce subjectivity is to compare the molecules in 3-D in the actual space of their physicochemical properties. The conformational flexibility of the molecules introduces another level of complication: due to the extreme difficulty of the problem, the modelist often relied in the past on visual judgements to control and modify manually the current conformations. Knowing the bioactive conformation of the ligand as it binds to the receptor or enzyme greatly facilitates the designing of new, potent and specific lead molecules. Experience shows that the bioactive conformation of a molecule is not necessarily the same as that found in solution or in the crystal. Rather, it is the specific conformation of lowest energy in the context of the receptor: during the process of complex formation, the protein and the ligand respond to each other's presence. In indirect design, this information is not available and the first step in this strategy therefore consists in generating such information by considering a molecule to be represented, as a set of energetically possible conformers.

Several computerized systems were recently developed for aligning molecules in 3-D (see table 1) and the most recent systems have incorporated automated procedures for the "molecular fitting" and "pharmacophore alignment" which are necessary procedures for revealing pharmacophores.

TABLE 1 . Computer programs for aligning molecules in 3-D
according to various molecular properties (35)

ASP	Oxford Molecular Ltd (1)
MEPCONF	Sanz *et al.* (2)
SPERM	Van Geerestein *et al.* (3)
OVID and SUPER	Hermann *et al.* (4)
RECEPS	Kato *et al.* (5)
AUTOFIT	Kato *et al.* (6)
DISCO	Martin *et al.*; Tripos Associates (7)
CoMFA	Cramer *et al.*; Tripos Associates (8)
BIOCES	Nokamura *et al.*; Kissei-Comtec (9)
YING	Chan *et al.* (10)
LOCKSMITH, LOCK & KEY	Kellog *et al.* (11)
SEAL	Kearsley *et al.* (12)
CATALYST	Sprague; Molecular Simulations (13)
APEX-3D	Golender *et al.*; Biosym (14)
LIPOT	Furet *et al.* (15)
YAK	Vedai *et al.* SIAT (16)

Automated computerized superimposition techniques are relatively new and very complex, especially when the multiple-conformation perspective is envisaged. Efforts of this type are current topics in modeling development and are going to extend the range of the tools already available.

Having the pharmacophore, the next step is to design new molecules carrying the desired molecular moieties in the proper 3-D orientation and in a conformation of reasonable energy. As already mentioned, this is not a trivial task and can be very subjective. The advantage of a computerized approach is, in the first place that the conformations which are going to be considered are seriously validated. Moreover the solutions are also expected to be free from any subjective bias. In current computer-aided strategies the candidate molecules are either extracted from a databank of known molecules (the database approach), or generated on the basis of construction algorithms (construction methods) that generate new molecules conforming to desired requirements. The differences between the two approaches are the same between standard and tailor-made solutions. One of the advantages of database searching is that it generally requires no synthetic effort for the validation of the first step. On the other hand, construction methods have the advantage of not being biased or dependent on a given databank, and provide the structural diversity necessary to allow a synthetic target to be chosen as a first lead. Both methods are outlined and discussed in this article.

THE CONCEPT OF A PHARMACOPHORE: A SWOT ANALYSIS

It is current practice to evaluate any project on the basis of the so-called "SWOT" (Strengths, Weaknesses, Opportunities and Threats) analysis. When applied to the concept of pharmacophore this leads to the features which are mentioned on the table 2.

TABLE 2. The concept of pharmacophore: a SWOT analysis (see text).

Strengths
- usable when the receptor is unknown
- provide means for superimposing chemically unrelated compounds
- reveals structural elements responsible for the biological activity
- systematic exploration of volume accessible to the receptor
- full exploitation of the S.A.R. observed

Weaknesses
- needs different structural types
- the receptor site is not known in 3-D
- indetermination if no molecules with high informational content are available
- choice in the multiplicity of solutions
- risk of incorrect identification of bioactive conformations

Opportunities
- allows the design of new leads
- increases chemical intuition
- rational improvement of poorly active series
- insight into mechanisms of actions

Threats
- attraction of "me too-ism"
- to forget that failure is part of the iterative scheme
- if the molecules compared bind at different regions of the receptor

The concept of a pharmacophore is useful because it can be used in drug design even when the receptor is unknown, however one should not forget that failure is also part of the iterative scheme and has to be integrated in the rational approach. When structure-activity relationships are carefully conducted and well unified in the strategy, they lead to reliable and useful pharmacophore models.

THE DATABASE APPROACH

In the database approach pharmacophores are used as queries for 3-D databases to search for known structures that match the targeted structural elements. In the first generation only rigid fit was considered. The database contained a very limited amount of conformations for a given molecule (from 1 to 10), and the risk of not identifying the relevant conformations was therefore extremely high. In more recent systems conformational flexible searching (CFS) is also considered (17,18,19). Presently the databank contains only a very reasonable amount of conformers of the molecule considered, however a large number of flexible forms are generated. Although this approach appears very promising, it is now too early to say whether it will actually lead to tools of great practical utility. Methods that use more chemical knowledge have been explored: for example, ALADDIN (20) and EMBED (21) use distance-geometry concepts that represent conformational multiplicity within the framework of a distance-bounds matrix. Along these lines, multiple interatomic distances ranges are used to implicitly define conformational flexibility. This approach has been shown to give encouraging results in matching molecules of a database to the requirements of a 3-D query (22).

In common practice the query consists of a 2-D diagram or a 3-D model described with the aid of a molecular-graphics system. Geometrical objects are then selected (points, lines, planes, centroids, spheres), and their desired 3-D relationships are defined (distances, angles, dihedral angles, etc.) with tolerance values associated to each. Such systems are generally easy to use, and when applied on relatively small databases they give very reasonable computing times.

The 3-D structure of small molecules can be obtained from various sources and a number of databases of this type are currently available (35). The molecular geometries generally derive either from experimental methods (X-ray crystallography, NMR spectroscopy), or from theoretical calculations (molecular modeling). Atomic coordinates from both sources can be stored in databanks for subsequent computerized treatments.

Current systems undergo continuous development and improvements, and were already used with some success in several projects. For example the group of DuPont published recently the spectacular design of non-peptide inhibitors of the HIV-PR protease (24). They adopted a database strategy, the pharmacophoric query was relatively simple (two key intramolecular distances between two hydrophobic groups) and the tolerance given for the distances was relatively high. The initial hits obtained were progressively modified, and eventually allowed the discovery of cyclic urea molecules active in the nanomolar range. Crystal structures of complexes with various cyclic urea-HIV-PR complexes proved the validity of the approach. Although oral bioavailability stymied the first clinical compound from this series, the research team is currently looking for analogues with improved properties.

A similar approach was used by Wang et al. (25) who were searching for inhibitors of protein-kinase C. The pharmacophore query consisted of the distances between three polar moieties. They used a database containing about more than 200,000 structures, and after a computerized treatment requiring approximately one week of computer time (on an Iris Indigo workstation) they obtained a number of hits which were subsequently analyzed and evaluated. The interesting aspect of this story is that by conducting additional molecular modeling studies they were able to define an improved pharmacophore model. This model was then validated by synthetic modifications of two inactive compounds which led to two active series of derivatives.

These studies illustrate the simplicity of the strategy based on 3-D database searching which has started to become very popular. Starting with a simple pharmacophore model the candidate molecules can be rapidly obtained and tested. The objective of such a method is to identify interesting hits showing activities at the micromolar range and then to optimize them. In order to increase the binding to the nanomolar range, it is necessary to undertake careful modeling and structure-activity relationship analyses, and introduce chemical groups that are expected to produce favourable attractions with the receptor as guided by the current pharmacophore model.

THE CONSTRUCTION APPROACH

Several programs based on construction approaches were recently developed in the field of pharmacophore-based drug design. Historically they started to appear at a time where the database approaches were more advanced. Such programs include the followings LINKOR (26), SPROUT(27,28), HOOK (29), SPLICE (30), RING-BRACING (31), NEWLEAD (32) and others (35). In our group we are continuously involved in the design of compounds mimicking some reference biological molecules such as peptides, coenzymes etc... In one of our early projects we devoted substantial effort to the conception of a new generation of nonpeptidic renin inhibitors by mimicking the bioactive conformation of known peptide ligands as predicted by previous modeling studies (33, 34, 36). The experience gained in such research projects allowed us to conceive a computer program named NEWLEAD which has proven to be a useful tool with wide applicability in rational drug design and particularly efficient in the areas of molecular mimicry and peptidomimetism (32).

In a construction approach, the treatment can be simply described as follows. Given a set of at least two disconnected fragments, the problem consists of finding appropriate spacers to assemble these moieties in a single molecule. Multiple solutions are likely to occur, because many connections with different combinations of spacers are possible. In construction approaches, the spacers are either generated atom by atom as in Linkor (26) or using elementary units such as library-spacers, single-atoms, and fuse-ring spacers as in Newlead (32). The conceptual frame used in construction approaches is visualized in Figure 2.

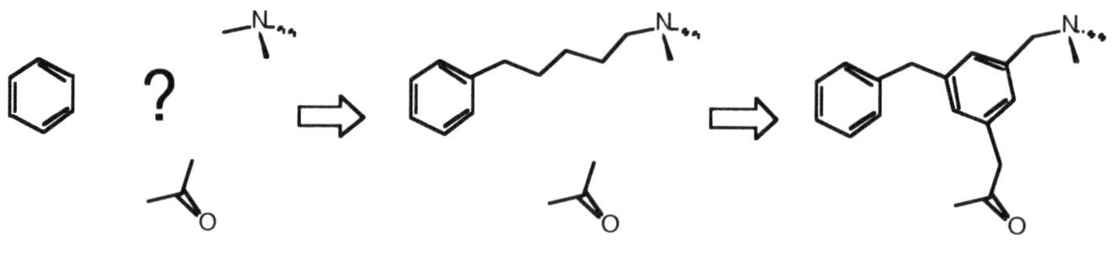

Figure 2. The conceptual frame used to create a structure by a construction approach. For clarity, the structures are represented in 2-D; in reality all the treatments are made in 3-D. Reproduced with permission from reference 23 (N.C. Cohen and V. Tschinke, Progress in Drug Research, vol. 45, in press).

The solutions are expected to conform to the following three criteria: they must be chemically plausible, geometrically valid, and energetically acceptable. The construction approach offers the advantage of not being biased or dependent on a given databank. The major benefit of these automatic linking programs is that they show many alternatives for positioning the same key fragments. For example on figures 2 and 3 are shown the chemical diversity of some of the solutions that can be associated with the actual 3-D query used.

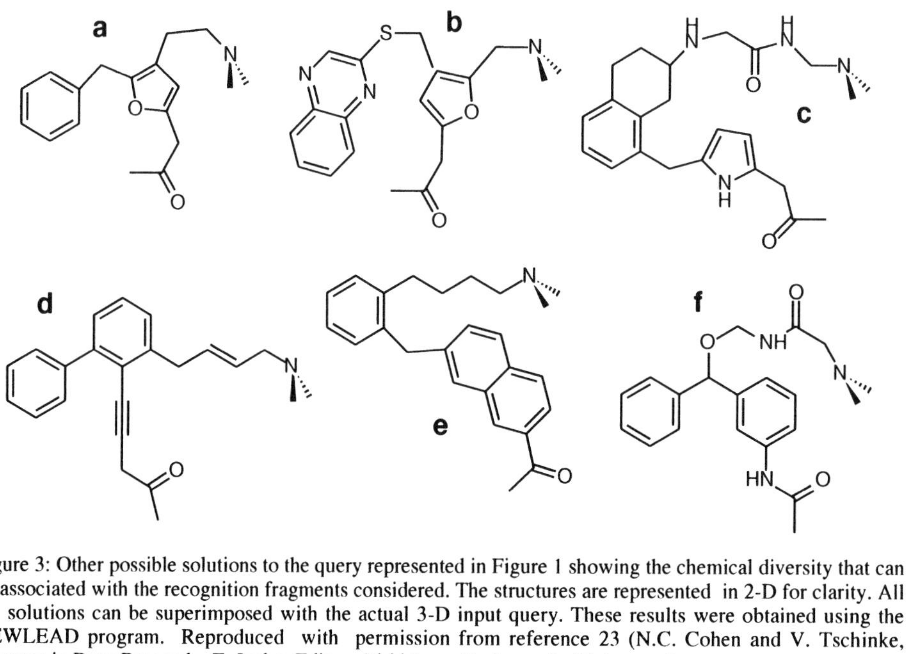

Figure 3: Other possible solutions to the query represented in Figure 1 showing the chemical diversity that can be associated with the recognition fragments considered. The structures are represented in 2-D for clarity. All the solutions can be superimposed with the actual 3-D input query. These results were obtained using the NEWLEAD program. Reproduced with permission from reference 23 (N.C. Cohen and V. Tschinke, Progress in Drug Research, E. Jucker Editor, Birkhäuser Verlag, vol. 45, in press).

PERSPECTIVES

The last decade has witnessed the development of useful approaches. However they were essentially of an analytical nature; they were oriented toward the modeling of known candidate structures, studied to verify how well they conformed to the desired requirements. The second generation has produced more creative methods whereby candidate new lead structures are now part of the results of the computerized approaches. Some of the current trends are indicated in the following table.

TABLE 3 . Some trends in pharmacophore-based drug design

- attempts to unify with QSAR
- automation of pharmacophore identification
- generalized pharmacophore language
- numerical coding and scoring
- commercial databanks of pharmacophores
- integrated design with databanks
- attempts to unify with combinatorial chemistry
- dynamic features of pharmacophores

The new developments and achievements in the field of pharmacophore-based drug design would have appeared incredible just a few years ago. Ten years ago the number of success stories published was very limited. Today no review of the field can ever include all of them (35). Currently, there is an intense activity for the development of new methods and tools in the area. Major efforts are directed towards integrated methods with even higher levels of computer automation, rationalization, quantification and, eventually, the design of useful novel molecules.

ACKNOWLEDGEMENTS

I am particularly grateful to Vincenzo Tschinke for the development of the NEWLEAD program, to my colleagues Pascal Furet, Marina Tintelnot-Blomley, Alex Sele and Alain Dietrich for stimulating discussions and pleasant teamwork.

REFERENCES

1.   Asp, distributed by Oxford Molecular Ltd., Oxford, England, U.K.
2.   F. Sanz, F. Manaut, J. Rodriguez, E. Lozoya and E. Lopez-de-Brinas. *J.Comput. Aided Mol. Des.*7, 337-347 (1993).
3.   (a) V.J. Van Geerestein, N.C. Perry, P.D.J. Grootenhuis and C.A.G. Haasnot. *Tetrahedron Comput. Methodol.* 3, 595-613 (1990). (b) J.R. Mellema and V.J. Van Geerestein. 12th Annual Conference of the Molecular Graphics Society, Interlaken, Switzerland, June 7-11, 1993; Abstract O29.
4.   R.B. Hermann and D.K. Herron. *J. Comput.-Aided Mol. Des.* 5, 511-524 (1991).
5.   Y. Kato, A. Itai and Y. Iitaka. *Tetrahedron* 43, 5229-5236 (1987).
6.   Y. Kato, A. Inoue, M. Yamada, N. Tomioka and A. Itai. *J.Comput.-Aided Mol.Des.* 6, 475-486 (1992).

7. (a) Y.C. Martin, M.G. Bures, E.A. Danaher, J. DeLazzer, I. Lico and P.A. Pavlik. *J.Comput.- Aided Mol. Des.* **7**, 83-102 (1993). (b) DISCO, distributed by Tripos Associates, St. Louis, MO, USA.

8. (a) R.D. Cramer III, D.E. Patterson and J.D. Bunce. *J. Amer.Chem.Soc.* **110**, 5959-5967 (1988). (b) J.J. Kaminski. *Adv. Drug Design Reviews* **14**, 331-337 (1994). (c) CoMFA, distributed by Tripos Associates, St. Louis, MO, USA.

9. (a) H. Nakamura, K. Komatsu, S. Nakagaura and H. Umayama . *J. Mol.Graphics* **3**, 2-11 (1985). (b) BIOCES, distributed by Kissei-Comtec, Matsumoto-City, Japan.

10. S.L. Chan, P.-L. Chau and J.M. Goodmann. *J.Comput.-Aided Mol.Des.* **6**, 461-474 (1992).

11. G.E. Kellog and D.J. Abraham. *J.Mol.Graphics* **10**, 212-217 (1992).

12. S.K. Kearsley and G.M. Smith . *Tetrahedron Comput. Methodol.* **3**, 615-633 (1990).

13. P.W. Sprague. *Recent Advances in Chemical Information*, Proceedings of the 1991 Chemical Information Conference (H. Coller, ed.), pp. 107-111, Royal Society of Chemistry, London (1991). (b) Catalyst, distributed by Molecular Simulations, Burlington, MA, USA.

14. (a) V. Golender and E.R. Vorpagel *3D QSAR in Drug Design :Theory, Methods and Applications* (H. Kubinyi, ed.), pp. 137-149, ESCOM, Leiden (1993). (b) Apex-3D, distributed by Biosym Technologies, Inc., CA, USA.

15. P. Furet, A. Sele and N.C. Cohen. *J.Mol.Graphics* **6**, 182-200 (1988).

16. (a) A. Vedani, P. Zbinden and J.P. Snyder. *J. Recept. Res.* **13**, 163-177 (1993). (b) A. Vedani, P. Zbinden, J.P. Snyder and P.A. Greenidge. *J.Amer.Chem.Soc.* **117**, 4987-4994 (1995). (c) Yak, distributed by Swiss Institute for Alternatives to Animal Testing (SIAT), Basel, Switzerland.

17. (a) N.W. Murral and E.K. Davies. *J. Chem. Inf.Comput.Sci.* **30**, 312-316 (1990). (b) ChemDBS, distributed by Chemical Design Ltd., Oxford, England, U.K.

18. UNITY, distributed by Tripos Associates, St. Louis, MO, USA.

19. (a) T.E. Moock, D.R. Henry, A.G. Ozkabak and M. Alamgir. *J. Chem. Inf. Comput. Sci.* **34**, 184-189 (1994). (a) Isis, distributed by Molecular Design Ltd., San Leandro, CA, USA.

20. (a) J.H. Van Drie, D. Weiniger and Y.C. Martin . *J. Comput.-Aided Mol. Des.* **3**, 225-251 (1989). (b) ALADDIN, distributed by Daylight Chemical Information Systems, Irvine, CA, USA.

21. I.D. Kuntz and G.M. Crippen : EMBED, Department of Pharmaceutical Chemistry, University of California, San Francisco, San Francisco CA, USA.

22. D.E. Clark, P. Willet and P.W. Kenny. *J.Mol.Graphics* **10**, 194-204 (1992).

23. N.C. Cohen and V. Tschinke. *Progress in Drug Research*, (E. Jucker, Editor), Birkhäuser Verlag, Basel, Switzerland, volume **45**, in press.

24. P.Y.S. Lam, P.K. Jadhav, C.J. Eyermann, C.N. Hodge, L.T. Bacheler, J.L. Meek, M.J. Otto,M.M. Rayner, Y.N. Wong, C.-H. Chang, P.C. Weber, D.A. Jackson, T.R. Sharpe and S. Erickson-Viitanen . *Science* **263**, 380-384 (1994).

25. (a) S. Wang, D.W. Zaharevitz, R. Sharma, V.E. Marquez, N.E. Lewin, L. Du,P.M. Blumberg and G.W.A. Milne. *J.Med.Chem.* **37**, 4479-4489 (1994). (b) P.A. Wender and C.M. Cribbs. *Advances in Medicinal Chemistry* 1, 1-53 (1992).

26. A. Inoue, N. Tomioka and A. Itai: unpublished.

27. V. Gillet, A.P. Johnson, P. Mata, S. Sike and William P. *J.Comput.-Aided Mol. Des.* **7**, 127-153 (1993).

28. P. Mata, V.J. Gillet, A.P. Johnson, J. Lampreia, G.J. Myatt, S. Sike and A.L. Stebbings . *J.Chem. Inf. Comput.Sci.* **35**, 479-493 (1995).

29. (a) M.B. Eisen, D.C. Wiley, M. Karplus and R.E. Hubbard . *Proteins* **19**, 199-221 (1994). (b) HOOK, distributed by Molecular Simulations, Burlington, MA, USA.

30. (a) C.M.W. Ho and G.R. Marshall. *J.Comput.-Aided Mol. Des.* **7**, 623-647 (1993). (b) SPLICE, distributed by Center for Molecular Design, Washington University, St. Louis, MO, USA.

31. A.R. Leach and R.A. Lewis. *J.Comput. Chem.* **15**, 233-240 (1994).

32. V. Tschinke and N.C. Cohen. *J.Med.Chem.* **36**, 3863-3870 (1993).

33. N.C. Cohen. *Trends in Medicinal Chemistry '88.* (H. van der Goot, G. Domany, L. Pallos and H. Timmermann, Editors). Elsevier Science Publishers, Amsterdam. pp.13-28 (1989).

34. J. Maibaum, V. Rasetti, H. Rüeger, R. Göschke, R. Mah, N.C. Cohen, M. Grütter, F. Cumin and J. Wood. The first paper of a series is presented at this meeting.

35. For a recent review see reference 23.

36. Papers in preparation.

Targeting delivery
of prostanoids and bioactive proteins

Yutaka Mizushima and Rie Igarashi

Institute of Medical Science, St. Marianna University, School of Medicine
2-16-1, Sugao, Miyamae-ku, Kawasaki, Japan 216

Abstract

Prostanoids and some bioactive proteins such as superoxide dismutase (SOD) are autacoids which are produced and act locally in the body as needed. The use of drug delivery systems in the treatment of diseases by these drugs may become necessary to achieve the needed pharmacokinetics. Lipid microspheres (LM) with an average diameter of 0.2μ, resembled liposomes in terms of tissue distribution, are more stable, easily mass produced and sterilized than liposome. In our studies, we have found that LM accumulate in the vascular lesions such as arteriosclerosis and vasculitis. Prostaglandin E1 was incorporated in LM (Lipo PGE1) and tested in animals and human. The clinical effect of Lipo PGE1 was excellent in the treatment of several diseases. These data in animals and human will be reported.

In order to increase the cellular affinity of SOD, a lecithin derivative was covalently bound to recombinant human SOD. The affinity of lecithinized SOD for several cell membranes was markedly increased as compared with unmodified SOD. Tissue distribution and accumulation of prostaglandins and lecithinized SOD in the diseased sites were studied using electron microscope, radioisotope technique and confocal laser scanning method. The pharmacological activity of lecithinized SOD was also increased up to 200 times. Phase I clinical trial of the SOD preparation is ongoing.

1. Introduction

Lipid microspheres containing prostanoids and bioactive proteins combined with lecithin are being studied in our institute for targeting delivery of drugs. We describe lipid microsphere preparation of prostaglandin E1 and lecithinized superoxide dismutase (SOD) in this paper.

2. Lipid microsphere preparation of prostaglandin E1

Prostaglandin E1 (PGE1) is being used for the treatment of various diseases as vasodilators and platelet aggregation inhibitors. PGE1 is synthesized, and act locally to regulate physiological responses and suppress the progression of disease. Therefore, the therapeutic use of PGE1 necessitates the targeting of specific sites in order to maximize local action and minimize systemic adverse effects. The authors have used lipid microspheres (LM) of 0.2 μm in diameter (O/W emulsion) as carriers of PGE1 for passive targeting of vascular lesions; this preparation is called lipo-PGE1. Favorable preclinical and clinical results have been achieved with these preparations. 1, 2, 3, 4)

The LM emulsion is prepared by emulsifying soybean oil containing lipid-soluble drugs using a Manton-Gaulin homogenizer with lecithin added as an emulsifying agent. PGE1 is considered to be located at the interface of LM. 5) This basic LM emulsion preparation has been marketed as a nutritional supplement ('Intralipid') and is given in doses of 300-500 ml. The base used in lipo preparations is 1-2 ml LM emulsion (drug carrier), the stability and safety of which have been established. The lipo preparation method is described elsewhere. 6)

After intravenous injection into spontaneously hypertensive rats (SHR) with vascular lesions, LM was shown to accumulate in the lesions, especially in the subendothelial spaces, 7) by electron microscopy in an affected SHR 5 min after LM injection. Many LM are present in the subendothelial space, whereas none were observed in the vessels of normal rats or unaffected SHR.

Table 1. Summary of phase III clinical results on lipo-PGE1

1. Arterial-duct-dependent congenital heart disease

	No. of patients	Efficacy	Dose	Adverse reaction
Lipo-PGE1	83	94.0%	5 ng/kg/min	30.1%
PGE1-CD	(historical)	50-70%	50-100 ng/kg/min	70-80%

2. Buerger's disease plus arteriosclerosis obliterans (double-blind trial)

	No. of patients	Improvement	Dose	Adverse reaction**
Lipo-PGE1	62	59.7%	10 µg/day/4 weeks i.v.	9.4%
IHN	62	40.3%	1200 mg/day/4 weeks p.o.	6.3%

IHN: Inositol Hexa Nicotinate $p < 0.05$, **N.S.

3. Diabetes-associated peripheral vascular and nervous disorders

	No. of patients	Improvement*	Dose	Adverse reaction**
Lipo-PGE1	84	60.7%	10 µg/day/4 weeks i.v.	12.9%
PGE1-CD	89	38.2%	40 µg/day/4 weeks i.v.	23.8%

*$p < 0.01$, **N.S.

4. Collagen disease-associated peripheral vascular disturbances (double-blind trial)

	No. of patients	Improvement*	Dose	Adverse reaction**
Lipo-PGE1	66	54.5%	10 µg/day/4 weeks	19.1%
Placebo	65	29.2%		16.4%

*$p < 0.001$, **N.S.

5. Vibration disease (double-blind trial)

	No. of patients	Improvement*	Dose	Adverse reaction*
Lipo-PGE1	49	85.7%	10 µg/day/4 weeks i.v.	36.8%
PGE1-CD	48	64.6%	40 µg/day/4 weeks i.v.	58.2%

*$p < 0.02$

After several phase II clinical studies were completed, controlled multi-centered trials of lipo-PGE1 were conducted (Table 1). In all diseases tested, lipo-PGE1 was significantly more beneficial than PGE1 CD (free PGE1) or other reference formulations.4) Coldness in the extremities, numbness, paresthesia, and pain were very responsive to lipo-PGE1. The size of the ulcers which could be assessed objectively was measured in a blind manner. Ulcer lesions associated with collagen diseases regressed significantly within a week of daily treatment with lipo-PGE1. After 4 weeks of treatment, the difference from the control group was significant. One double-blind study demonstrated a better safety profile for lipo-PGE1 than for placebo. Lipo-PGE1 was reported to be 10 to 20 times more effective than PGE1 in the treatment of ductus-dependent congenital heart diseases. It can be used at much lower doses, and accordingly, adverse reactions are reduced. The drug is approved for this indication, and has already been used safely in many patients.4)

Lipo-PGE1 was introduced onto the Japanese market in October, 1988, and since then, it has been widely used in clinical practice. There have been many sporadic reports of its effectiveness in the treatment of fulminant hepatitis 7), neuralgia associated with herpes zoster, multiple spinal canal stenosis, cerebral infarction, myocardial infarction, chronic renal failure, and bed sores as well as for its approved indications.

3. Problems of lipo-PGE1 and development of a better lipo-PGE1

As described above, lipo-PGE1 is very effective in the treatment of many vascular diseases. However, this preparation has two disadvantages. (1) In LM, PGE1 is chemically unstable and largely converted to PGA1 within 6 months of storage at room temperature, and (2) PGE1 is rapidly released from LM on incubation with human blood or serum. 8) Since LM are intended to target PGE1 distribution, PGE1 must be retained in LM until its delivery. To develop a better lipo-PGE1 system and eliminate the disadvantages of lipo-PGE1, prodrugs of PGE1 that are chemically stable, show little leakage from LM, and are readily hydrolyzed to PGE1 at the target site were investigated. 8)

The carbonyl group of PGE1 at C-9 undergoes tautomerization to its enol form under basic and even acidic conditions, which leads to inactivation by hydrolysis to form PGA1. 9) In view of the ready chemical conversion of PGE1, the isomer must be made more chemically stable. This was accomplished by acylation at C-9 and the introduction of a double bond between C-8 and C-9. It was also anticipated that esterification at C-1 and/or acylation at C-9, -11, and -15 would increase the solubility of PGE1 derivatives in LM and thereby decrease their *in vivo* leakage from LM preparations. Several PGE1 prodrugs were therefore synthesized with these objectives.

Then, AS-013, whose chemical structure is shown in Fig. 1, was chosen as a candidate. A small amount of AS-013 is readily hydrolyzed to PGE1 in human serum. The substance is stable in an LM preparation, with 80% recovery of unchanged compound, even after storage as lipo-AS013 for 4 weeks at 40°C. In

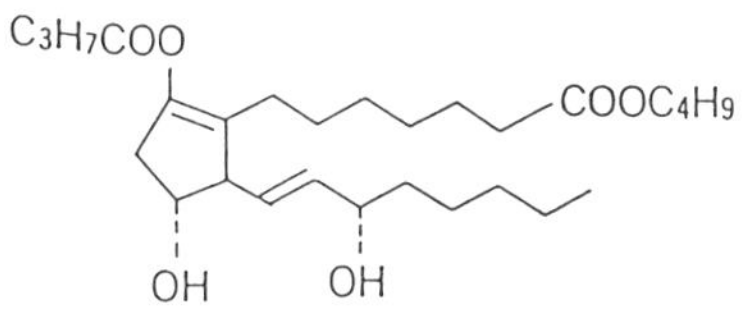

Fig. 1 Chemical structure of AS013

contrast, only 5% of PGE1 was recovered from lipo-PGE1 under the same conditions. Moreover, the new lipo-PGE1 (lipo-AS013) upon incubation in human serum was over eightfold more effective in LM compared with the current lipo-PG. Lipo-AS013 may be expected to exhibit very high potency in various clinical applications.

4. Lecithinized SOD

Active oxygen species such as O_2^-, have been reported to be involved in the pathogenesis of myocardial infarction, tissue injury related to ischemia-reperfusion, Crohn's disease, allergic disease and oncogenesis. Hereditary insufficiency of superoxide dismutase (SOD) activity has also been suggested to be involved in amyotrophic lateral sclerosis. Since SOD, an enzyme scavenging O_2^-, was discovered by McCord & Friedvich in 1969, various attempts have been made to use it for the treatment of oxygen radical-related diseases. However, the clinical application of SOD has been unsuccessful probably because of its low affinity for cells or tissue and its rapid excretion. Therefore, it is necessary to introduce SOD into a drug delivery system. It is important to deliver SOD to the cell membrane since O_2^-, which causes tissue damage, is mainly produced by NADPH oxidase on neutrophil's membranes or by xanthine oxidase in endothelial cells. To develop a more effective drug delivery system for SOD, we synthesized a lecithinized SOD.

In the lecithinized SOD preparation, 4 molecules of a phosphatidylcholine derivative (Fig. 2) are covalently bound to each dimer of recombinant human Cu, Zn SOD, to increase the affinity of SOD for the cell membrane. Lecithinized SOD shows a high cell membrane affinity, prolonged elimination half-life, and remarkable enhancement of pharmacological activity, as described below.

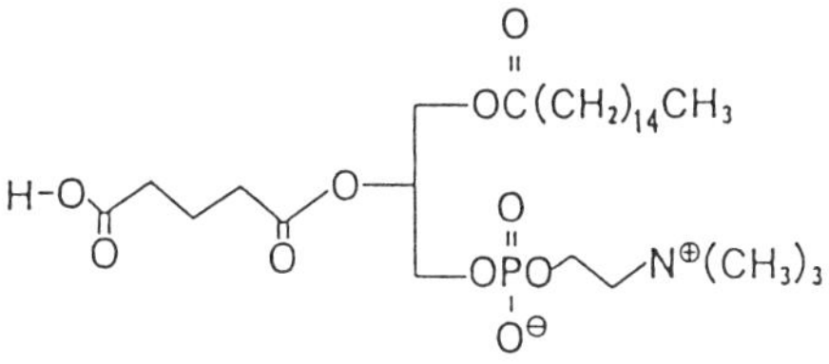

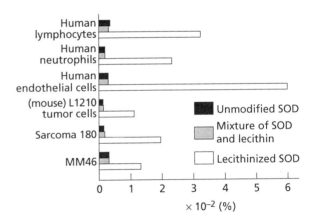

Fig. 2 Chemical structure of a phosphatidylcholine derivative bound to rhSOD

Fig. 3   Affinity for various cells of SOD preparations

The higher cell affinity of lecithinized SOD when compared to unmodified SOD was confirmed by the studies using [3H]-SOD. 3H-lecithinized SOD was added to various cells and, after 3 hours of incubation, the [3H]SOD binding to each cell type was determined. Lecithinized SOD showed higher affinity for every cell type as shown in Fig. 3. 10) Laser confocal imaging human vascular endothelial cells incubated with lecithinized SOD or unmodified SOD for 3 hr and stained with fluorescein isothiocyanate (FITC)-labeled anti-SOD antibodies demonstrated cells incubated with lecithinized SOD show prominent fluorescence, indicating SOD uptake, whereas those incubated with unmodified SOD show only trace fluorescence. 10)

Lecithinized SOD (5 mg/kg) was administered intravenously to rats and plasma SOD levels were determined by enzyme-linked immunosorbent assay. Lecithinized SOD showed slower clearance compared with unmodified SOD (Fig. 4) Protective effect of lecithinized SOD against vascular endothelial cell injury caused by PMA-stimulated neutrophils was studied. Human unblical vascular endothelial cells were cultured in 24-well culture dishes. At confluence, human neutrophils were added plus either lecithinized SOD or unmodified SOD. After incubation, PMA and FITC-labeled albumin were added. The fluorescence leaking out into the lower layer of the culture inserts was measured and used as an index of endothelial cell injury, and the inhibitory effect of each SOD preparation was determined. The result shown in Fig. 5 clearly indicates a remarkable increase in pharmacological effect following lecithinization. 10)

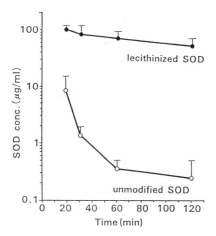

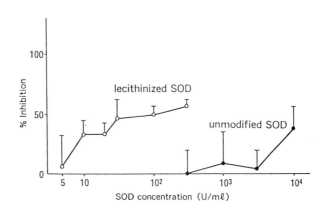

Fig. 4 Plasma elimination curve of SOD preparations in rats

Fig. 5 Protective effect of lecithinized SOD against vascular endothelial cell injury caused by PMA-stimulated neutrophilis

The effect of lecithinized SOD on Forssman antiserum-induced respiratory resistance in guinea pigs was studied. The Forssman antiserum was administered 15 min after an intravenous administration of lecithinized SOD, and the respiratory resistance was measured. It was shown that the protective effect of lecithinized SOD was over 200 times more potent than unmodified SOD, according to the ED50 values measured 30 min after challenge. 11)

5. Conclusion

The use of drug delivery systems in the treatment of diseases with prostanoids and some bioactive proteins may be a highly useful means to achieve the needed pharmacokinetics. As described in this paper, lipid microspheres (LM) with an average diameter of 0.2μ, resembled liposomes in terms of tissue distribution, are more stable, and are easily mass produced. It was found that LM accumulate in the vascular lesions such as those seen in arteriosclerosis and vasculitis. Prostaglandin E1 was incorporated in LM (Lipo-PGE1) and tested in animals and humans. The remarkable clinical effects of Lipo PGE1 were proved in the treatment of several diseases. In order to increase the cellular affinity of SOD, a lecithin derivative was covalently bound to recombinant human SOD. The affinity of lecithinized SOD for several cell membranes was markedly increased as compared with unmodified SOD. The tissue distribution and accumulation of prostaglandins and lecithinized SOD in the diseased sites were studied using electron microscopy, a radioisotope technique, and confocal laser scanning. Furthermore, the pharmacological activity of lecithinized SOD was increased up to 200 times. These results indicate that drug delivery systems are very useful for these drugs.

References

1) Mizushima, Y., Yanagawa, A., Hoshi, K. (1983) Prostaglandin E1 is more effective, when incorporated in lipid microspheres, for treatment of peripheral vascular diseases in man. J. Pharm. Pharmacol., 35, 666-667

2) Mizushima, Y., Igarashi, R., Hoshi. K., Sim, A. K., Cleland, M. E., Hayashi, H., Goto, J. (1987) Marked enhancement in antithrombotic activity of isocarbacycline following its incorporation into lipid microspheres. Prostaglandins, 33, 161-168

3) Mizushima, Y., Shiokawa, Y., Homma, M., Kashiwazaki, S., Ichikawa, Y., Hashimoto, H., Sakuma, A. (1987) A multicenter double blind controlled study of lipo-PGE1, PGE1 incorporated in lipid microspheres, in peripheral vascular disease secondary to connective tissue disorders. J. Rheumatol., 14, 97-101.

4) Mizushima, Y. (1991) Lipo-prostaglandin preparations (review). Prostagl. Leukotr. Essent. Fatty Acids, 42, 1-6.

5) Teagarden, D. L., Anderson, B. D., Petre, W. J. (1989) Dehydration Kinetics of prostaglandin E1 in a lipid emulsion. Pharmaceut. Res., 6, 210-215

6) Mizushima, Y., Hamano, T., Yokoyama, K. (1982) Tissue distribution and anti-inflammatory activity of corticosteroids incorporated in lipid emulsion. Ann. Rheum. Dis. 41, 263-267.

7) Mizoguchi, Y., Kuboi, H., Kawada, N., Kobayashi, K., Morisawa, S. (1990) Effects of lipo prostaglandin E1 in an experimentally-induced acute hepatic failure model. Jpn. J. Inflam., 10, 115-118

8) Igarashi, R., Mizushima, Y., Takenaga, M., Matsumoto, K., Morizawa, Y., Yasuda, A. (1992) A stable PGE1 prodrug for targeting therapy. J. Cont. Rel., 20, 37-46

9) Samuelsson, B. (1965) The prostaglandins. Angew. Chem. Int. Ed. Engl., 4, 410-416.

10) Igarashi, R., Hoshino, J., Ochiai, Y., Morizawa, Y., Mizushima, Y (1994) Lecithinized Superoxide Dismutase Enhances Its Pharmacologic Potency by Increasing Its Cell Membrane Affinity. J. Pharmacol. Exp. Ther., 271, 1672-1677

11) Igarashi, R., Hoshino, M., Takenaga, M., Kawai, S., Morizawa, Y., Yasuda, A., Otani, M., Mizushima, Y (1992) Lecithinization of Superoxide Dismutase Potentiates Its Protective Effect against Forssman Antiserum-Induced Elevation in Guinea Pig Airway Resistance. J. Pharmacol. Exp. Ther., 262, 1214-1219

Cytokine receptors as a new tool for therapy of inflammation

G. Schulz[1], K. Enßle[2], R. Kurrle[2], L. Lauffer[2], F. Seiler[2]

[1] CellGenix AG, Elsässer Str. 2n, 79110 Freiburg, FRG

[2] Behringwerke AG, Postfach 1140, 35001 Marburg, FRG

Abstract: Recombinant, soluble interleukin-4 receptors are able to neutralize circulating interleukin-4. This results in a down-regulation of TH2 cells which are dependent on the availability of this cytokine. As TH2 cells play a crucial role in the development of allergies, this may lead to new therapeutic strategies for treatment of these and other chronic diseases, e.g. autoimmune and parasitic diseases.

Abbreviations

SLE	systemic lupus erythematosus
cGvHD	chronic graft versus host disease
IL-4	interleukin-4
IL-4R	interleukin-4 receptor
sIL-4R	soluble interleukin-4 receptor
rhuIL-4R	recombinant human interleukin-4 receptor
rmuIL-4R	recombinant murine interleukin-4 receptor
rhuIL-4R/Fc	recombinant human interleukin-4 receptor - Fc - fusion protein
rmuIL-4R/Fc	recombinant murine interleukin-4 receptor - Fc - fusion protein

Introduction

Molecular cloning of membrane receptors has opened the way to the engineering of recombinant versions of these proteins with potentially valuable therapeutical properties. Immune and inflammatory reactions are mediated at the cellular level by a large number of specialized membrane proteins, many of which have been characterized in the recent past. Even though these reactions are natural and necessary responses to injury or infection, unregulated activity can be detrimental and is of clinical significance in many human diseases. The receptors for the inflammatory cytokines Interleukin-1 (IL-1) and Tumor Necrosis Factor (TNF) are an example of such proteins. IL-1 and TNF play a central role in mediating toxic shock syndrome. If it would be possible to specifically intercept these cytokines before they bound to their cognate receptors, one could possibly alleviate the symptoms of this dangerous pathologic condition. An elegant way of neutralizing the action of these cytokines is to use their specific receptors as blocking agents. For this purpose, one has to generate soluble forms of these normally membrane-associated proteins which retain binding affinity for their ligands. Many of the cytokine receptors are type I membrane proteins consisting of an aminoterminal extracellular region, a transmembrane region and a carboxyterminal cytoplasmic region involved in signal transduction. The simplest procedure to generate a soluble form is to introduce a translational stop signal in the receptor gene after the coding region for the extracellular domain(s) (Fig. 1). As the recombinant protein thus generated still retains its signal sequence which directs it into the secretory pathway, it will be transported into the extracellular medium like any "normal" secretory protein. Alternatively, the soluble receptor can be fused genetically to an additional protein moiety providing it with new properties. This has been successfully achieved using the receptor portion as the aminoterminal partner in a fusion to regions of immunoglobulin heavy chains (Fig. 2). By virtue of disulphide-bridging between cystein residues in the hinge region, the resulting fusion protein dimerizes. This can result in an increase of avidity to the receptor ligand as well as of serum half life. A soluble receptor then can be used to bind and thereby neutralize its ligand. Most interestingly, the soluble-receptor-concept is probably used by the body itself to regulate the cytokine network. In this review we will focus on the potential therapeutical use of recombinant Interleukin-4 receptors (IL-4R) in allergic and autoimmune diseases.

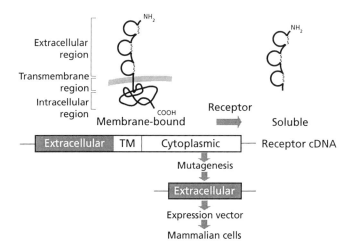

Fig. 1:   Generation of recombinant soluble interleukin receptors.

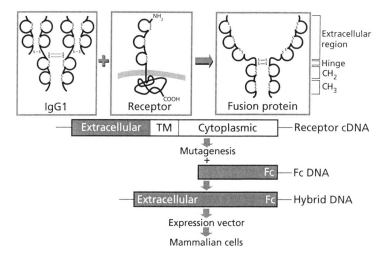

Fig. 2:   Generation of recombinant soluble interleukin receptor - immunoglobulin fusion proteins.

TH2 Cells and IL-4 in Allergy

Dependent from several factors T-helper cells can differentiate from "TH0" cells with a broad pattern of cytokine secretion to cells with a more restricted pattern (1). It has been demonstrated in man that allergen-specific T cells secreting mainly the interleukin-4, interleukin-5 and GM-CSF ("TH2" cells) are preferentially expanded at the site of allergy, but not in peripheral blood (2, 3, 4, 5, 6). Cytokines like IL-4, IL-5 and IL-13 are to a great extend responsible for clinical observations as eosinophilia and the switch of B cells to synthesis of IgE in allergic diseases, infections with e.g. parasites, or some autoimmune diseases like systemic lupus erythematosus (SLE) or chronic graft versus host disease (cGvHD) (7, 8, 9, 10, 11, 12). This dysregulated system stabilizes itself, because there is strong evidence, that IL-4 is a skewing factor towards TH2 cells (13, 14, 15). Therefore, cells of the TH2 phenotype and IL-4 play a crucial role in e.g. type-I allergies and other diseases with a marked TH2 dysregulation (16). Inhibition of IL-4 may therefore normalize the altered TH1/TH2 ratio and prevent the outcome of symptoms in such diseases.

Physiological Role of IL-4 Receptors

For several cytokine receptors and adhesion molecules soluble forms had been identified. As e.g. the receptors for IL-2, IL-6, IL-7, TNF, or the growth hormone receptor, Interleukin-4 receptors (IL-4R) exist in a membrane-bound and a soluble form (17). Whereas binding of IL-4 to a complex of the membrane-bound IL-4R and the gamma chain of the Interleukin-2 receptor leads to signal transduction (18, 19), the function of the soluble IL-4R is still a matter of speculation. Whereas a soluble IL-6 receptor/IL-6 complex leads to signal transduction via binding to gp130, sIL-4R will inhibit signalling by binding to the ligand and therefore compete with the membrane bound receptor. SIL-4R may have a dual physiological role by (a) preventing binding of IL-4 to the membrane protein and (b) acting as a carrier protein for IL-4 (20, 21). Murine spleen cells can be stimulated by e.g. IL-4 itself or mitogens to produce sIL-4R (22). As Interleukin-4 is produced predominantly in local tissue (23, 24, 25, 26), it may thus be speculated, that sIL-4R may act as a natural buffering protein to counteract locally high concentrations of IL-4. It may be therefore highly attractive to support this natural system of IL-4 regulation by use of a recombinant soluble Interleukin-4 receptor (rIL-4R) (Fig. 3).

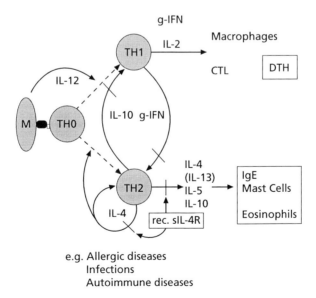

Fig. 3: Down-regulation of IL-4 production and generation of TH2 cells by recombinant soluble IL-4 receptor.

Recombinant IL-4 Receptors and Chimeric Fusion Proteins

For this purpose, using cDNA for the extracellular region of human and murine IL-4R (27, 28), soluble recombinant monomeric IL-4R (rhuIL-4R/rmuIL-4R) and dimeric fusion proteins with Fc-parts of human IgG1 (rhuIL-4R/Fc fusion protein) or IgG2b (rmuIL4R/Fc fusion protein) had been constructed (29). The murine and human forms of rIL-4R or rIL-4R/Fc were expressed in animal cells (BHK and CHO) and are therefore glycosylated. The proteins were purified by affinity columns using monoclonal antibodies (human or murine forms of rIL-4R) or Protein A (human or murine forms of rIL-4R/Fc). Bioactivity, glycosylation and pharmacokinetics in mice were controlled for each lot of the molecules to get a uniform reactivity in in-vitro and in-vivo experiments. Both murine and human rIL-4R and rIL-4R/Fc bind IL-4 in a species specific manner with high affinity and are neutralizing IL-4 in bioassays with nearly the same activity (Fig. 4). Pharmacokinetic studies in Balb/c-mice revealed for both the monomeric form as the dimeric fusion protein a distribution half life of about 30 minutes and an elimination half life of about 7 hours.

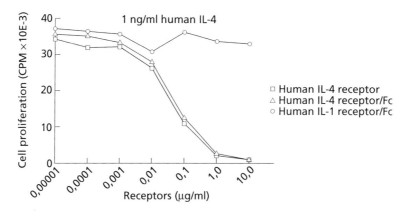

Fig. 4:   Specific inhibition of IL-4 dependent cell proliferation by soluble IL-4 receptor and IL-4 receptor/Fc fusion protein.

Recombinant IL-4 Receptors in Allergic Diseases

Allergen-specific T cell clones are present in lesional skin from atopic patients. In order to mimic such a pathological situation, Müller et al. have injected activated murine clones of the TH2- and TH1-phenotype into skin of mice. Both cell types can induce inflammation, but differ in their kinetics of swelling. Whereas TH2 induced swelling peaks at about 6 hours after injection, TH1 clones induced delayed swelling at 24 to 48 hours. The TH2 induced inflammation, but not swelling induced by TH1 cells, is very efficiently blocked with rmuIL-4R. IL-4 seems to be of importance in the early stages of this inflammation, because swelling is only affected by rmuIL-4R if injected at the same time or before, but not after application of the clones. In addition to IL-4, involvement of TNF produced by TH2 clones has been demonstrated in H-2D mice (30, 31).

In order to get more information about the role of IL-4 in allergy and the regulation of the complex immune response during sensitization a murine model for allergic responses after pulmonale sensitization with allergen has been used. In a first approach, mice were pulmonally sensitized with ragweed or ovalbumin using an ultrasonic nebulizer. These mice developed increased allergen specific IgE and IgG1 levels, positive skin test responses and increased airway responsiveness (AR). Splenic mononuclear cells were restimulated in vitro with the same antigen and cultured in the presence or absence of rmuIL-4Rs (rmuIL-4R or rmuIL-4R/Fc). Both rmuIL-4R and rmuIL-4R/Fc, but not rhuIL-4Rs inhibit in vitro allergen-induced proliferation and production of total IgE and IgG1 and of allergen-specific IgE and IgG1. This

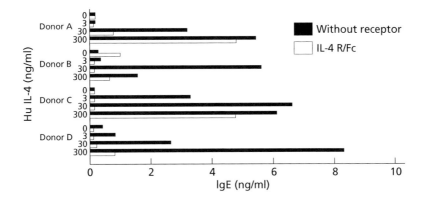

Fig. 5:   Inhibition of IgE-synthesis in vitro (14 days culture) by peripheral blood mononuclear cells from donors with asthma by addition of hu IL-4R Fc.

inhibition seems to be dependent on the early presence of the rmuIL-4Rs in culture (days 0-6). Addition of rmuIL-4Rs in late phases of culture after day 6 can not inhibit the production of antibodies, reflecting the necessity of IL-4 for the class-switch of murine cells to IgE or IgG1 and the independence of synthesis of IgE from IL-4 under these experimental conditions if B cells were already switched (29).

Influence of rhuIL-4R/Fc on the IL-4 induced human IgE-synthesis in vitro had been proved in a first set of experiments with peripheral blood mononuclear cells (PBMNC) from donors with symptoms of asthma and increased levels of serum IgE in their plasma. The induction of IgE in the culture supernatants after 14 days in vitro by addition of 3 to 30 ng/ml IL-4 could be efficiently blocked if huIL-4R/Fc is present (Fig. 5). The effect of the receptor was counter-acted if high concentrations of IL-4 (300 ng/ml) were added. In parallel, the IL-4 induced proliferation of these cells was blocked by rhuIL-4R/Fc, whereas control proteins had no effect (32). These results were consistent with results of Garrone et al. using cells from healthy donors (33).

Allergen specific TH2 clones from skin biopsies of patients with atopic dermatitis secrete in vitro a "TH2" like pattern of cytokines, including IL-4 (2). Such clones can support synthesis of IgE in vitro in a co-cultivation system with autologous, freshly isolated B cells and allergen. Addition of rhuIL-4R/Fc can suppress at least partially this allergen induced synthesis of total IgE (34).

Recombinant IL-4 Receptors in Autoimmune Diseases and Infections

Besides these experiments with an focus on allergic diseases rmuIL-4R has been tested in other animal models with a TH2 dysregulation. In murine models which correspond to the human autoimmune disease systemic lupus erythematosus (SLE) and chronic graft versus host disease (cGvHD), rmuIL-4R has been found to increase survival of mice and to decrease several disease parameters. Whereas serum levels of IgE and IgG1 were reduced by rmuIL-4R treatment, IgG2a antibody levels were not affected in both experimental systems. Therefore, in contrast to immunosuppressive agents like cyclosporin, a specific immunomodulation is observed, which still allows antibody production with changed isotype patterns (35, 36).

This immunomodulating capacity of rIL-4R is supported by results from tests of rmuIL-4R in murine models for infections with C.albicans (37) or Leishmania major (38). In both systems, rmuIL-4R reduces the load of infective particels. In the groups treated with rmuIL-4R this is accompanied in both models by a switch of the cytokine profile of T cells ex vivo. Application of rmuIL-4R results in a change of the phenotype of T-cells from a TH2 profile (production of IL-4, IL-5) to a TH1 profile (expression of IL-2, IFN-γ) of cytokine secretion.

Conclusion

Taken together, the results point out, that rIL-4R can specifically modulate the self-per-petuating system of dysregulated "TH2-helper" cells and may be therefore attractive to treat various diseases with a pathological role of TH2 cells, e.g. allergy. This approach may be more physiological and safe to balance the dysregulation of the immune system in allergic patients than the use of classical immunosuppressive agents or anti-inflammatory drugs like corticosteroids which induce severe adverse events and damage the normal immune system. Clinical trials with rhu-IL-4R will have to prove if positive effects of animal studies can be repeated in patients.

Acknowledgement

We thank Dr. Leander Lauffer for manuscript review and helpful suggestions and Ms Martina Körner and Ms Stephanie Stadler for typing the manuscript.

Literature

1. A. Sher and R.L. Coffmann. Annu. Rev. Immunol. 10, 385 (1992).
2. N. Sager et al. J. Allergy Clin. Immunol. 89/4, 801 (1992).
3. A.B.Kay et al. J. Exp. Med. 173, 775 (1991).
4. D.S. Robinson et al. N. Engl. J. Med. 326, 298 (1992).
5. E. Maggi et al. J. Immunol. 146/4, 1169 (1991).
6. M.L. Kapsenberg et al. Immunol. Today 12/11, 392 (1991).
7. R. Moser et al. J. Immunol. 149/4, 1432 (1992).
8. H.L. Spiegelberg. Springer Semin Immunopathol. 12, 365 (1990).
9. G. Zurawski and J.E. de Vries. Immunol. Today 15/1, 19 (1994).
10. P. Scott and S.H.E. Kaufmann. Immunol. Today 12/10, 346 (1991).
11. D.B. Magilavy et al. Eur. J. Immunol. 22, 2359 (1992).
12. R.D. Allen et al. Eur. J. Immunol. 23, 333 (1993).
13. R.A. Seder and W.E. Paul. Annu. Rev. Immunol. 12, 635 (1994).
14. T. van der Pouw-Kraan et al. Eur. J. Immunol. 23, 1 (1993).
15. C.Y. Wu et al. J. Immunol. 1141 (1994).
16. S. Romagnani. Annu. Rev. Immunol. 12, 227 (1994).
17. W.C. Fanslow et al. Cytokine 2/6, 398 (1990).
18. K. Izuhara et al. Biochem. Biophys. Res. Commun. 190/3, 992 (1993).
19. S.M. Zurawski et al. EMBO Journal 12/7, 2663 (1993).
20. R. Fernandez-Botran and E.S. Vitetta. J. Exp. Med. 174, 673 (1991).
21. T. Sato et al. J. Immunol. 150/7, 2717 (1993).
22. P.M. Chilton and R. Fernandez-Botran. J. Immunol. 151/11, 5907 (1993).
23. H. Secrist et al. J. Immunol. 152/3, 1120 (1994).
24. W.W. Hancock et al. Transplant. Proc. 24, 5 (1992).
25. D.A. Cunningham et al. Transplant. 57/9, 1333 (1994).
26. A. Kupfer et al. Proc. Natl. Acad. Sci. USA 88/3, 775 (1991).
27. R.L. Idzerda et al. J. Exp. Med. 171, 861 (1990).
28. B. Mosley et al. Cell 59, 335 (1989).
29. H. Renz et al. Int. Arch. Allergy Immunol. 316 (1994).
30. K.M. Müller et al. J. of Immunol. 150/12, 5576 (1993).
31. K.M. Müller et al. J. Immunol. 316 (1994).
32. K. Enssle et al. Eur. J Allergy Clin Immunol 12/47 (1992).
33. P. Garrone et al. Eur. J. Immunol. 21, 1365 (1991).
34. K. Enssle et al. Immunobiol. 189, 1 (1993).
35. H.U. Schorlemmer et al. Special Conference Issue. C180 (1994).
36. H.U. Schorlemmer et al. Submitted to Agents & Actions.
37. P. Puccetti et al. J. Infect. Dis. 169, 1325 (1994).
38. A. Gessner et al. Infect. Immun. 62/10, 4112 (1994).

Functional analysis of glutamate receptors by drug-based molecular approaches

Shigetada Nakanishi

Institute for Immunology, Kyoto University Faculty of Medicine, Yoshida, Sakyo-ku, Kyoto 606, Japan

Abstract: Glutamate is a major excitatory neurotransmitter in the central nervous system. Our molecular cloning study has disclosed the existence of diverse members of both NMDA receptors and metabotropic glutamate receptors (mGluRs). Functional analysis of mGluR subtypes has demonstrated that the specific subtypes of the mGluR family play critical roles in information processing in visual and olfactory transmissions as well as in olfactory recognition memory.

Introduction

Glutamate is a major excitatory neurotransmitter in the central nervous system and plays a critical role in neural plasticity, neural development and neurodegeneration (1-3). Glutamate transmission is mediated by a variety of glutamate receptors that are classified into two distinct groups termed ionotropic and metabotropic glutamate receptors (mGluRs). The ionotropic receptors are subdivided into NMDA receptors and AMPA/kainate receptors, both of which contain glutamate-gated, cation-specific ion channels. The mGluRs are coupled to intracellular signal transduction via G proteins and modulate the production of second messengers. This article first summarizes our molecular cloning study that indicates the existence of diverse members of both NMDA receptors and mGluRs and second discusses the specialized and critical roles of mGluRs in segregation/discrimination of information processing in the visual and olfactory systems as well as in olfactory recognition memory.

Molecular diversity of NMDA receptors and mGluRs

We have cloned and characterized the NMDA receptors and mGluRs with the aid of our functional cloning strategy that combines electrophysiology and a *Xenopus laevis* oocyte expression system (4). Both receptors form a family consisting of distinct members of receptor subunits or subtypes.

The NMDA receptors are composed of two distinct types of subunits, NMDAR1 and NMDAR2A-2D (1-3)(*Figure 1*). NMDAR1 is a key subunit that possesses all properties characteristic of the NMDA receptor-channel complex. These include a high Ca^{2+} permeability, agonist and antagonist selectivity, glycine activation, voltage-dependent Mg^{2+} blockade, and Zn^{2+} inhibition (5). The NMDAR2A-2D subunits show no such NMDA receptor activity but potentiate the NMDA receptor activity by forming heteromeric assemblies with NMDAR1 and confer the functional variability by different heteromeric formations (6). The NMDAR1 mRNA is ubiquitously expressed throughout the brain regions, whereas different members of NMDAR2 mRNAs are distinctly distributed in different brain regions (6). The functional and anatomical differences of the NMDAR2 subunits thus contribute to generation of the NMDA receptor heterogeneity in different neuronal cells.

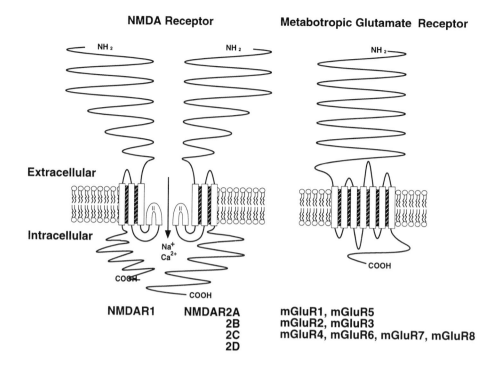

NMDA Receptor   **Metabotropic Glutamate Receptor**

NMDAR1   NMDAR2A   mGluR1, mGluR5
2B   mGluR2, mGluR3
2C   mGluR4, mGluR6, mGluR7, mGluR8
2D

Figure 1.

Model structures of NMDA receptor and mGluR

The NMDA receptor subunits and the mGluR subtypes can be classified into two and three subgroups, respectively, as indicated.

The second transmembrane segments (TM II) of NMDA receptor subunits resemble the channel structure of K^+ channels and are thought to form a folded configuration to permeate cationic ions. A series of mutations with single amino acid substitutions in the TM II of NMDAR1 indicated that the asparagine at the TM II is critical in determining a high Ca^{2+} permeability and the actions of voltage-dependent Mg^{2+} blockade and other channel blockers such as MK801, Zn^{2+} and the antidepressant despramine (7). This study also indicated that mutational analysis is useful for defining the sites of actions of inhibitors and blockers.

The mGluRs forms a large receptor family consisting of at least eight different subtypes termed mGluR1-mGluR8 (1-3, 8)(*Figure 1*). The mGluR family shows no sequence similarity to other G protein-coupled receptors (9), and their large amino-terminal extracellular domain serves as a glutamate binding site (10). The mGluR family thus represents a novel family of G protein-coupled receptors. We characterized the signal transduction mechanisms and agonist selectivity of individual mGluRs subtypes after DNA transfection in CHO cells (11-16). mGluR1 and mGluR5 are coupled to IP_3/Ca^{2+} signal transduction and potently respond to quisqualate. mGluR2 and mGluR3 are linked to the inhibitory cAMP cascade and effectively react with L-(2S, 1'S, 2'S)-(carboxycyclopropyl)glycine (L-CCG-1)(17) and (1S, 3R)-1-aminocyclopentane-1,3-dicarboxylate (ACPD). mGluR4, mGluR6, mGluR7 and mGluR8 are also linked to the inhibitory cAMP cascade but potently react with L-2-amino-4-phosphonobutyrate (L-AP4). The eight subtypes of the mGluR family are thus classified into three subgroups according to their sequence similarities, signal transduction mechanisms and agonist selectivities (2, 8). All but mGluR6 mRNA are widely but distinctly expressed in various brain regions (11-21). These findings strongly indicate that individual mGluR subtypes have their own functions by specializing the signal transduction and expression patterns in various brain regions.

Role of mGluR6 in visual transmission

On the basis of our knowledge of the molecular diversity of the glutamate receptors, it is important to explore physiological roles of individual receptor subtypes or subunits in brain functions and neural development. We have investigated roles of specific mGluR subtypes in visual transmission in the retina and olfactory transmission in the accessory olfactory bulb. In these synaptic transmissions, glutamate plays an essential role in synaptic operation and regulation. Furthermore, these sensory systems are relatively simple in their synaptic organizations, but their synaptic operation is fundamentally similar to those of the central parts of the brain.

Visual signals are transmitted from photoreceptors to bipolar cells. A key processing of visual transmission is detecting visual contrasts at the level of bipolar cells, and this processing occurs by segregating the visual signals into ON-center and OFF-center pathways (22). Interestingly, electrophysiological evidence has suggested that a putative mGluR sensitive to L-AP4 mediates postsynaptic responses of ON-bipolar cells by enhancing cGMP hydrolysis similar to that of signal transduction in photoreceptors (23). We cloned mGluR6 from a retinal cDNA library and indicated that this mGluR subtype selectively responds to L-AP4 after DNA transfection in CHO cells and is also exclusively expressed in the bipolar cell layer (15). Immunohistochemical and immunoelectron-microscopic analysis disclosed that mGluR6 is restrictedly located at the postsynaptic site of rod (ON-type) bipolar cells in the adult rat retina (24). In addition, mGluR6 is localized from cell bodies to the postsynaptic site during retinal development, and this developmental change in mGluR6 distribution accords well with the synaptic formation between photoreceptors and bipolar cells (24). Furthermore, targeted disruption of the mGluR6 gene indicated that the mGluR6 deficiency results in abolishment of ON responses without any change of OFF responses (25). Thus, all these findings demonstrate that mGluR6 is essential in synaptic transmission from photoreceptors to ON-bipolar cells.

On the basis of the above characterization of mGluR6, together with the electrophysiological study of the mGluR-mediated synaptic transmission (23), the synaptic mechanism in visual transmission from photoreceptors to ON-bipolar cells can be summarized as follows (22): In response to light exposure, photoreceptors lower cGMP concentrations through the stimulation of phosphodiesterase via transducin. This decrease in cGMP hyperpolarizes photoreceptors by closing the cGMP-gated ion channel and reduces glutamate release. The mGluR6-G protein-phosphodiesterase system in ON-bipolar cells thus becomes inactive, and a high concentration of cGMP stimulates the cGMP-gated ion channel of ON-bipolar cells and depolarizes these cells. Glutamate release is then augmented in ON-bipolar cells and in turn excites the subsequent ON pathway.

The key role of mGluR6 in ON-bipolar cells is thus to mediate sign inversion between photoreceptors and bipolar cells, so that the presynaptic hyperpolarization (depolarization) is converted to depolarization

VISUAL SENSORY SYSTEM   **OLFACTORY SENSORY SYSTEM**

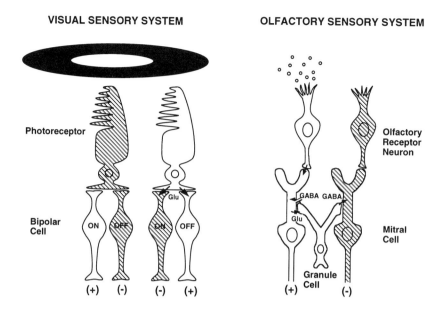

Figure 2.

Visual and olfactory Sensory Transmissions

Models for segregation/discrimination of Visual and Olfactory transmissions at the level of second-order neurons and the modulation of olfactory transmission in combination with the regulatory function of the granule cell are illustrated.

(hyperpolarization) at the postsynaptic bipolar cells (*Figure 2*). In contrast, OFF-bipolar cells preserve the sign of photoreceptors by using the postsynaptic AMPA receptors. Thus, two distinct types of glutamate receptors effectively operate for the ON and OFF segregation in response to the common glutamate transmitter. In addition, the two G protein-coupled receptors, rhodopsin and mGluR6, at the initial steps of visual transmission play an important role in the amplification of visual inputs through a second messenger system. Our investigation has provided the first and compelling evidence that a specific mGluR subtype plays an essential role in synaptic transmission.

Role of mGluR2 in olfactory transmission

In the olfactory system, olfactory receptor neurons make synaptic connections with mitral cells (22, 26)(*Figure 2*). Granule cells are inhibitory interneurons that form typical dendrodendritic synapses with mitral cells, and these synapses undergo reciprocal regulation (22, 26); the mitral cell excites the granule cell via glutamate and in turn receives inhibition of γ-aminobutyrate (GABA) from the granule cell. The granule cell also forms divergent synaptic contacts with a large number of neighboring mitral cells and is thus thought to contribute to both recurrent and lateral inhibition via inhibitory GABA transmission.

Our functional analysis of mGluR2 in the reciprocal synapses of the accessory olfactory bulb (AOB) has revealed a novel mechanism of the olfactory transmission modulation. In situ hybridization and immunoelectron-microscopic analysis indicated that mGluR2 is highly expressed and selectively localized at the presynaptic site of dendrodendritic connections of granule cells in the AOB (27). We also identified a glutamate analogue, (2S, 1'R, 2'R, 3'R)-2-(2,3-dicarboxycyclopropyl)glycine (DCG-IV), as a potent and selective agonist for mGluR2 (27). We thus investigated the role of mGluR2 in olfactory transmission in the AOB by examining the effect of the DCG-IV-induced activation of mGluR2 on GABA transmission from the granule cell to the mitral cell. This analysis indicated that the mGluR2 activation reduces inhibitory GABA transmission from granule cells to mitral cells. These observations led to the conclusion that glutamate released from an excited mitral cell activates mGluR2 at the presynaptic site of granule cells and relieves the mitral cell from the GABA-mediated inhibition (27). Because the activation of mGluR2 is thought to be confined to the synapses of excited mitral cells, the modulation by mGluR2 would relieve the recurrent inhibition of excited mitral cells but keep lateral inhibition to unexcited neighboring mitral cells.

The mechanism discussed above would evidently enhance the signal-to-noise ratio between excited and unexcited mitral cells and would contribute to discrimination of olfactory transmission. The AOB is also well known a crucial site for olfactory memory formation. It is thus plausible that the persistent activation of mGluR2 by glutamate results in prolonged excitation of the olfactory sensory system and may enhance synaptic efficacy responsible for olfactory memory induction. We investigated a potential role of mGluR2 in olfactory memory formation by animal behavioral analysis (28).

Female mice form an olfactory memory at mating, and this memory maintains pregnancy during exposure to familiar pheromones of the stud male but evokes pregnancy block after exposure to unfamiliar pheromones of a different male strain (29). This olfactory block of pregnancy, known as the Bruce effect, is caused by enhancement of norepinephrine in the AOB through coital stimulation (29). Norepinephrine enhanced in the AOB after mating persistently excites mitral cells by reducing GABA transmission from granule cells, thus forming olfactory memory specific to pheromones exposed during mating. The activation of mGluR2 similarly reduces inhibitory GABA transmission and may thus create olfactory memory formation. To examine a possible role of mGluR2 in olfactory memory formation, we conducted the following animal behavioral analysis (28). mGluR2 agonists were infused into the AOB of females during exposure to male pheromones without mating. At the next oesterus, the females were mated with a different strain and re-exposed to test pheromones of the original strain. This protocol allows us to evaluate drug-induced memory formation by measuring the protection of pregnancy block evoked by the test pheromonal exposure. The results of this analysis explicitly demonstrated that the activation of mGluR2 induces a specific olfactory memory formation (28). Thus, our study demonstrates that a specific mGluR subtype plays an important role in synaptic modulation and neural plasticity responsible for olfactory recognition memory formation.

Concluding remarks

Our molecular studies demonstrate that both NMDA receptors and mGluRs consist of diverse members of the receptor subunits and subtypes. In combination with drug-based approaches, we also indicate that some specific mGluR subtypes have specialized and crucial roles in the segregation or discrimination of sensory stimuli in the visual and olfactory systems. The peripheral sensory organs receive graded external signals, and these graded signals must be converted into action potentials in an early step of signal transmission (*Figure 2*). Thus, the fundamental function of the second-order neurons (bipolar cell and mitral cell) is to discriminate the signal-to-noise ratio of external stimuli prior to transmission of these signals to higher centers of the brain. Our studies demonstrate not only that the glutamate receptors play distinct and important roles in segregating or discriminating external sensory signals in the visual and olfactory systems but also that different glutamate receptors are effectively utilized in information processing to transmit external signals to the central parts of the brain. Furthermore, the modulation by the regulatory function of the specified mGluR subtype is important in evoking neural plasticity for an olfactory recognition memory. Our understanding of roles of different glutamate receptors is still fragmentary, and further studies with drug-based molecular approaches to the diverse members of glutamate receptors will thus be fruitful for understanding the mechanisms of brain functions.

REFERENCES

1. S. Nakanishi. Science **258**, 597-603 (1992).
2. S. Nakanishi. Neuron **13**, 1031-1037 (1994).
3. S. Nakanishi, M. Masu. Annu. Rev. Biophys. Biomol. Struct. **23**, 319-348 (1994).
4. Y. Masu, K. Nakayama, H. Tamaki, Y. Harada, M. Kuno and S. Nakanishi. Nature **329**, 836-838 (1987).
5. K. Moriyoshi, M. Masu, T. Ishii, R. Shigemoto, N. Mizuno and S. Nakanishi. Nature **354**, 31-37 (1991).
6. T. Ishii, K. Moriyoshi, H. Sugihara, K. Sakurada, H. Kadotani, Y. Yokoi, C. Akazawa, R. Shigemoto, N. Mizuno, M. Masu and S. Nakanishi. J. Biol. Chem. **268**, 2836-2843 (1993).
7. K. Sakurada, M. Masu and S. Nakanishi. J. Biol. Chem. **268**, 410-415 (1993).
8. R. M. Duvoisin, C. Zhang and K. Ramonell. J. Neurosci. **15**, 3075-3083 (1995).
9. M. Masu, Y. Tanabe, K. Tsuchida, R. Shigemoto and S. Nakanishi. Nature **349**, 760-765 (1991).
10. K. Takahashi, K. Tsuchida, Y. Tanabe, M. Masu and S. Nakanishi. J. Biol. Chem. **268**, 19341-19345 (1993).
11. Y. Tanabe, M. Masu, T. Ishii, R. Shigemoto and S. Nakanishi. Neuron **8**, 169-179 (1992).
12. I. Aramori and S. Nakanishi. Neuron **8**, 757-765 (1992).
13. T. Abe, H. Sugihara, H. Nawa, R. Shigemoto, N. Mizuno and S. Nakanishi. J. Biol. Chem. **267**, 11361-11368 (1992).
14. Y. Tanabe, A. Nomura, M. Masu, R. Shigemoto, N. Mizuno and S. Nakanishi. J. Neurosci. **13**, 1372-1378 (1993).
15. Y. Nakajima, H. Iwakabe, C. Akazawa, H. Nawa, R. Shigemoto, N. Mizuno and S. Nakanishi. J. Biol. Chem. **268**, 11868-11873 (1993).
16. N. Okamoto, S. Hori, C. Akazawa, Y. Hayashi, R. Shigemoto, N. Mizuno and S. Nakanishi. J. Biol. Chem. **269**, 1231-1236 (1994).
17. Y. Hayashi, Y. Tanabe, I. Aramori, M. Masu, K. Shimamoto, Y. Ohfune and S. Nakanishi. Brit. J. Pharmacol. **107**, 539-543 (1992).
18. R. Shigemoto, S. Nakanishi and N. Mizuno. J. Comp. Neurol. **322**, 121-135 (1992).
19. H. Ohishi, R. Shigemoto, S. Nakanishi and N. Mizuno. J. Comp. Neurol. **335**, 252-266 (1993).
20. R. Shigemoto, T. Abe, S. Nomura, S. Nakanishi and T. Hirano. Neuron **12**, 1245-1255 (1994).
21. H. Ohishi, R. Ogawa-Meguro, R. Shigemoto, T. Kaneko, S. Nakanishi and N. Mizuno. Neuron **13**, 55-66 (1994).
22. S. Nakanishi. Trends Neurosci. **18**, 359-364 (1995).
23. R. Shiells. Current Biology **4**, 917-918 (1994).
24. A. Nomura, R. Shigemoto, Y. Nakamura, N. Okamoto, N. Mizuno and S. Nakanishi. Cell **77**, 361-369 (1994).
25. M. Masu, H. Iwakabe, Y. Tagawa, T. Miyoshi, M. Yamashita, Y. Fukuda, H. Sasaki, K. Hiroi, Y. Nakamura, R.Shigemoto, M. Takada, K. Nakamura, K. Nakao, M. Katsuki, M and S. Nakanishi. Cell **80**, 757-765 (1995).
26. H. Kaba and S. Nakanishi. Rev. Neurosci. **6**, 125-141 (1995).
27. Y. Hayashi, A. Momiyama, T. Takahashi, H. Ohishi, R. Ogawa-Meguro, R. Shigemoto, N. Mizuno and S. Nakanishi. Nature **366**, 687-690 (1993).
28. H. Kaba, Y. Hayashi, T. Higuchi and S. Nakanishi. Science **265**, 262-264 (1994).
29. P. Brennan, H. Kaba and E. B. Keverne. Science **250**, 1223-1226 (1990).

Biologically active microbial metabolites

Satoshi Ōmura

The Kitasato Institute, Minato-ku, Tokyo 108, Japan

Abstract: We have searched for biologically active substances of microbial origin for over 25 years. Consequently, we discovered several useful compounds for human and animals and in agriculture. Furthermore, we found a number of compounds, including enzyme inhibitors important for studies in the fields of biochemistry, molecular biology, pharmacology and so on. We have also performed chemical and biochemical studies on these compounds and basic research such as taxonomy and gene manipulation of the producing organisms. This paper describes about my expectation to create medicines of microbial metabolites by introducing outline of our results on four biologically active microbial metabolites that we discovered.

1. Motilide (Macrolide with motilin mimic activity)[1]

It is known that some of antibiotics have so-called pharmacological activities, including hypotensive, anesthesia competeting and cardiotonic activities. For instance, cyclosporine A and bledenin are clinically used owing to their side activity as potent immunosuppressants, though they were firstly discovered as antifungal antibiotics. And they have been used as possessing intrinsic antifungal activity. We paid attention to the gastrointestinal motor stimulating (GMS) activity of erythromycin, a side effect of the antibiotic discovered by Itoh *et al.* By collaboration with them, we synthesized over 250 erythromycin derivatives and estimated their GMS activity. GMS activity was measured by means of chronically implanted force transducers in consious dogs. Test samples were given intravenously as single bolus injection during the initial period of quiescence of the interdigestive state. Gastric contractile activity in the gastric antrum was measured with integrators connected to tranducer amplifiers. We selected splendid derivatives, EM-523 and EM-574[2] for further evaluation *in vivo*. These two compounds lacked antibacterial activity, but increased in GMS activity (Table 1) and could advance to the phase II trials on a new drug for the treatment of gastrointestinal disorder. EM-574 displaced specifically ^{125}I-motilin bound to smooth muscle tissue of human gastric antrum with a Kd value of 7.8×10^{-9} M compared with 4.5×10^{-9} M for cold motilin (Fig.1)[3]. Therefore, a series of macrolide with motilin like activity were named "motilide". This example gives new expectation that some of known antibiotics may be renewed as useful medicines by re-evaluating their secondary effects. Another fundamentally different activity of erythromycin A, efficacy against diffuse panbronchiolitis (DPB) found by S. Kudoh, *et al.*, is postulated to depend on its immunosuppressive activity.

Table 1. Antibacterial activity and gastrointestinal motor-stimulating (GMS) activity of motilides

Compound	Antibacterial activity (MIC, μg/ml) [1]	GMS activity [2]
EM A	0.2	1
EM 201	50	10
EM 523	>100	18
EM 574	>100	248
EM 536	>100	2890

1) *Staphylococcus aureus* ATCC 6538P, agar dilution method.
2) GMS activity was estimated by 2 x 2 points parallel line assay.

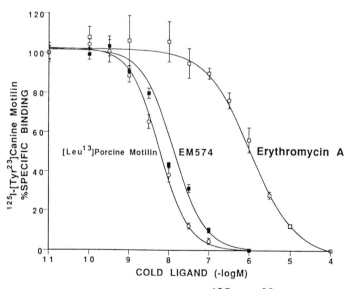

Erythromycin A (EM A)

EM 201 R=CH₃
EM 523 R=CH₂CH₃
EM 574 R=CH(CH₃)₂

EM 536

Fig. 1. Inhibition of specific binding of ^{125}I-[Tyr^{23}]canine motilin by motilin and motilides in human antral smooth muscle tissue

Increasing concentrations of [Leu^{13}]porcine motilin, EM574, and EM-A were added to the incubation medium at 30°C for 90 min. Mean ± S.E. of triplicate determinations.

2. Lactacystin (Inducer of neuritogenesis)

During the course of our screening for microbial metabolites which induce differentiation of the mouse neuroblastoma cell line Neuro 2a, a novel compound named lactacystin was isolated from the cultured broth of *Streptomyces* sp. OM-6519.[4] A unique γ-lactam structure for lactacystin was determined by NMR spectroscopy and X-ray crystallographic analysis.[5] We also reported the biosynthetic origin and stereochemical assignment of methyl group, C-11 and C-12 of lactacystin by feeding experiments with ^{13}C-enriched compounds.[6] Corey and Reichard have recently reported the first total synthesis of lactacystin.[7] We have reported concise alternative approach (Scheme 1).[8] The key steps in the elaboration of lactam moiety include stereoselective hydroxymethylation of oxazoline **3** and an asymmetric allylboration of **5**, which introduces the hydroxyl and methyl substituents at C(6) and C(7), respectively (Scheme 1).

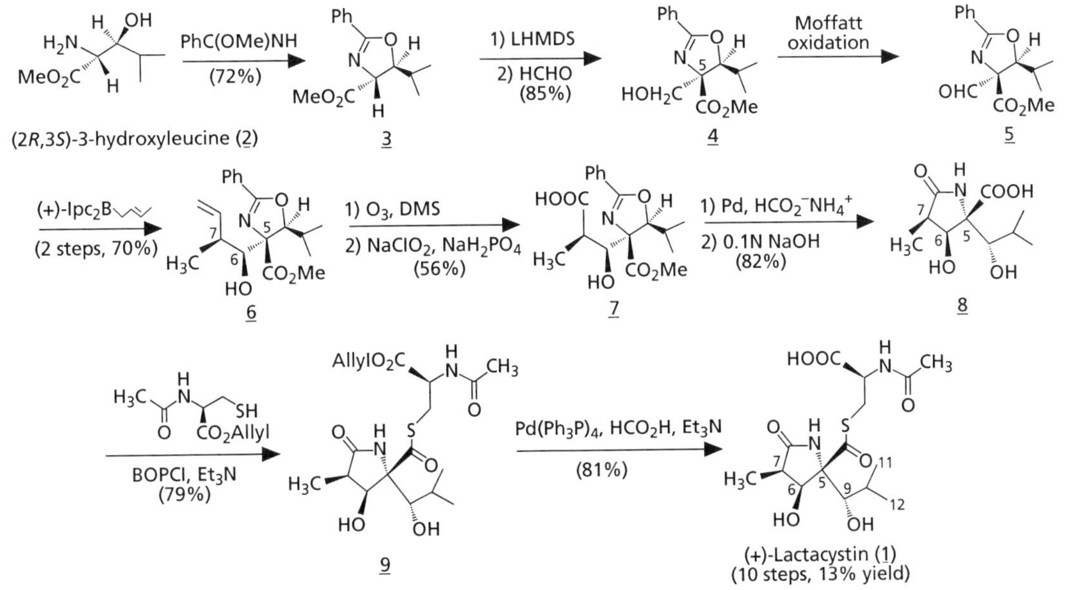

Scheme 1

Several derivatives have been synthesized from *clasto*-lactacystin β-lactone. Minimum effective dose on neuritogenesis in Neuro 2a cells of decarboxy-lactacystin is 15 times lower than that of lactacystin (Table 2).[9] Neuritogenesis of Neuro 2a cells induced by lactacystin was confirmed by electron-microscopy, and immunofluorescence staining of 200 KD neurofilament, as well as by the induction of acetylcholinesterase. When neuritogenesis is induced by lactacystin, proliferation of Neuro 2a cells is inhibited.[4] So it was expected that lactacystin inhibits cell cycle of Neuro 2a cells. Fenteany *et al.*, reported that lactacystin inhibits progression of M phase of Neuro 2a cells and MG-63 osteosarcoma cells beyond the G1 phase of the cell cycle.[10] We found that lactacystin arrests the cell cycle at both G0/G1 and G2 phases in Neuro 2a cells (Fig. 2)[11].   We examined the effect of lactacystin on enzymes, cdc 2 kinase, casein kinase II, mitogen-activated protein kinase and histone deacetylase, which are known as target enzymes of cell cycle inhibitors arresting at both G1 and G2 phases.   Lactacystin did not affect these enzymes even at 100 μM, but weakly inhibited histone deacetylase.[11]

Table 2   The minimum effective dose on neuritogenesis and cytotoxicity of lactacystin derivatives

Compound	A minimum effective dose μM	B cytotoxicity μM	B / A	Compound	A minimum effective dose μM	B cytotoxicity μM	B / A
(structure)	not effective	—	—	(structure, COOEt)	0.20	3.13	1.6
(structure) *clasto*-lactacystin β-lactone	1.56	12.5	8	(structure) decarboxyllactacystin	0.10	12.5	125
(structure, AllylO₂C)	0.20	6.25	31	(structure) Lactacystin	1.56	12.5	8

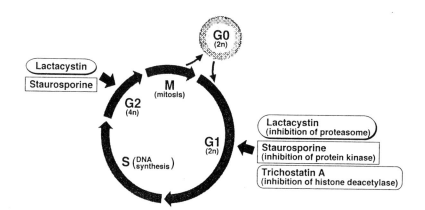

Fig. 2.   Cell cycle and its inhibitors with neuritogenesis activity in Neuro 2a cells

Fenteany *et al.* found ^{3}H-lactacystin to bind specifically to bovine 20S proteasome subunit X[12). They reported that lactacystin appears to modify covalently the highly conserved amino-terminal threonine of the mammalian 20S proteasome subunit X which has a catalytic role of 20S proteasome. Lactacystin is the most selective proteasome inhibitor known at present and may become a useful reagent for examining the role of the proteasome in the proteolytic processes such as antigen processing and regulation of cell cycle in the cell.[13)

3. Pyripyropenes (ACAT inhibitor)

Hypercholesterolemia is a serious risk factor of atherosclerosis developing into coronary heart disease or myocardial infarction. Several therapeutic means have been proposed to lower cholesterol levels in plasma. One successful example is to inhibit *de novo* cholesterol biosynthesis. Lovastatin and its analogs (pravastatin and simvastatin), potent inhibitors of HMG-CoA reductase which is the rate-limiting enzyme in the biosynthetic pathway, are now clinically used. Another possible target is acyl-CoA:cholesterol acyltransferase (ACAT). ACAT utilizes long-chain fatty acyl-CoA and cholesterol as substrates to catalize the intracellular formation of cholesteryl ester. ACAT plays important roles in cholesterol metabolism in human body; 1) Dietary cholesterol is absorbed at intestines as cholesteryl ester by the reaction of ACAT, 2) cholesterol in liver supplied by both factors of taking diet and *de novo* synthesis is acylated by ACAT to be secreted as VLDL, and 3) macrophages and smooth muscle cells in arterial wall accumulate cholesterol as cholesteryl ester in oil droplets by the reaction of ACAT, leading to foam cells in atherosclerotic lesions. Therefore, ACAT inhibitors are expected not only to lower plasma cholesterol levels but also to have a direct effect on the arterial wall.

A large number of synthetic ACAT inhibitors have been reported, but nothing used clinically. Microbial ACAT inhibitors had been rarely known at that time when we started screening for ACAT inhibitors in 1988. Culture broths of over 10,000 microorganisms, including actinomycetes, fungi and bacteria were screened with an *in vitro* assay using rat liver microsomes as the enzyme source. Consequently, we have discovered five kinds of new ACAT inhibitors, all of which were produced by fungal strains. Among them, glisoprenins,[14) pyripyropenes[15) and terpendoles[16) have isoprene units in their structures. In this paper, structure, biological activity and total synthesis of pyripyropenes are described.

Highly potent ACAT inhibitors were found to be produced by *Aspergillus fumigatus* FO-1289. A series of active compounds, consisting of pyridine, α-pyrone and sesquiterpene moieties in common, were isolated and named pyripyropenes A~R. Structures[17-19) and IC$_{50}$ values of ACAT inhibition of pyripyropenes are shown in Table 3. Pyripyropenes A and C are the most potent inhibitors of microbial origin. We have succeeded in the first total synthesis of pyripyropene A[20). The effective and concise approach was designed to afford easy access to pyripyropene and a variety of its analogs (Scheme 2).

Table 3. Structure and ACAT inhibitory activities of pyripyropenes

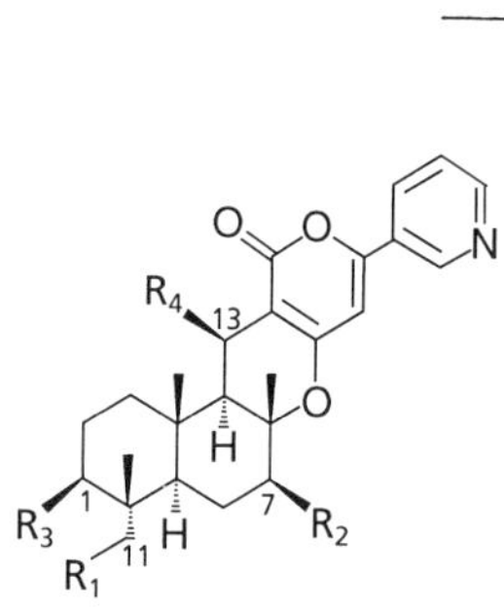

Pyripyropene	R_1	R_2	R_3	R_4	ACAT inhibition $IC_{50}(\mu M)$
A	-OCOCH$_3$	-OCOCH$_3$	-OCOCH$_3$	-OH	0.16
B	-OCOCH$_2$CH$_3$	-OCOCH$_3$	-OCOCH$_3$	-OH	0.32
C	-OCOCH$_3$	-OCOCH$_2$CH$_3$	-OCOCH$_3$	-OH	0.15
D	-OCOCH$_3$	-OCOCH$_3$	-OCOCH$_2$CH$_3$	-OH	0.74
E*	-H	-H	-OCOCH$_3$	-H	399
F	-H	-H	-OCOCH$_2$CH$_3$	-H	559
G	-H	-H	-OCOCH$_3$	-OH	221
H	-H	-H	-OCOCH$_2$CH$_3$	-OH	270
I	-OCOCH$_2$CH$_3$	-OCOCH$_2$CH$_3$	-OCOCH$_2$CH$_3$	-OH	2.45
J	-OCOCH$_3$	-OCOCH$_2$CH$_3$	-OCOCH$_2$CH$_3$	-OH	0.85
K	-OCOCH$_2$CH$_3$	-OCOCH$_3$	-OCOCH$_2$CH$_3$	-OH	2.65
L	-OCOCH$_2$CH$_3$	-OCOCH$_2$CH$_3$	-OCOCH$_3$	-OH	0.27
M	-OCOCH$_3$	-OCOCH$_2$CH$_3$	-OCOCH$_3$	-H	3.80
N	-OCOCH$_2$CH$_3$	-H	-OCOCH$_2$CH$_3$	-OH	48.0
O	-OCOCH$_3$	-H	-OCOCH$_3$	-H	11.0
P	-OCOCH$_2$CH$_3$	-H	-OCOCH$_3$	-H	40.0
Q	-OCOCH$_2$CH$_3$	-H	-OCOCH$_3$	-OH	40.0
R	-OCOCH$_3$	-H	-OCOCH$_2$CH$_3$	-H	78.0
CL-283,546					1.30

*Identical with GERI-BP001 (Jeong T.S. *et al.* Tetrahedron Lett. <u>35</u> 3569 (1994))

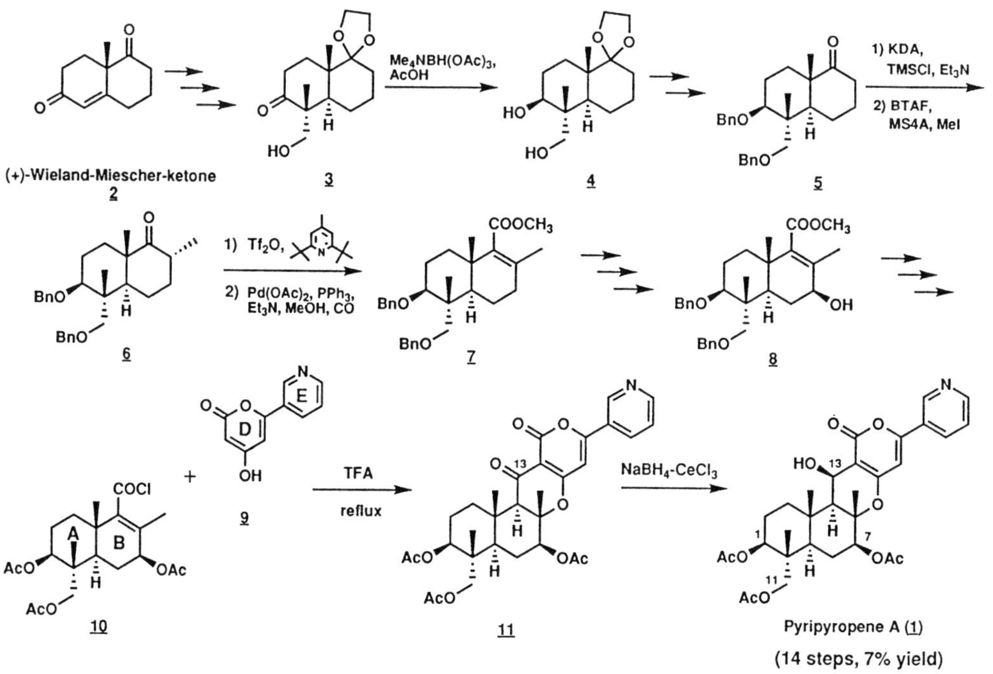

Scheme 2

The key step is the coupling reaction between α-pyrone-pyridine moiety (DE ring) (**9**) and the acid chloride of sesquiterpene moiety (AB ring) (**10**) in the presence of Lewis acid to construct ketone **11**. The sesquiterpene moiety was synthesized *via* stereoselective and reductive formylation, palladium associated carbonylation, and allylic oxidation started from (+)-Wieland-Miescher ketone

(2). The evaluation of pyripyropene and its derivatives *in vivo* is in progress using an animal model.

4. Staurosporine (Microbial alkaloid with inhibitory activity against protein kinases) [21]

The search for biologically active substances from microorganisms is based on the diversity in structure and activity of microbial metabolites and on the expectation that various substance(s) may exist in them. A conventional approach to explore novel and useful metabolites is activity-guided biological screening. An alternative approach is chemical screening. It is at first to find new compounds by looking only at the chemical nature, next to seek for its biological activity by evaluating for various activities as much as possible, and finally to find practical applications.

We started the screening based on such a concept in 1970 and have discovered several compounds, including staurosporine[22] pyrindicin,[23] herquline[24] etc., whose biological activities were later clarified. Staurosporine possesses varied activities, including increase of blood current and hypotensive effect on the basis of its relaxation activity for blood vessel, inhibition of platelet aggregation and cell proliferation. We studied on the relationship of the structure and inhibitory activity against platelet aggregation and could obtain some interesting compounds such as AGL103 and AZ32101. These staurosporine derivatives have specific inhibitory activity against platelet aggregation (Table 4).

On the other hand, since staurosporine has been reported to inhibit protein kinase C in 1986[25], and later known to be the most potent non-specific inhibitor of protein kinases (Table 5), it has attracted attention as one of the most important reagents for studies on cellular signal transduction.

Table 4. Structures and inhibitory activities of staurosporine and its derivatives

	Staurosporine	AGL-103	AZ-32101	L-103-(OH)$_2$
Platelet aggregation (Guinea pig, PRP) IC$_{50}$, μM (A)	7.5	10	6.8	0.4
Vasoconstriction (Rabbit aorta) IC$_{50}$, μM (B)	0.025	35	30	3
A / B	300	0.28	0.22	0.13

Staurosporine   AGL-103   AZ-32101   L-103-(OH)$_2$

Table 5. Inhibitory effects on protein kinases

Kinase	K_i value (nM)
PKC	2.7
PKA	22.0
PKG	8.5
PTK	6.4
MLCK	5.0
cdc2kinase	4.0

Acknowledgements: I am grateful to Drs. H. Tomoda, H. Takeshima, Y. Iwai of The Kitasato Institute, Prof. H. Tanaka and Dr. T. Sunazuka of Kitasato University, Prof. Amos B. Smith, III of University of Pennsylvania, Dr. Y. Sasaki of Asahi Chemical Industry, Co., Ltd., Prof. Z. Itoh of Gunma University, and Dr. N. Inatomi of Takeda Chemical Industries for their long, extensive and friendly collaboration in the field of microbial metabolites. I also thank Dr. H. Takeshima for his critical reading of the manuscript.

References:
1) S. Ōmura et al., In "Motilin", ed. by Z. Itoh, p. 245, Academic Press (1990). 2) K. Tsuzuki et al., Chem. Pharm. Bull., 37, 2687 (1989). 3) M. Satoh et al., J. Pharmacol. Exp. Ther., 271, 574 (1994). 4) S. Ōmura et al., J. Antibiotics, 44, 113 (1991). 5) S. Ōmura et al., J. Antibiotics, 44, 117 (1991). 6) A. Nakagawa et al., Tetrahedron Lett., 35, 3009 (1994). 7) E. J. Corey et al., J. Am. Chem. Soc., 114, 10677 (1992). 8) T. Sunazuka et al., J. Am. Chem. Soc., 115, 5302 (1993). 9) T. Nagamitsu et al., J. Antibiot., 48, 747 (1995). 10) G. Fenteany et al., Proc. Natl. Acad. Sci., USA, 91, 3358 (1994). 11) M. Katagiri et al., J. Antibiotics, 48, 344 (1995). 12) G. Fenteany et al., Science, 268, 726 (1995). 13) A. Ichihara et al., Mol. Biol. Reports, 21, 49 (1995). 14) H. Tomoda et al., J. Antibiotics, 45, 1202 (1992). 15) S. Ōmura et al., J. Antibiotics, 46, 1168 (1993). 16) X.-H. Huang et al., J. Antibiotics, 48, 1 (1995). 17) H. Tomoda et al., J. Antibiotics, 47, 148 (1994). 18) Y. K. Kim et al, J. Antibiotics, 46, 154 (1994). 19) H. Tomoda et al., J. Am. Chem. Soc., 116, 12097 (1994). 20) T. Nagamitsu et al., in preparation. 21) S. Ōmura et al., J. Antibiot., 48, 535 (1995). 22) S. Ōmura et al., J. Antibiot., 30, 275 (1977). 23) S. Ōmura et al., Agric. Biol. Chem., 38, 899 (1974). 24) S. Ōmura et al., J. Antibiot., 32, 786 (1979). 25) T. Tamaoki et al., Biochem Biophys. Res. Commun., 135, 397 (1986).

Total synthesis of natural products and designed molecules: brevetoxin B

K. C. Nicolaou* and E. A. Theodorakis

Department of Chemistry, The Scripps Research Institute, 10666 North Torrey Pines Road, La Jolla, California 92037, and Department of Chemistry and Biochemistry, University of California, San Diego, 9500 Gilman Drive, La Jolla, California 92093

Abstract: Natural products with novel molecular structures and important biological properties provide exciting challenges and opportunties to synthetic chemists, biologists, and clinicians. The pursuit of such target molecules by total synthesis is often accompanied by the discovery and development of new synthetic technology and strategies, molecular design studies, and biological investigations with the natural and designed molecules. In this article this philosophy of research is exemplified with brevetoxin B (**1**) as a target molecule.

Introduction

Nature has been amply generous to chemists, biologists and clinicians by providing a wealth of compounds with unusual molecular structures that find utility as powerful biological tools and life-saving medicines (1). Within the realm of chemistry, such unusually complex and challenging structures help shape the art and science of total synthesis (2,3), by offering opportunities for the discovery and development of new synthetic methods and strategies (4). In addition, naturally occurring substances with important biological actions provide opportunities for molecular design by serving as prototype structures for fine-tuning in order to arrive at desirable biological tools and clinical agents (5).

In the past few years, research in our laboratories has been focused on the total synthesis of natural products with special emphasis on the following aspects: (a) the selection and pursuit of target molecules with unique molecular architectures offering opportunities of discovery not only in chemistry but also in biology and medicine; (b) the systematic development of new synthetic methods and new strategies to face the challenges at hand; (c) the molecular design and chemical synthesis of novel biological mimics of the natural products; and (d) the biological investigation of designed molecules in search for useful applications to biology and medicine (5).

Some of the natural products that have served admirably as synthetic targets and provided a number of opportunities for discovery and invention in our laboratories are shown in Fig. 1 (4,5). They include compounds from the families of the eicosanoid hormones (6,7), the ionophore (8) and macrolide antibiotics (9), the endiandric acids (10), sialyl Lewis^x (11) and NodRm-IV factor oligosaccharides (12), rapamycin (13), the taxoids (14), the balanoids (15), the zaragozic acids (16), the enediynes (17), and the brevetoxins (18). The molecules shown in Fig. 2 were, in general, inspired by the natural products shown in Fig. 1. The chemical synthesis of these designed molecules allowed some important contributions to be made in biology and medicine.

In this article we exemplify these principles using brevetoxin B (**1**) as the naturally occurring target molecule.

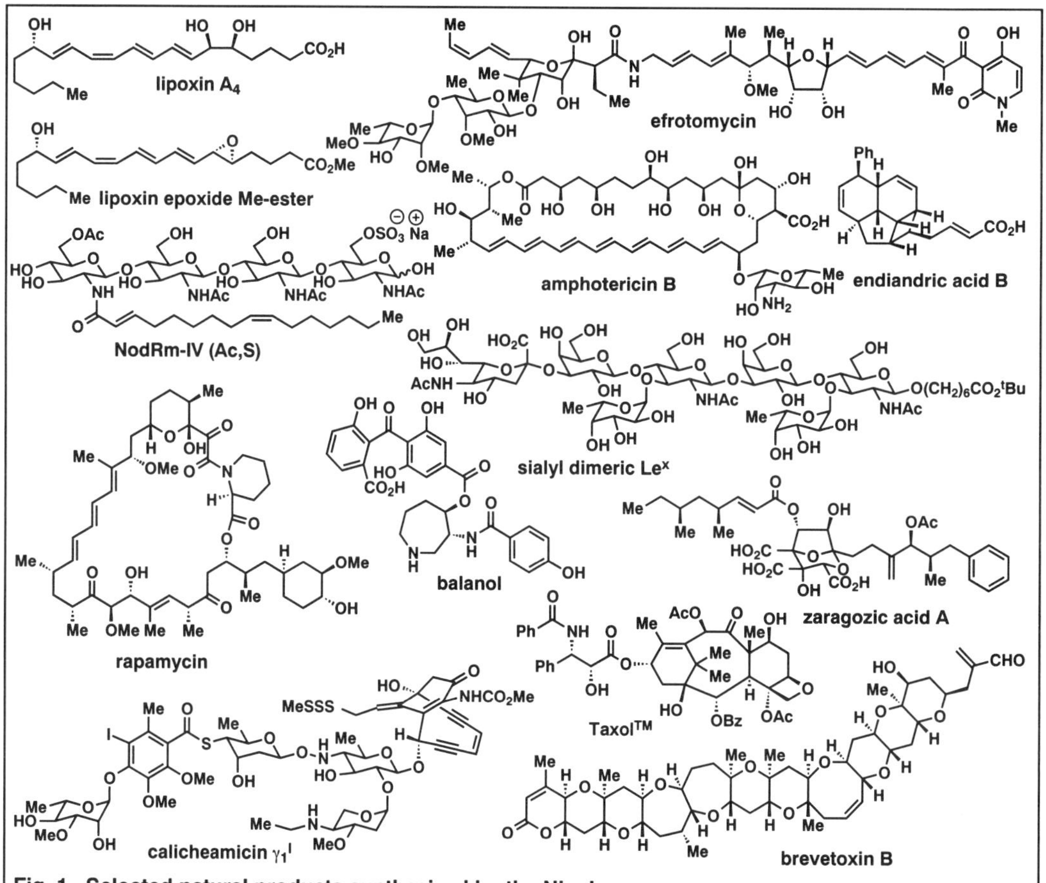

Fig. 1. Selected natural products synthesized by the Nicolaou group.

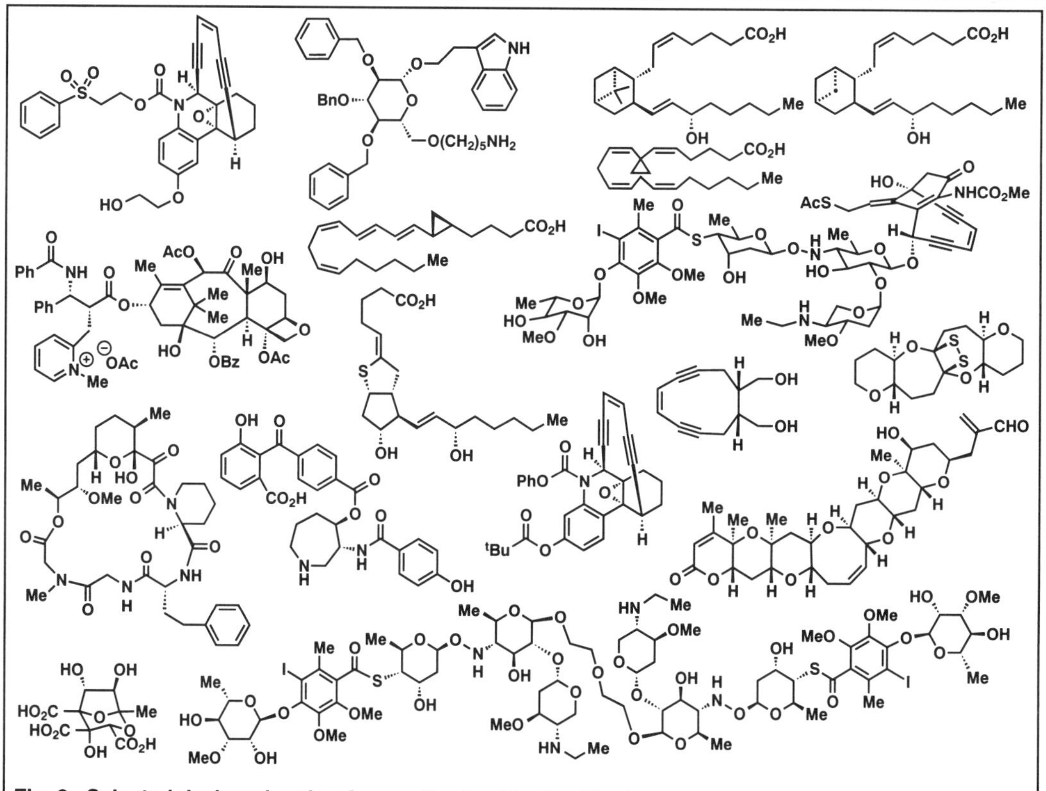

Fig. 2. Selected designed molecules synthesized by the Nicolaou group.

The Brevetoxin B project

Occurrence and structure.

Brevetoxin B (**1**) (19) (Fig. 3) is one of the toxic compounds responsible for the massive poisoning of marine life that occurs during certain "red tide" phenomena (20). With its eleven fused oxygen rings and twenty-three stereogenic centers, the structure of brevetoxin B is undeniably impressive. As a synthetic target it offers a highly

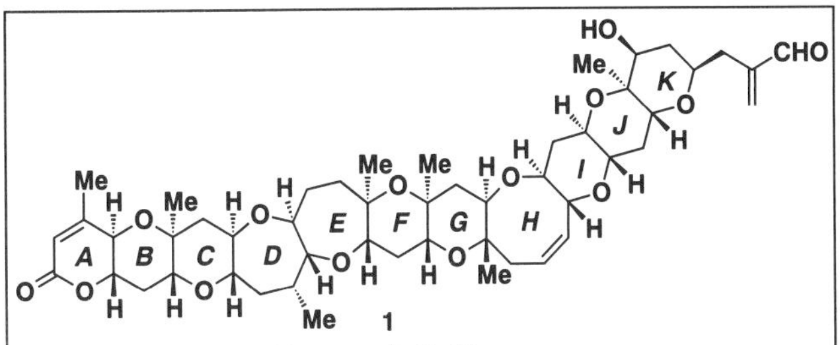

Fig. 3. Structure of brevetoxin B (**1**).

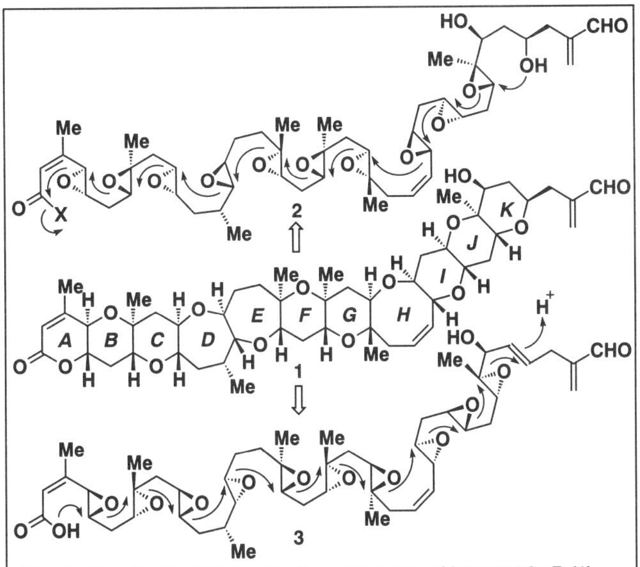

Fig. 4. Hypothetical biosynthetic precursors of brevetoxin B (**1**).

attractive "playground" for the development of new synthetic methods and novel strategies (4,21). The ability of brevetoxin B to interfere with the function of sodium channels in membranes provides opportunities for molecular design studies (22). With these attractive features adding up to an irresistible challenge, brevetoxin B (**1**) became the subject of our investigations soon after its structural elucidation (19).

It was at first tempting to consider a possible "biomimetic" scheme (Fig. 4) even though no evidence existed for the shown postulated intermediates **2** and **3** (23). The anticipated difficulties, however, with both the construction of the required polyepoxide precursors **2** and **3**, and their regioselective "zipping" to the brevetoxin skeleton, prompted us to seek alternative pathways.

The first retrosynthetic analysis and strategy.   Fig. 5 outlines a more rational retrosynthetic analysis of brevetoxin B (**1**) which requires the development of a number of new methods to construct the various rings of the molecule. This first generation retrosynthetic analysis, therefore, defined for us a set of initial goals, namely the development of efficient and practical means to build the six-, seven-, and eight-membered rings of brevetoxin B (**1**). The strategy for the total synthesis entailed: (a) construction of three fragments representing ring systems ABC, FG, and IJK; (b) coupling of FG with IJK and construction of the oxocene ring H; (c) attaching fragment ABC and construction of the bisoxepane ring system DE *via* bridging of the 12-membered ring dithioxolactone (**4**); and (d) oxygenation and side chain extension to complete the synthesis, as illustrated in Fig. 5.

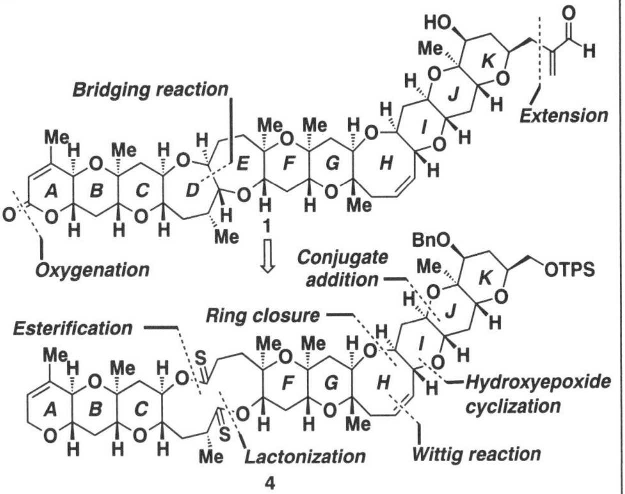

Fig. 5. First retrosynthetic analysis of brevetoxin B (**1**).

Figs 6-9 depict the chemistry developed to address the problems of the total synthesis of brevetoxin B.

path a: Exclusive with unsaturated chain. The developing electron deficient orbital (δ +) is stabilized by conjugation with the π-orbital.
path b: Exclusive with saturated chain.

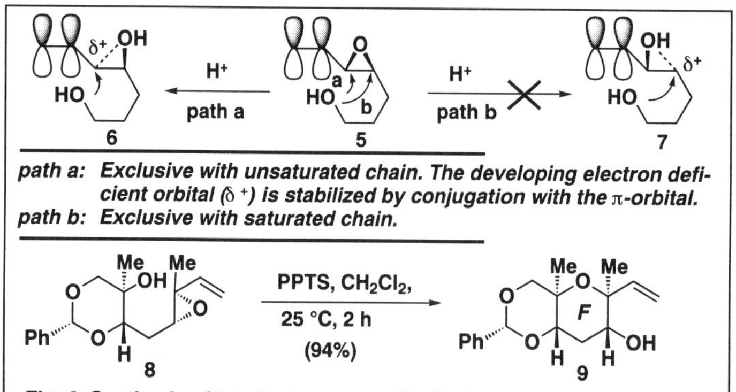

Fig. 6. Synthesis of tetrahydropyrans : the hydroxyepoxide cyclization.

For the construction of tetrahydropyrans, the methods developed include: (a) the regioselective and stereospecific hydroxyepoxide cyclization (5→6) (Fig. 6), and (b) the Michael-type intramolecular addition of an alkoxide ion to an α,β-unsaturated ester (10→11) (Fig. 7).

For the construction of the oxocene ring systems, the technology is based on the hydroxy-dithioketal cyclization, followed by reductive desulfurization (12→14→15) as shown in Fig. 8.

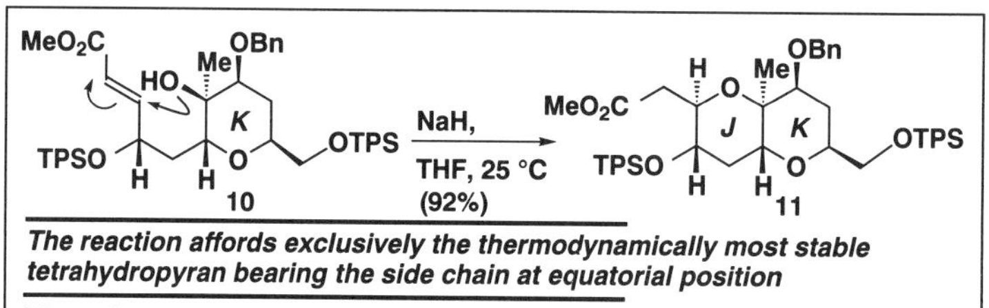

The formation of the bisoxepane region of brevetoxin B was based on the photo-induced bridging of dithioxolactones and the subsequent reduction of the dienol ether as illustrated in

The reaction affords exclusively the thermodynamically most stable
tetrahydropyran bearing the side chain at equatorial position

Fig. 7. Synthesis of tetrahydropyrans: the intramolecular conjugate addition.

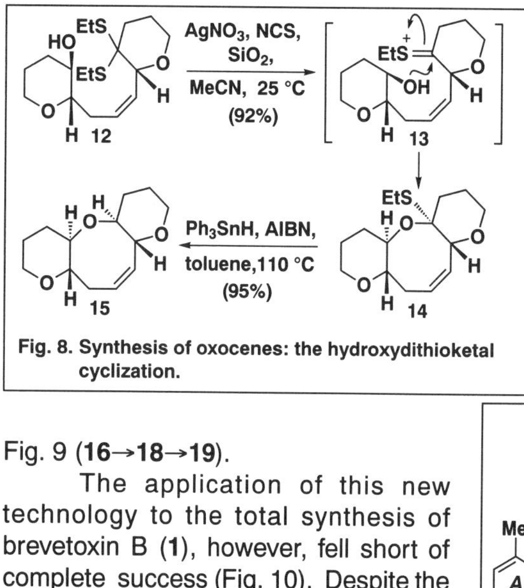

Fig. 8. Synthesis of oxocenes: the hydroxydithioketal
cyclization.

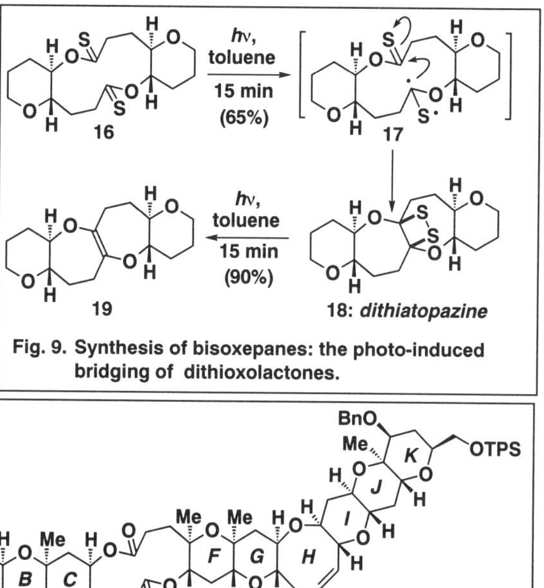

Fig. 9. Synthesis of bisoxepanes: the photo-induced
bridging of dithioxolactones.

Fig. 9 (16→18→19).

The application of this new technology to the total synthesis of brevetoxin B (1), however, fell short of complete success (Fig. 10). Despite the efficient assembly of the advanced precursor 20 (Fig. 10), attempts at the thionation and the bridging reactions (20→21→22) were both rewarded only with low yields and stereochemical problems. The methyl group on ring D was held responsible for these failures. This setback stimulated a second generation retrosynthetic analysis and a new wave of developments in synthetic methodology.

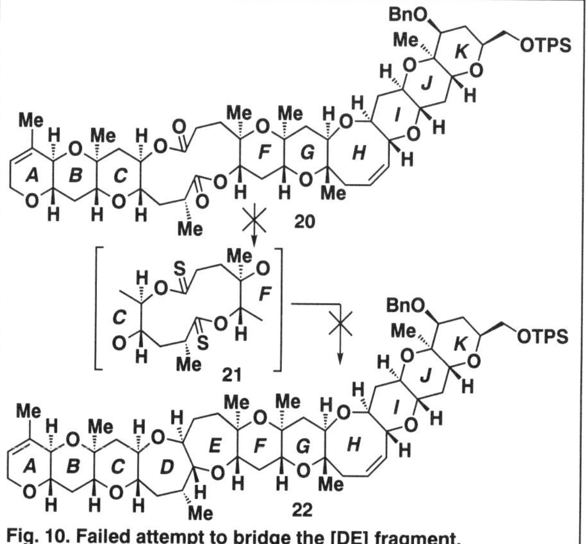

Fig. 10. Failed attempt to bridge the [DE] fragment.

The second retrosynthetic analysis and strategy. As outlined in Fig. 11, the new approach towards brevetoxin B required the stepwise construction of the two oxepane rings, each formed by a different method. Figs 12 and 13 summarize the two methods developed specifically for this purpose.

The first method involves photo-induced bridging of dithioxoesters such as **27**. This bridging is accompanied by extrusion of disulfur, thus

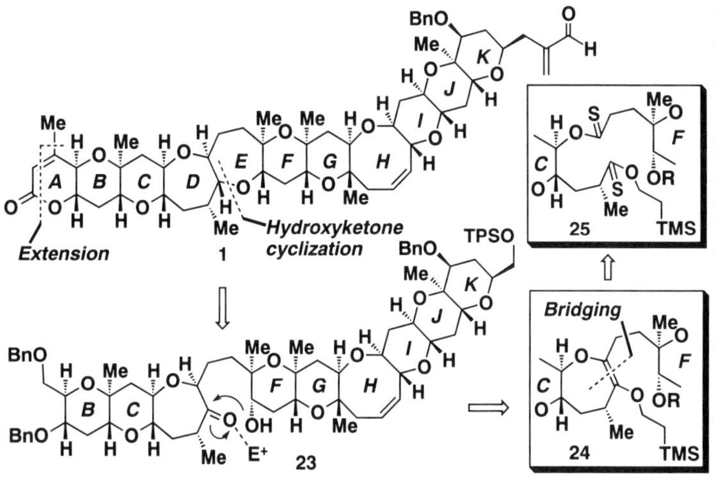

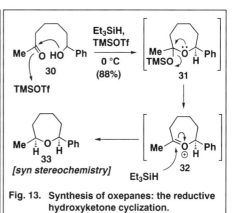

Fig. 11. Second retrosynthetic analysis of brevetoxin B (1).

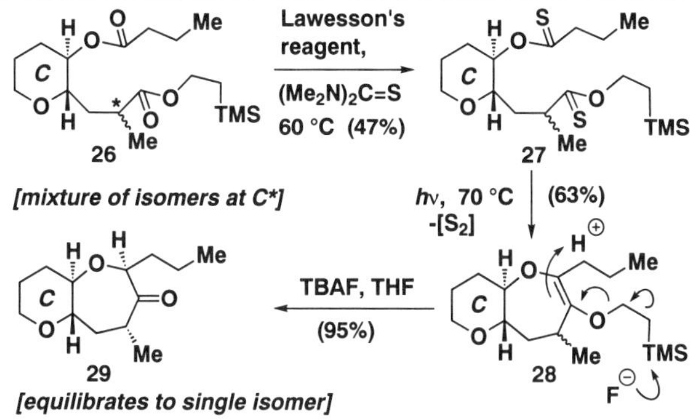

Fig. 12. Synthesis of oxepanes: photo-induced ring closure of dithioxoesters.

Fig. 13. Synthesis of oxepanes: the reductive hydroxyketone cyclization.

forming the oxepene system **28**, which serves as the precursor of the desired oxepanone **29** (Fig. 12). The second piece of new methodology accomplishes the construction of simple oxepanes by reductive cyclization of hydroxyketones (**30→33**) (Fig. 13).

The photo-induced ring closure of dithioxoesters was successfully applied to the preparation of the advanced intermediate **34** (Fig. 14). Again, however, and despite the demonstrated power of the reductive hydroxyketone cyclization in model systems, the final ring closure to form ring E was unsuccessful. The only identifiable product during these attempts was the diol resulting from the reduction of the carbonyl group in **34** (Fig. 14). The methyl groups on rings D and F were again the prime suspects for this unfortunate turn of events. Yet another approach was clearly needed.

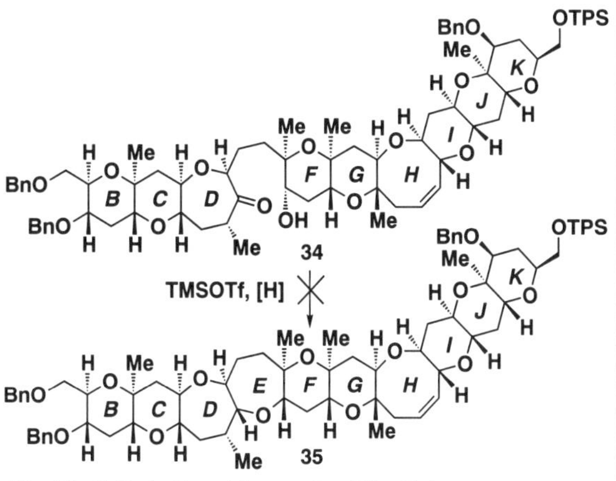

Fig. 14. Failed attempt to construct the E ring.

The final retrosynthetic analysis and strategy. Fig. 15 summarizes the final and

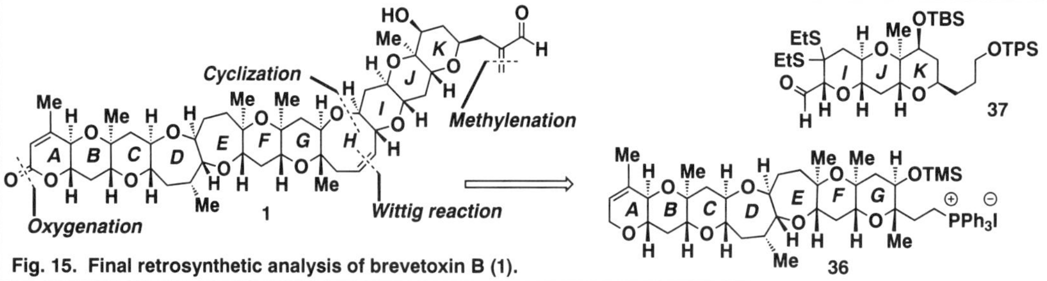

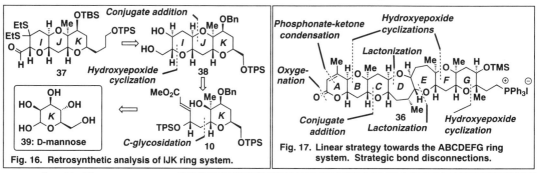

Fig. 15. Final retrosynthetic analysis of brevetoxin B (1).

successful strategy that led to the total synthesis of brevetoxin B (1) (24). According to this strategy, intermediates **36** and **37** were defined as the requisite building blocks. Their constructions

Fig. 16. Retrosynthetic analysis of IJK ring system.

Fig. 17. Linear strategy towards the ABCDEFG ring system. Strategic bond disconnections.

are outlined in Figs 16 and 17, respectively.

The final stages of this convergent sequence to brevetoxin B (1) are shown in detail in Fig. 18. Thus, Wittig coupling of phosphonium salt **36** and aldehyde **37**, furnished, after removal of the trimethylsilyl group, olefin **40**. Ring closure of the hydroxydithioketal **40** followed by reductive desulfurization of the resulting O,S-acetal **42** led to **41**, a compound that contains the entire brevetoxin B skeleton. Compound **41** was then smoothly converted to the natural product (**1**) by the sequence indicated in Fig. 18.

Brevetoxin B (**1**) was obtained as a crystalline compound exhibiting identical properties with the natural substance. Its total synthesis required 83 steps (longest linear sequence) starting from 2-deoxy-D-ribose and the overall yield was 0.043%.

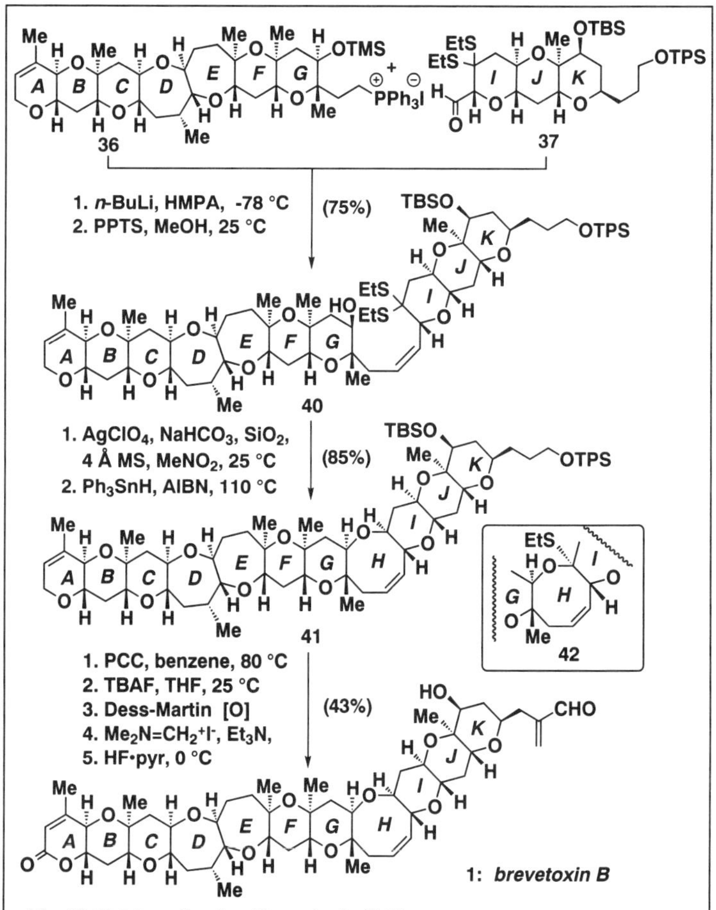

1. *n*-BuLi, HMPA, -78 °C
2. PPTS, MeOH, 25 °C (75%)

40

1. AgClO₄, NaHCO₃, SiO₂, 4 Å MS, MeNO₂, 25 °C (85%)
2. Ph₃SnH, AIBN, 110 °C

41

42

1. PCC, benzene, 80 °C
2. TBAF, THF, 25 °C
3. Dess-Martin [O]
4. Me₂N=CH₂⁺I⁻, Et₃N,
5. HF·pyr, 0 °C (43%)

1: *brevetoxin B*

Fig. 18. Total synthesis of brevetoxin B (1).

Biological studies. Brevetoxin B belongs to a family of marine neurotoxins with unique molecular structure and biological function. Its biological mode of action is presumed to involve binding to and activation of the voltage sensitive sodium channels, resulting in enhanced cellular sodium ion influx, as represented schematically in Fig. 19 (25). Due to such an important biological profile, brevetoxin B and related molecules are considered as powerful tools for the study of ion channels (26).

During our molecular design studies, we have focused on the preparation and biological

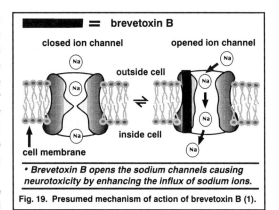

Fig. 19. Presumed mechanism of action of brevetoxin B (1).

• Brevetoxin B opens the sodium channels causing neurotoxicity by enhancing the influx of sodium ions.

evaluation of a variety of brevetoxin B analogs, possessing different ring skeletons and overall lengths. Particularly important was the "length hypothesis" put forward by Baden and Gawley (26), regarding the mode of action of brevetoxin B. According to this hypothesis the overall length of the structure (*ca.* 30 Å) is necessary for binding of the toxin to its receptor (26).

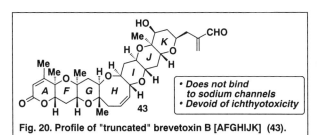

• Does not bind to sodium channels
• Devoid of ichthyotoxicity

Fig. 20. Profile of "truncated" brevetoxin B [AFGHIJK] (43).

Using chemistry developed during the total synthesis of brevetoxin B, "truncated" brevetoxin B [AFGHIJK] (**43**, Fig, 20) was synthesized (28) and subjected to biological studies (29). Indeed, with an overall length of *ca.* 20 Å, "truncated" brevetoxin B (**43**) shows a decreased binding affinity to the brevetoxin B receptor site and is devoid of ichthyotoxicity, as predicted by the Baden-Gawley hypothesis. However, it activates the sodium channels, thus partially mimicking the natural agonist.

According to the Baden-Gawley presumed model, the "tail" of brevetoxin B (K ring side chain) resides in the vicinity of the S5-S6 extracellular loop of domain IV of the channel protein. Its "cigar-shaped" skeleton is positioned, at a nearly vertical manner, at the interface of domains IV and I, thus positioning the "head" (A ring) in close proximity to the inactivation gate (IIIS6-IVS1 intracellular loop). In fact, the A-ring carbonyl oxygen is now explicitly implicated in the obstruction of the inactivation gate,

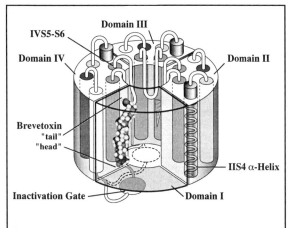

Fig. 21. Presumed model of sodium channel hosting brevetoxin B. (reprinted with permission from *Chemistry & Biology*, ref. 29).

which in the absence of brevetoxin B closes the channel. A schematic representation of the sodium channel serving as a host to brevetoxin B (**1**) is shown in Fig. 21 (29).

Ongoing structure-activity studies, are expected to shine further light on the mechanism of action of these fascinating marine neurotoxins.

Conclusion

The brief summary of the brevetoxin B (**1**) project given in this article is representative of several of our past and present total synthesis projects. The value of such programs should be measured not only by eventual arrival at the target molecule but also in terms of new contributions to synthetic organic chemistry and to biology and medicine. While the present program is still evolving, its potential value to these disciplines is already quite evident (30).

Acknowledgments. We would like to express our gratitude and appreciation to all our co-workers, whose names appear in the references, for their contributions to the research described in this article. Our programs have been financially supported by the National Institutes of Health, Merck, Sharp and Dohme, Glaxo, Inc., Schering-Plough, Pfizer, Hoffman-La Roche, and the ALSAM Foundation.

References

1. J. Mann, *Murder, Magic, and Medicine*, Oxford University Press, Oxford, 1994.
2. E. J. Corey and X.-M. Cheng, *The Logic of Chemical Synthesis*, J. Wiley, N. Y., 1989.
3. K. C. Nicolaou and E. J. Sorensen, *Classics in Total Synthesis*, VCH, Weinheim, 1995.
4. K. C. Nicolaou, *Aldr. Acta* **26**, 63 (1993).
5. K. C. Nicolaou, E. A. Theodorakis and C. F. Claiborne, *Pure & Appl. Chem.* (1995) in press.
6. K. C. Nicolaou, G. P. Gasic and W. E. Barnette, *Angew. Chem. Int. Ed. Engl.* **17**, 293 (1978).
7. K. C. Nicolaou, J. Y. Ramphal, N. A. Petasis and C. N. Serhan, *Angew. Chem. Int. Ed. Engl.* **30**, 1100 (1991).
8. R. E. Dolle and K. C. Nicolaou, *J. Am. Chem. Soc.* **107**, 1695 (1985).
9. K. C. Nicolaou and W. W. Ogilvie, *ChemTracts* **3**, 327 (1990).
10. K. C. Nicolaou, N. A. Petasis and R. E. Zipkin, *J. Am. Chem. Soc.* **104**, 5560 (1982).
11. K. C. Nicolaou, C. W. Hummel and Y. Iwabuchi, *J. Am. Chem. Soc.* **114**, 3126 (1992).
12. K. C. Nicolaou, N. J. Bockovich, D. R. Carcanague, C. W. Hummel and L. F. Even, *J. Am. Chem. Soc.* **114**, 8701 (1992).
13. K. C. Nicolaou, T. K. Chakraborty, A. D. Piscopio, N. Minowa and P. Bertinato, *J. Am. Chem. Soc.* **115**, 4419 (1993).
14. K. C. Nicolaou, Z. Yang, J. J. Liu, H. Ueno, P. G. Nantermet, R. K. Guy, C. F. Claiborne, J. Renaud, E. A. Couladouros, K. Paulvannan and E. J. Sorensen, *Nature* **367**, 630 (1994).
15. K. C. Nicolaou, M. E. Bunnage and K. Koide, *J. Am. Chem. Soc.* **116**, 8402 (1994).
16. K. C. Nicolaou, E. W. Yue, S. LaGreca, A. Nadin, Z. Yang, J. E. Lereche, T. Tsuri, Y. Naniwa and F. De Riccardis, *Chem. Eur. J.*, (1995) in press.
17 K. C. Nicolaou and W.-M. Dai, *Angew. Chem. Int. Ed. Engl.* **30**, 1387 (1991). K. C. Nicolaou, A. L. Smith and E. W. Yue, *Proc. Natl. Acad. Sci. U.S.A.* **90**, 5881 (1993).
18. K. C. Nicolaou, *Angew. Chem. Int. Ed. Engl.* (1996) in press.
19. Y.-Y. Lin, M. Risk, S. M. Ray, D. Van Engen, J. Clardy, J. Golik, J. C. James and K. Nakanishi, *J. Am. Chem. Soc.* **103**, 6773 (1981).
20. For a comprehensive review of the "red tides" see: D. M. Anderson, *Scientific American*, **271**, 62 (1994) and refs cited therein. T. Yasumoto and M. Murata, *Chem. Rev.* **93**, 1897 (1993).
21. K. R. Reddy, G. Skokotas and K. C. Nicolaou, *Gazz. Chim. Ital.* **123**, 337 (1993).
22. V. L. Trainer, R. A. Edwards, A. M. Szmant, A. M. Stuart, T. J. Mende and D. G. Baden, *Marine Toxins: Origin, Structure and Molecular Pharmacology* (Eds.: H. Sherwood and G. Strichartz), *ACS Symposium Series*, **418**, 167 (1990).
23. H.-N. Chou and Y. Shimizu, *J. Am. Chem. Soc.* **109**, 2184 (1987). M. S. Lee, G.-W. Qin, K. Nakanishi and M. G. Zagorski, *J. Am. Chem. Soc.* **111**, 6234 (1989). See also: M. J. Garson *Chem. Rev.* **93**, 1699 (1993).
24. K. C. Nicolaou, E. A. Theodorakis, F. P. J. T. Rutjes, J. Tiebes, M. Sato, E. Untersteller and X.-Y. Xiao, *J. Am. Chem. Soc.* **117**, 1171 (1995). K. C. Nicolaou, F. P. J. T. Rutjes, E. A. Theodorakis, J. Tiebes, M. Sato and E. Untersteller, *J. Am. Chem. Soc.* **117**, 1173 (1995).
25. V. L. Trainer, W. J. Thomsen, W. A. Catterall and D. G. Baden, *Mol. Pharmacol.* **40**, 988 (1991). D. G. Baden, *Int. Rev. Cytology* **82**, 99 (1983). V. L. Trainer, D. G. Baden and W. A. Catterall, *J. Biol. Chem.* **40**, 988 (1991).
26. W. A. Catterall, *Annu. Rev. Biochem.* **55**, 953 (1986). S. A. Cohen and R. L. Barchi, *Intern. Rev. Cytology*, **137C**, 55 (1993).
27. K. C. Nicolaou, J. Tiebes, E. A. Theodorakis, F. P. J. T. Rutjes, K. Koide, M. Sato and E. Untersteller, *J. Am. Chem. Soc.* **116**, 9371 (1995).
28. K. S. Rein, D. G. Baden and R. E. Gawley, *J. Org. Chem.* **59**, 2101 (1994). K. S. Rein, B. Lynn, R. E. Gawley and D. G. Baden, *J. Org. Chem.* **59**, 2107 (1994). D. G. Baden, K. S. Rein, R. E. Gawley, G. Jeglitsch and D. J. Adams, *Nat. Toxins* **2**, 212 (1994).
29. R. E. Gawley, K. S. Rein, G. Jeglitsch, D. J. Adams, E. A. Theodorakis, J. Tiebes, K. C. Nicolaou and D. G. Baden, *Chemistry & Biology*, **2** 533 (1995).
30. For a full account of the brevetoxin B project see ref 3, chapter 37 and ref. 18.

Enzyme inhibitors for the treatment of AIDS

Paul A. Aristoff

Discovery Chemistry, Upjohn Laboratories
The Upjohn Company, Kalamazoo, Michigan 49001

Abstract: Extensive research on the human immunodeficiency virus (HIV), the causative agent of AIDS, has identified a number of critical targets for potential therapeutic intervention. Particularly promising in this regard are the HIV reverse transcriptase (RT) and the HIV protease enzymes. A computer-assisted broad screening strategy was utilized to identify novel templates with selective activity against HIV RT and protease. These templates were then optimized so as to enhance both their potency and their pharmacokinetics. An interesting class of non-nucleoside RT inhibitors, the BHAPs, was discovered, and two compounds, ateviridine mesylate (U-87201E) and delavirdine mesylate (U-90152S), have undergone clinical evaluation. Similarly, a non-peptide HIV protease inhibitor lead, phenprocoumon, was developed into a series of clinical candidates, as exemplified by U-96988 and U-103017, that combine good enzyme inhibitor properties with superior pharmacokinetic behavior. A common characteristic of both the RT and protease compounds in development is their relative ease of synthesis.

The expanding medical, social, and economic problem of AIDS worldwide emphasizes the need for rapid development of therapy that is both efficacious and cost-effective. Prevailing opinion is that the disease is best treated by a combination of agents so as to prevent or delay the onset of resistance. Furthermore, these agents should have the appropriate combination of potency and pharmacokinetic properties such that resistance development is discouraged.   Finally, chemical complexity of the agent under development is also an issue, since a drug which is very difficult to synthesize and has either low potency or poor pharmacokinetics is very difficult to develop.

Two promising points of therapeutic intervention in the life cycle of HIV are represented by the viral enzymes, HIV reverse transcriptase and HIV protease. Both enzymes are essential for effective viral replication, and both are sufficiently different from human enzymes that selective inhibition of the viral enzymes is possible. In terms of HIV RT inhibitors, a number of nucleoside inhibitors have reached the marketplace; however, ultimately their efficacy is limited by toxicity or development of resistance. Thus, an opportunity exists if non-toxic, potent non-nucleoside inhibitors can be identified which are synthetically accessible, have good pharmacokinetics (e.g. good oral availability and duration of action), and can be combined with the nucleoside agents.

At Upjohn a computer-assisted broad screening approach was used to identify agents which inhibited the RT enzyme in a cell-free assay. A computer algorithm was utilized to separate the chemical collection into 1500 dissimilar classes, and a compound from each class was evaluated in the initial RT assay. When a hit was found, chemical similarity methods were then utilized to select related compounds for screening. Using this approach over 100 compounds were found which had some degree of enzyme inhibition. Follow-up assays to determine enzyme specificity (e.g. by comparing RT activity versus inhibition of human polymerases) and selective activity in cell culture against HIV, revealed U-80493E (1) as a promising template. (1) While U-80493E has only weak enzyme inhibitory activity (IC_{50} = 20 μM) and modest antiviral potency (ED_{50} = 2 μM), it had at least a ten-fold selectivity index both at the enzyme level and in cell culture. Furthermore, a wide variety of analogues could be readily prepared in just a few simple chemical steps. (2)

1

2 R_1 = H; R_2 = Et
3 R_1 = OMe; R_2 = Et
4 R_1 = NHSO$_2$Me; R_2 = i-Pr

An important milestone in the program was the discovery that replacement of the benzene ring in 1 with an indole gave a derivative, U-85961 (2), which was 10 - 100 times more potent and selective than 1. Further studies established that 2 and similar compounds, which are referred to as BHAPs (bis-heteroaryl piperazines), are non-competitive and selective inhibitors of HIV-1, acting at an allosteric binding site where a number of non-nucleoside inhibitors appear to interact with the enzyme. (1) Manipulation of the indole moiety in 2 so as to enhance pharmaceutical properties (especially oral bioavailability and half-life) provided the methoxy derivative 3 (U-87201E, atevirdine mesylate). Atevirdine mesylate had potency comparable to AZT, but was less toxic, having greater than a 10,000-fold selectivity index. It was synergistic with the nucleoside RT inhibitors, and was well tolerated in rats and dogs where it was orally bioavailable in excess of 60%. (1) Thus, it was selected as the first generation BHAP for clinical development and has progressed to the stage of evaluation in HIV-infected individuals.

Because of concerns about the possible development of resistance with atevirdine mesylate, a search for analogues with improved potency and pharmacokinetics, yet still maintaining the favorable safety profile of 3, was undertaken. The resulting compound 4, delavirdine mesylate (U-90152S), is one of the most potent RT inhibitors described, being more potent than AZT in a number of assays. (3) Relative to atevirdine mesylate, delavirdine mesylate was five-fold more potent at the enzyme level and at least ten-fold more potent against the virus in cell culture. In addition to excellent antiviral potency (average ED_{50} = 70 nM against a panel of clinical isolates), it has superior pharmacokinetic properties, being 60-100% bioavailable in rats and dogs with high micromolar blood levels being readily achieved, maintained, and tolerated. (4) Furthermore, the compound is readily synthesized as shown in Fig. 1. (5) Thus, delavirdine mesylate was selected as the second generation RT clinical candidate and has

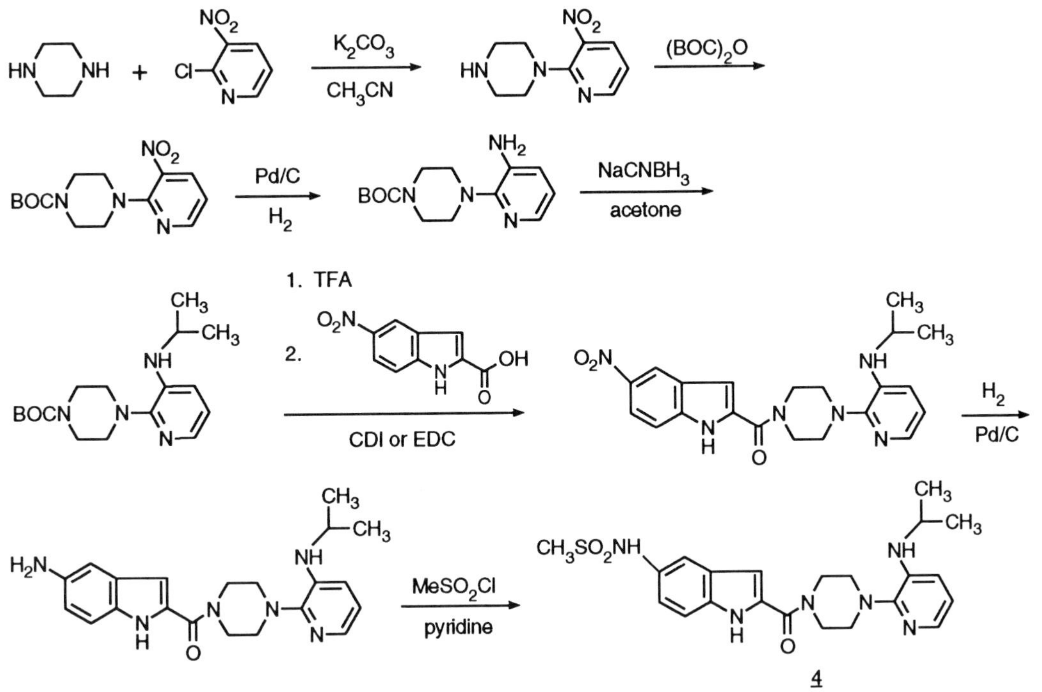

Fig. 1 Synthesis of U-90152S (3).

progressed to an advanced stage of clinical evaluation in AIDS patients. Like atevirdine mesylate, the compound is well tolerated in man with steady state blood levels of the drug in excess of ten micromolar easily achievable. In the clinic the most common side effects were occasional headache and skin rash.

Studies on resistance development with the BHAP class of RT inhibitors have produced some surprising findings. In particular, when laboratory strains of HIV resistant to delavirdine mesylate were developed and characterized, a new mutant RT containing a proline to leucine mutation at amino acid position 236 was obtained. (6) Delavirdine mesylate was roughly 70 times less potent against the P236L mutant RT than the wild-type enzyme. This was clearly different from the tyrosine to cysteine mutation at 180 which had been previously identified as a single point mutation causing marked resistance to a number of other non-nucleoside RT inhibitors, such as nevirapine and L-697661. Even more unexpected was the finding that the P236L mutant was now sensitized to the other non-nucleoside inhibitors, with nevirapine and L-697661 being about seven to ten fold more potent against the delavirdine resistant mutant than the wild-type RT. (6) Thus, this raises the intriguing possibility of combination of delavirdine mesylate in the clinic not only with nucleoside inhibitors but also non-nucleoside inhibitors of HIV RT.

Finally, recent studies with a new class of BHAPs wherein the piperazine moiety is replaced with an N-alkyl piperidine have revealed that some of these analogues, such as compound 5, are not only quite active against the wild-type virus, but are also active against many of the common mutant strains, including the Y181C (ED_{50} = 100 nM) and P236L (ED_{50} = 250 nM) variants. On the other hand, compound 5 produces yet another mutant virus in the laboratory, this time with a glycine to glutamate mutation at position

190 in RT. What is interesting about this G190E mutation is that it appears to be associated with reduced replication kinetics, thus suggesting an impaired or attenuated HIV. Fortunately, the G190E mutant enzyme retains its sensitivity to inhibition by delavirdine mesylate.

5   **6**

Initially, Upjohn's program to discover inhibitors of HIV protease centered on optimizing leads derived from a previous renin (another aspartyl protease) program. While extremely potent compounds such as U-75875 (6, K_i < 1 nM) were prepared, typically these peptidic compounds had poor pharmacokinetics (low oral bioavailability and very rapid clearance) and were rather complicated to synthesize on large scale because of the presence of so many chiral centers. Thus, a broad screening program was initiated to try to identify a non-peptide lead template. Relatively few templates emerged from this screening program; however, one compound, phenprocoumon (7) was identified as a weakly potent inhibitor (K_i = 1 μM) of the enzyme. (7) While the antiviral activity of phenprocoumon in cell culture was rather poor (ED_{50} = 100 - 300 μM), the known excellent pharmacokinetics and relatively simple structure of the compound looked promising. Furthermore, x-ray crystallographic studies of 7 and closely related agents revealed that the compound bound at the active site of HIV protease.

7   **8**

Thus, an iterative process of crystallography, molecular modelling, and synthesis could be effectively utilized in the design of compounds with improved potency. This eventually lead to the first generation protease candidate 8 (U-96988), which has a substituted phenethyl moiety in place of the fused phenyl ring in the coumarin 7. (7) U-96988 has a K_i of 38 nM against HIV-1 and 32 nM against HIV-2. It is not active against human protease enzymes (such as pepsin, renin, cathepsin, and gastricsin) below 10 μM. It has modest activity against the virus in cell culture (ED_{50} = 4 μM against clinical isolates), and is synergistic in cell culture with AZT. Importantly, the compound demonstrated the desired pharmacokinetics in rats (oral bioavailability of 75 - 90%) and dogs (oral bioavailability of 45%) with a six hour half-life in each species. The compound was well tolerated in animals, with very high micromolar blood levels safely maintained. Finally, the synthesis of U-96988 was achieved in just a few steps (Fig. 2). U-96988 is a mixture of four compounds, and an efficient chiral synthesis was developed (8); however, all four isomers proved to be inhibitors of the protease. Thus U-96988 entered clinical trials as a first generation non-peptidic protease inhibitor. In studies in healthy patients good blood levels were achieved and no significant toxicity was observed.

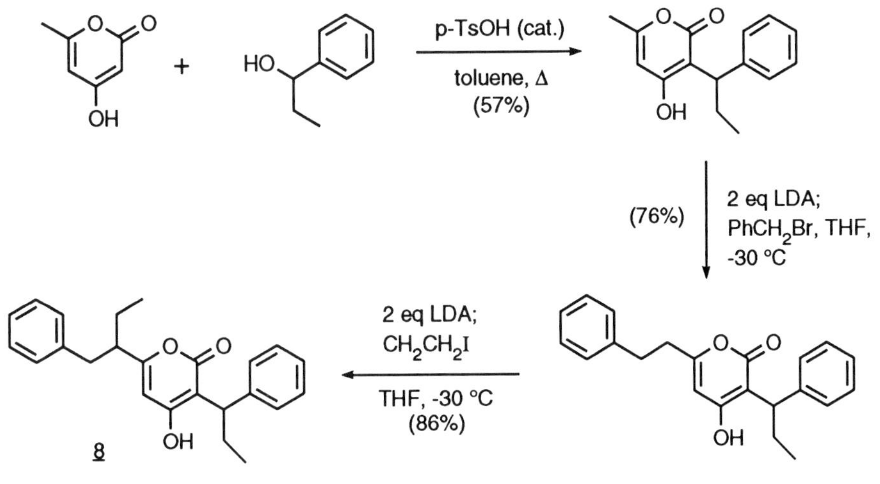

Fig. 2 Synthesis of U-96988 (8).

Hoping to improve on U-96988, the phenyl ring of the coumarin ring in 7 was reduced, and as predicted by modelling experiments this lead to a marked enhancement of the enzyme inhibitory activity. For example the cyclooctylpyranone 9 (U-99499) had a K_i of about 15 nM against both HIV-1 and HIV-2 protease. (9) The compound also maintained excellent pharmacokinetics with a five hour half-life and greater than 90% oral bioavailability in rats. Unfortunately, however, compound 9 had only weak antiviral activity (ED_{50} = 60 μM).

A more modest improvement in enzyme inhibitory activity (but again not antiviral potency) which was predicted by molecular modeling involved adding an amide side chain to the coumarin 7 so as to give compounds such as 10 which had a K_i of 280 nM. (10) Additional modelling suggested that replacing the amide linkage with a sulfonamide linkage would provide additional bonding interactions with the enzyme. This eventually lead to a series of extremely potent arylsulfonamide derivatives as exemplified by the p-cyanophenyl analog 11 (U-103017). Compound 11 has K_i of 0.8 nM against HIV-1 and 3.2 nM against HIV-2 protease, but does not significantly inhibit other human proteases. Another salient feature of the arylsulfonamides is that they imparted significant antiviral activity to a large number of compounds. In particular, U-103017 had an average ED_{50} of 5 μM against a panel of clinical isolates. Again, it also maintained the good pharmacokinetics of the cyclooctylpyranones in both rats and dogs. For example, it had 77% oral bioavailability and a six hour half-life in dogs. Impressively large serum concentrations (> 100 μM) could safely be achieved and maintained in animals. Finally, a relatively simple synthesis of U-103017 could be utilized to prepare large amounts of

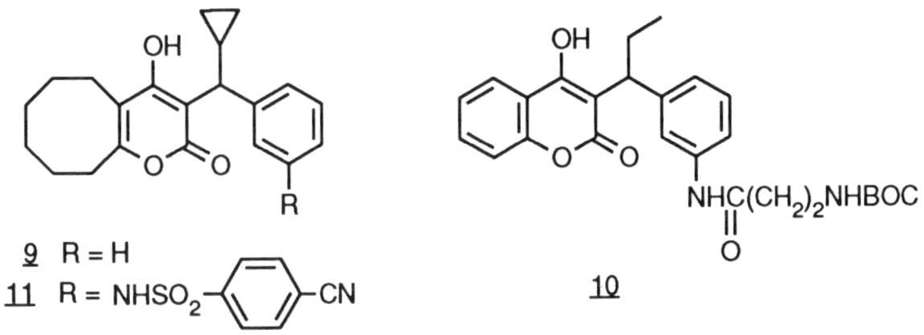

9 R = H
11 R = NHSO₂—⟨ ⟩—CN

10

the material (Fig. 3).   Compound 11 has a single chiral center. Testing of the individual enantiomers showed at most a two-fold difference in enzyme inhibitory activity or antiviral acivity between the two isomers.   Thus, U-103017 was selected as the second generation protease inhibitor clinical candidate.

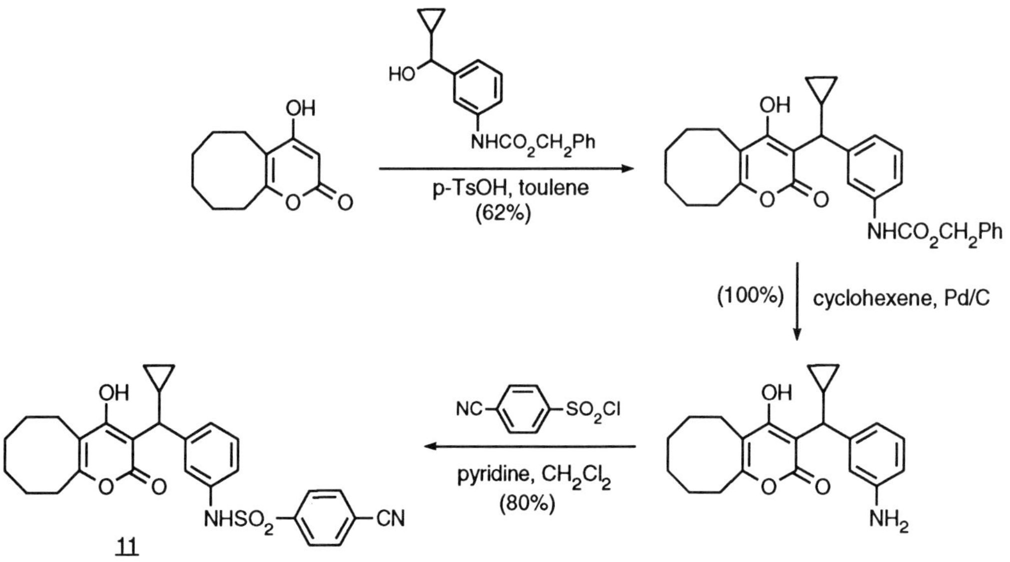

Fig. 3 Synthesis of U-103017 (11).

In conclusion, a strategy to provide potent enzyme inhibitors with good pharmacokinetics and which can be readily prepared on large scale has been successfully realized with the discovery of a variety of clinical candidates. In particular, the RT inhibitor delavirdine mesylate is especially promising in that serum concentrations in great excess of in vitro ED_{90}'s can be safely maintained in HIV patients. This agent, along with some of the new and encouraging HIV protease inhibitors, will hopefully provide additional weapons to meaningfully attack the HIV virus and provide a clinical benefit to the patient.

Acknowledgement   The results summarized herein represent the combined efforts of the many dedicated and talented individuals on Upjohn's HIV RT and HIV Protease Program Teams.

References

1.   D.L. Romero *et al. Proc. Nat. Acad. Sci. USA* **88**, 8806 (1991).
2.   D.L. Romero *et al. J. Med. Chem.* **37**, 999 (1994).
3.   T.J. Dueweke *et al. Antimicrob. Agents Chemother.* **37**, 1127 (1993).
4.   D.C. Romero *Drugs Fut.* **19**, 238 (1994).
5.   D.L. Romero *et al. J. Med. Chem.* **36**, 1505 (1993).
6.   T.J. Dueweke *et al. Proc. Nat. Acad. Sci. USA* **90**, 4713 (1993).
7.   S. Thaisrivongs *et al. J. Med. Chem.* **37**, 3200 (1994).
8.   R.B. Gammill *et al. J. Amer. Chem. Soc.* **116**, 12113 (1994).
9.   K.R. Romines *et al. J. Med. Chem.* **38**, 1884 (1995).
10.   S. Thaisrivongs *et al. J. Med. Chem.*, in press.

New approaches to the synthetic problems of *myo*-inositol phosphates and their biological utilities

Sung-Kee Chung,* Young-Tae Chang and Kwang-Hoon Sohn

Department of Chemistry, Pohang University of Science & Technology
Pohang 790-784, Korea

Abstract:
A number of *myo*-inositol phosphates (IP$_n$) have been implicated as either second messengers or key metabolic intermediates in the intracellular signal transduction pathways. Syntheses of all possible 9 and 12 regioisomers of *myo*-inositol tetrakis- and tris-phosphates(IP$_4$ and IP$_3$), respectively were accomplished from *myo*-inositol via its di- and tri-benzoate derivatives as the key intermediates. The requisite sets of *myo*-inositol benzoate intermediates were most conveniently produced by base-catalyzed migration of the benzoate group in partially benzoylated inositols, followed by suitable separation methods. The biological properties of the synthetic IP$_n$ samples have been examined in terms of their abilities for the receptor binding, Ca^{2+} release and its interference, and for the possible inhibition of the metabolic enzymes such as kinases.

The phospholipase C catalyzed cleavage of membrane bound phosphatidylinositol bisphosphate (PIP$_2$) into *myo*-inositol-1,4,5-trisphosphate(IP$_3$) and diacylglycerol is a crucially important means of cellular signalling processes. A large fraction of receptors identified to date are linked to the increase of PIP$_2$ turnover following the receptor activation (ref. 1). More recently, a number of *myo*-inositol phosphates including IP$_4$ and IP$_5$ have also been implicated as either second messengers or key metabolic intermediates in the intracellular signal transduction pathways (ref. 2 and 3). Systematic research on the relationship of the structure of IP$_n$ and their biological functions has been hampered by the limited availability of IP$_n$ regioisomers.

As a part of our efforts to understand the molecular recognition aspect of the *myo*-inositol phosphate dependent signal transduction, we set out to synthesize all the possible regioisomers of IP$_n$. There are 39 racemic or meso regioisomers (enantiomerically 63) available for IP$_n$: IP$_1$, 4(6); IP$_2$, 9(15); IP$_3$, 12(20); IP$_4$, 9(15); IP$_5$, 4(6); IP$_6$, 1. One of the key problems in the synthesis of inositol phosphates is to prepare suitable, selectively protected inositol intermediates for phosphorylation. We have envisaged that the synthetic challenge may be reduced to a facile generation by some means of all possible regioisomers of *myo*-inositol benzoate(IBz$_{6-n}$) intermediates, which are expected to be readily phosphorylated to provide the target structures(IP$_n$) after deprotection. It was also envisioned that the simplest way of generating all possible regioisomers of *myo*-inositol benzoate intermediates would be the acyl migrations of partially acylated inositols.

The acyl migrations among vicinally related diol systems are well precedented in the literature. For example, the initial observation that the acyl groups in mono- and di-acylglycerols migrate from the oxygen to which they are attached to an adjacent free hydroxyl group by a transesterification process catalyzed by acid, base or heat goes back to Emil Fischer. In fact, the facile acyl group migration presents a persistent major difficulty in the synthesis of pure acylglycerols (ref. 4). The acyl group migrations in the glucopyranoside and glucofuranoside systems have frequently been observed under the conditions employing silver oxide or a mild base (ref. 5). In the aminoacyl-tRNA, the critical component in the biosynthesis of proteins, the aminoacylation occurs on the 2' or 3'-OH group of the adenosine at the 3' end of tRNA, and the ester formed initially equilibrates very rapidly (t$_{1/2}$ of ca. 0.2 ms) between the two vicinal hydroxyl groups (ref. 6). The acetyl group migrations in partially acetylated derivatives of *myo*-inositols are also known (ref. 7).

Synthesis of 9 Regioisomers of *myo*-Inositol Tetrakisphosphates.

Up until our recent synthesis (ref. 8 and 9), only 4 out of possible 9 (enatiomerically 15) IP$_4$ isomers, I(1,2,4,5)P$_4$, I(1,3,4,5)P$_4$, I(1,3,4,6)P$_4$, I(1,4,5,6)P$_4$ have been synthesized via a variety of synthetic routes (ref. 10 and 11). Our key synthetic strategy for the synthesis of IP$_4$ regioisomers is based on the facile generation of all 9 regioisomers of *myo*-inositol dibenzoate(IBz$_2$) as the intermediates, which are expected to be amenable to phosphorylation to give the target structures (Scheme 1). Since *myo*-inositol acetates and benzoates are known to isomerize upon base treatment (ref. 7 and 12), we have examined this method as a quick way of generating IBz_2 isomers. Thus, compound **4**, prepared from *myo*-inositol (ref. 13), was hydrolyzed in 80 % aqueous acetic acid at reflux to give $I(1,4)Bz_2(\mathbf{1c})$. The desired complete benzoyl migration in **1c** could successfully be effected by treatment with 60 % aq. pyridine at elevated temperatures. The HPLC analysis conditions were found, which allowed a complete separation of the 9 isomers(**1a-i**) present in the reaction mixture. In practice the 9 isomers of $IBz_2(\mathbf{1})$were separated in pure forms by combinations of silica gel column chromatography, fractional crystallization and preparative HPLC, and they were fully characterized by ^1H and ^{13}C NMR spectroscopy including ^1H-^1H COSY. It was determined that the increasing order of the HPLC retention time of these isomers is 1,4(**1c**), 2,4(**1f**), 2,5(**1g**), 1,5(**1d**), 1,2(**1a**), 4,6(**1i**), 1,3(**1b**), 4,5(**1h**) and 1,6(**1e**), and the increasing order of Rf values on SiO_2 is **1a**, **1h**, **1e**, **1f**, **1b**, **1g**, **1i**, **1d**, and **1c**. There is no obvious correlation between these two sequences.

Scheme 1

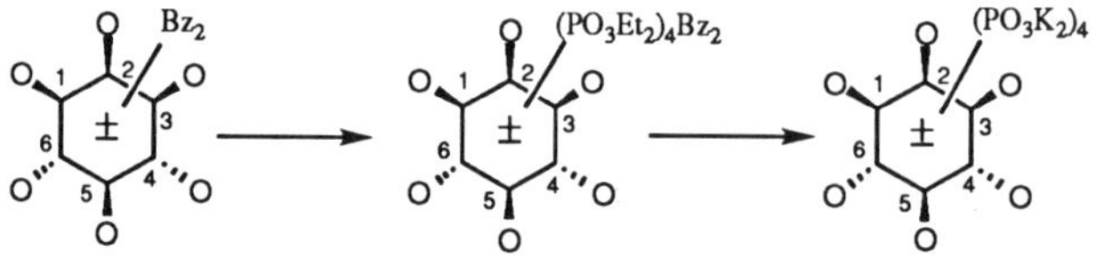

1
Dibenzoate

2

3
Inositol-tetrakisphosphate

a. $I(1,2)Bz_2 = I(2,3)Bz_2$
b. $I(1,3)Bz_2$
c. $I(1,4)Bz_2 = I(3,6)Bz_2$
d. $I(1,5)Bz_2 = I(3,5)Bz_2$
e. $I(1,6)Bz_2 = I(3,4)Bz_2$
f. $I(2,4)Bz_2 = I(2,6)Bz_2$
g. $I(2,5)Bz_2$
h. $I(4,5)Bz_2 = I(5,6)Bz_2$
i. $I(4,6)Bz_2$

a. $I(3,4,5,6)P_4 = I(1,4,5,6)P_4$
b. $I(2,4,5,6)P_4$
c. $I(2,3,5,6)P_4 = I(1,2,4,5)P_4$
d. $I(2,3,4,6)P_4 = I(1,2,4,6)P_4$
e. $I(2,3,4,5)P_4 = I(1,2,5,6)P_4$
f. $I(1,3,5,6)P_4 = I(1,3,4,5)P_4$
g. $I(1,3,4,6)P_4$
h. $I(1,2,3,6)P_4 = I(1,2,3,4)P_4$
i. $I(1,2,3,5)P_4$

The separational difficulties could be substantially ameliorated by carrying out the benzoyl group migration in partially protected derivatives of $I(1,4)Bz_2$. Thus, compound **5c** was prepared from **4** by selective hydrolysis, and compound **6c** from **1c** by monoacetalization (Scheme 2). When **5c** and **6c** were subjected to 60 % aqueous pyridine conditions and then 80 % aqueous acetic acid at reflux, two sets of 5 isomers of IBz_2 were obtained from the limited benzoyl group migrations.

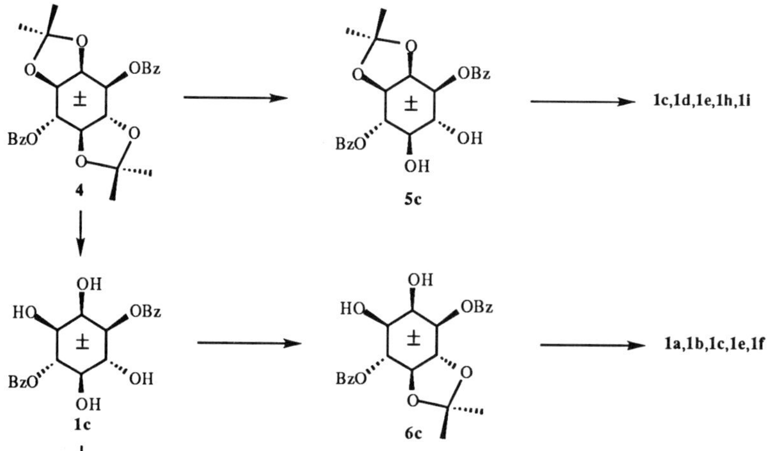

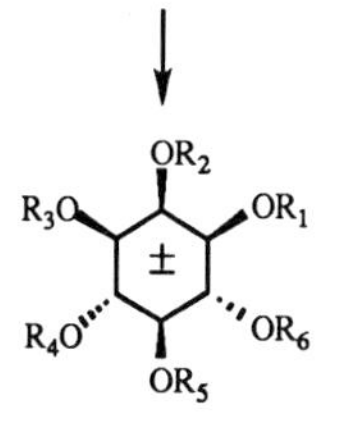

4

5c

1c,1d,1e,1h,1i

1c

6c

1a,1b,1c,1e,1f

1a $R_1, R_2 = Bz, R_3, R_4, R_5, R_6 = H$
1b $R_1, R_3 = Bz, R_2, R_4, R_5, R_6 = H$
1c $R_1, R_4 = Bz, R_2, R_3, R_5, R_6 = H$
1d $R_1, R_5 = Bz, R_2, R_3, R_4, R_6 = H$
1e $R_1, R_6 = Bz, R_2, R_3, R_4, R_5 = H$
1f $R_2, R_4 = Bz, R_1, R_3, R_5, R_6 = H$
1g $R_2, R_5 = Bz, R_1, R_3, R_4, R_6 = H$
1h $R_4, R_5 = Bz, R_1, R_2, R_3, R_6 = H$
1i $R_4, R_6 = Bz, R_1, R_2, R_3, R_5 = H$

Scheme 2

Each of the IBz$_2$ regioisomers thus obtained was separately phosphorylated by successive treatments with diethyl chlorophosphite and N,N-diisopropylethylamine in DMF, and then 30 % hydrogen peroxide to yield all the 9 isomers of compound **2**, which were thoroughly characterized by ^{1}H, ^{13}C and ^{31}P NMR. In the final steps, the protecting groups of **2** were removed by successive reactions with trimethylsilyl bromide and then KOH. Cleavage of the ethyl phosphate esters was monitored by ^{31}P-NMR, which clearly showed upfield chemical shift changes of 10-20 ppm due to the conversion of the ethyl ester to the silyl ester (ref. 14). The product(**3**) was obtained after chromatography on Dowex 50x8-100(H$^+$ form), pH adjustment to 10 with KOH, and lyophilization.

Synthesis of 12 Regioisomers of *myo*-Inositol Trisphosphates.

D-*myo*-inositol-1,4,5-trisphosphate(I(1,4,5)P$_3$) receptors are shown to be a homologous family of tetrameric ligand-gated Ca^{2+} channels, which allow mobilization of intracellular Ca^{2+} stores in response to activation of cell-surface receptors linked to I(1,4,5)P$_3$ generation. Although the ligand recognition site of the IP$_3$ receptors appears to be in the N-terminal region, the exact nature of the molecular interactions has not been understood. Several other IP$_3$s are also found in living systems, and studies to elucidate the receptor binding as well as metabolism are in progress (ref. 15-17). Until now 9 out of possible 12(enantiomerically 20) IP$_3$ regioisomers have been synthesized by various synthetic routes: I(1,2,3)P$_3$, I(1,2,6)P$_3$, I(1,3,4)P$_3$, I(1,3,5)P$_3$, I(1,4,5)P$_3$, I(1,4,6)P$_3$, I(1,5,6)P$_3$, I(2,4,5)P$_3$, I(2,4,6)P$_3$ (ref. 10 and 11).

Based on the same synthetic strategy as described above for IP$_4$, now the benzoyl migration technique was applied to the facile generation of all 12 regioisomers of *myo*-inositol tribenzoate(IBz$_3$), **7** as the key synthetic intermediates, which were expected to be readily phosphorylated to provide the target IP$_3$ structures, **8** (Scheme 3). Thus, the random, divergent benzoyl group migration in compound **7g**, prepared from *myo*-inositol (ref. 18), was successfully effected under the conditions of 60 % aq. pyridine at elevated temperatures. The HPLC analysis allowed clean separation of 10 isomers out of the total 12 regioisomers(**7a-l**) present in the reaction mixture. Structures of the isolated regioisomers of **1** could be readily elucidated by ^{1}H-NMR. The increasing order of the HPLC retention time of these regioisomers was found to be **1f**, **1g**, **1k**, **1b**, **1e**, **1h**, **1c**, **1j**, **1i**, (**1a+1l**), **1d** (Figure 1).

Scheme 3

7	**8a-l**	**9**
Tribenzoate		Inositol-trisphophate

a I(4,5,6)Bz$_3$	**a** I(1,2,3)P$_3$
b I(3,5,6)Bz$_3$ = I(1,4,5)Bz$_3$	**b** I(1,2,4)P$_3$ = I(2,3,6)P$_3$
c I(3,4,6)Bz$_3$ = I(1,4,6)Bz$_3$	**c** I(1,2,5)P$_3$ = I(2,3,5)P$_3$
d I(3,4,5)Bz$_3$ = I(1,5,6)Bz$_3$	**d** I(1,2,6)P$_3$ = I(2,3,4)P$_3$
e I(2,5,6)Bz$_3$ = I(2,4,5)Bz$_3$	**e** I(1,3,4)P$_3$ = I(1,3,6)P$_3$
f I(2,4,6)Bz$_3$	**f** I(1,3,5)P$_3$
g I(2,3,6)Bz$_3$ = I(1,2,4)Bz$_3$	**g** I(1,4,5)P$_3$ = I(3,5,6)P$_3$
h I(2,3,5)Bz$_3$ = I(1,2,5)Bz$_3$	**h** I(1,4,6)P$_3$ = I(3,4,6)P$_3$
i I(2,3,4)Bz$_3$ = I(1,2,6)Bz$_3$	**i** I(1,5,6)P$_3$ = I(3,4,5)P$_3$
j I(1,3,6)Bz$_3$ = I(1,3,4)Bz$_3$	**j** I(2,4,5)P$_3$ = I(2,5,6)P$_3$
k I(1,3,5)Bz$_3$	**k** I(2,4,6)P$_3$
l I(1,2,3)Bz$_3$	**l** I(4,5,6)P$_3$

In practice, the separational problems dealing with the mixture of the 12 regioisomers could be substantially helped by carrying out the benzoyl group migrations in partially protected derivatives of IBz$_3$. In the event, compounds **10** and **11**, prepared from *myo*-inositol (ref. 19), were hydrolysed in acid, and then selectively benzoylated on the axial 2-OH via the orthoester intermediates to give two IBz$_3$, **7g** and **7i** (ref. 20). Each IBz$_3$ was monoacetalized under kinetic conditions to give the 4,5-acetal **6** and the 1,6-acetal **7** (Scheme 4). On the other hand, compound **14**, prepared from **10** by selective hydrolysis (ref. 8), was monobenzoylated to give a mixture of the 1,2-acetals **15** and **16** (Scheme 5).

Figure 1

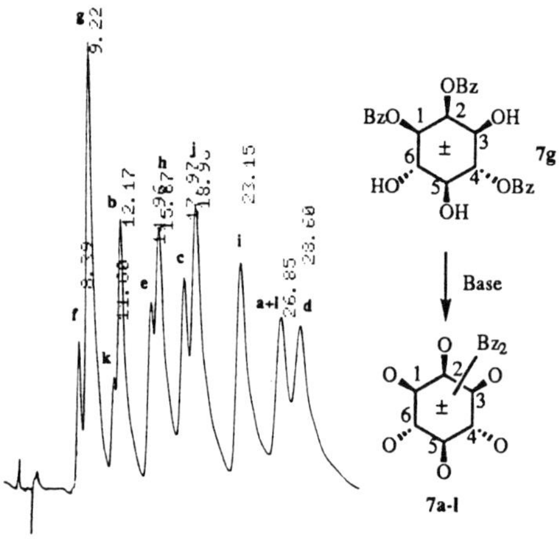

Scheme 4

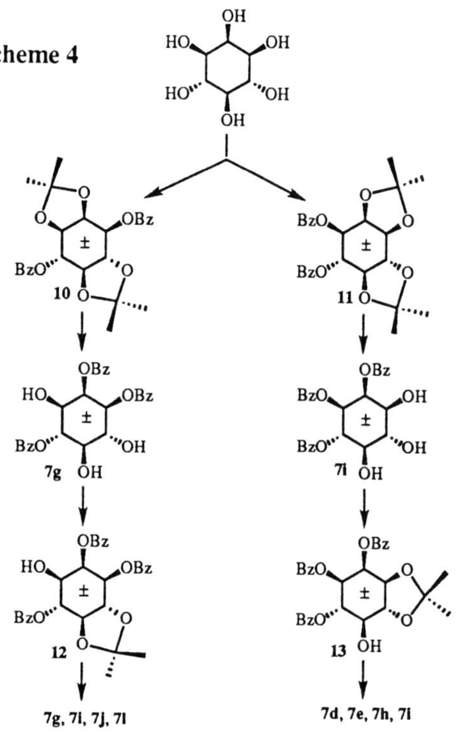

Scheme 5

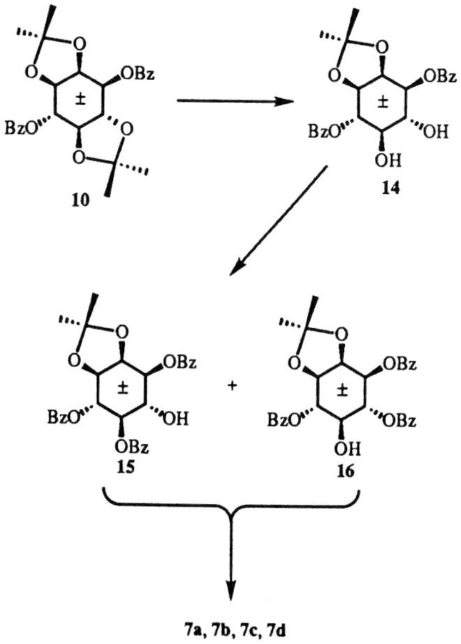

Scheme 6

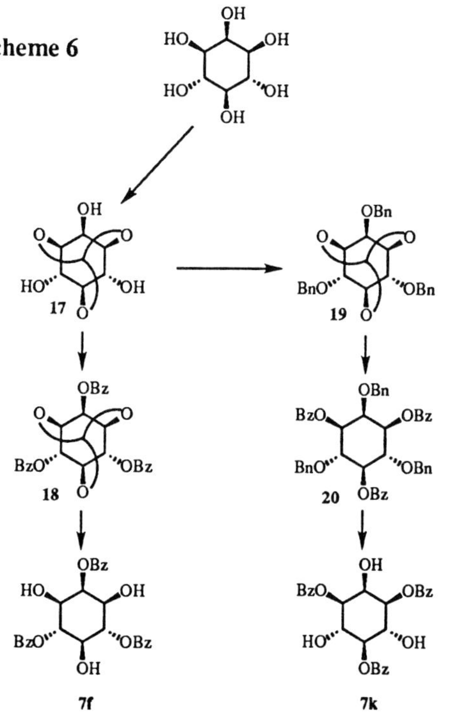

When **12, 13** and the mixture of **15** and **16** were subjected to the 60 % aqueous pyridine conditions and then 80 % aqueous acetic acid at reflux, three sets of 4 regioisomers providing a total of 10 different IBz$_3$ were obtained from the limited benzoyl group migrations. The kinetic behaviors of the benzoyl migration in each set could be conveniently monitored by HPLC at various temperatures. More importantly, in these cases, the individual regioisomer was easily separated by silica gel chromatography at two stages, i.e. with or without the acetonide group. However, the remaining two IBz$_3$ intermediates **7f** and **7k** could not be obtained in this way, and had to be separately prepared from *myo*-inositol orthoformate **17** (Scheme 6). Benzoylation of compound **17**, derived from *myo*-inositol (ref. 21), followed by acid-catalyzed hydrolysis gave **7f**. Alternatively, benzylation of **17**, acid-catalyzed hydrolysis, benzoylation, and then hydrogenolysis provided **7k**. Each of the 12 regioisomers thus obtained was fully characterised by ^{1}H, ^{13}C NMR including H-H COSY and mass(FAB) spectroscopy.

Each of the pure IBz$_3$ isomers(**7**) was phosphorylated as previously described for compound **1** to yield all 12 regioisomers of compound **8**, which were fully characterized by ^{1}H, ^{13}C, ^{31}P NMR including H-H COSY and mass(FAB) spectroscopy. In the final steps, the protecting groups of **8** were removed by successive reactions with trimethylsilyl bromide and then LiOH. The 12 regioisomers of IP$_3$(**9**) was obtained after chromatography on Dowex 50x8-100(H$^+$ form), pH adjustment to 10 with LiOH, and lyophilization (ref. 22).

It is now clear that all 39 regioisomers of *myo*-inositol phosphates(IP$_n$) could be conveniently synthesized by the acyl group migration method as demonstrated for IP$_4$ and IP$_3$ (ref. 8, 9 and 22). Furthermore, it is suggested that the group migration method represents a way of generating molecular diversities in small molecules, and in particular might prove to be a very useful and general synthetic strategy to generate a diverse molecular array of carbohydrate isomers, which would be necessary for the determination of structural specificities in their reactions with biological macromolecules such as receptors, enzymes and antibodies. We are continuing our efforts toward syntheses of the other IP$_n$ isomers and optically active versions of IP$_4$ and IP$_3$ isomers, and exploring the possible generation of small molecule libraries based on inositols and carbohydrates by the group migration method.

Utilities of IP$_4$ and IP$_3$ Regioisomers in Biological Studies.

The issue of whether or not I(1,3,4,5)P$_4$ is a second messenger, and if so what its functions may be, is still controversial. The putative IP$_4$ receptor has recently been purified to homogeneity from porcine platelets, which showed a high affinity and specificity for I(1,3,4,5)P$_4$ (ref. 23 and 24). The specificity of this I(1,3,4,5)P$_4$ binding protein has been examined using all nine synthetic regioisomers of IP$_4$ (Table 1). From the relative potencies of these compounds, three features are readily apparent. First, D/L-I(1,3,4,5)P$_4$ is half as potent as D-enantiomer, consistent with L-I(1,3,4,5)P$_4$ being weaker than the D-enantiomer. Second, any IP$_4$ with a phosphate at the 2 position is essentially inactive. The third observation is that D/L-I(1,4,5,6)P$_4$ is the only isomer that shows any significant ability to displace I(1,3,4,5)P$_4$; K$_d$s for I(1,3,4,5)P$_4$ and L-I(1,4,5,6)P$_4$ are 6.4 nM and 793 nM, respectively, whereas K$_d$ for D/L-I(1,4,5,6)P$_4$ is 390 nM, indicating that the D-I(1,4,5,6) is slightly more potent than the L-enantiomer. In terms of the structure-activity relationship, it appears that phosphorylation of the 1-, 3- and 5- hydroxyl groups is essential for high affinity binding, that there is some tolerance of phosphorylation of the 6-hydroxyl group, but little of a phosphate at the 2-hydroxyl, and that phosphorylation at the 4-hydroxyl has a minor influence (ref. 25).

Table 1

Inositol phosphates (D/L isomer)	IC$_{50}$ (μM)
Ins(1,2,4,5)P$_4$	8.1 ± 0.1
Ins(1,2,4,6)P$_4$	6.3 ± 0.1
Ins(1,2,3,5)P$_4$	> 10
Ins(1,3,4,6)P$_4$	8.2 ± 0.4
Ins(2,4,5,6)P$_4$	5.8 ± 0.1
Ins(1,3,4,5)P$_4$	0.098 ± 0.01
Ins(1,2,5,6)P$_4$	≥ 10
Ins(1,2,3,4)P$_4$	7.7 ± 0.1
Ins(1,4,5,6)P$_4$	1.7 ± 0.2
D-Ins(1,3,4,5)P$_4$	0.056 ± 0.001

myo-Inositol hexakisphosphate(phytic acid) is widely distributed in all eucaryotic cells and in soil. It has been suggested to have such biological roles as phosphate or inositol store, metal chelator, and anti-oxidant. The anti-oxidant properties of various IP_n moleculs including all 9 regiosomers of IP_4 have recently been investigated in terms of their ability to inhibit the hydroxyl radical in the iron-catalyzed Haber-Weiss reaction. It has been demonstrated that the 1,2,3-grouping of phosphates in *myo*-inositol was necessary for the inhibition, thus explaining the powerful anti-oxidant property of phytic acid, D/L-I(1,2,3,4)P_4 and I(1,2,3)P_3 (ref. 26).

All the synthetic regioisomers of IP_3 and IP_4 are under examination for their abilities for binding to the I(1,4,5)P_3 receptors (ref. 27-29), for Ca^{2+} release and its interference either as agonist or antagonist, and for the possible inhibition of the metabolic enzymes such as kinases and phosphatases (ref. 30). These studies are beginning to generate some significant results.

Acknowledgments

We thank Professors Robert H. Michell (Birmingham Univ.), Robin F. Irvine (Cambridge Univ.), David R. Poyner (Aston Univ.), Colin W. Taylor (Cambridge Univ.) and Stefan. R. Nahorski (Univ. of Leicester) for the collaborations involving biological studies. Financial support from the Korea Science & Engineering Foundation/Center for Biofunctional Molecules, and The Ministry of Education/Basic Science Research Fund is gratefully acknowledged.

References and Notes

1.   M. J. Berridge, *Nature (London)*, **1993**, *361*, 315.
2.   R. F. Irvine and P. J. Cullen, *Current Biology*, **1993**, *3*, 540.
3.   S. B. Shears, *Biochem. J.*, **1989**, *260*, 313
4.   F. D. Gunstone, "*Compre. Org. Chem.*", Pergamon Press, London, **1979**, vol. *6*, p 633.
5.   R. M. Rowell, *Carbohydr. Res.*, **1972**, *23*, 417.
6.   S. M. Hecht, *Acc. Chem. Res.*, **1977**, *10*, 239.
7.   V. I. Shvets, *Russian Chem. Rev.*, **1974**, *43*, 488.
8.   S.-K. Chung and Y.-T. Chang, *J. Chem. Soc. Chem. Commun.*, **1995**, 11.
9.   S.-K. Chung and Y.-T. Chang, *J. Chem. Soc. Chem. Commun.*, **1995**, 13.
10. D. C. Billington, "*The Inositol Phosphates, Chemical Synthesis and Biological Significance*", VCH, Weinheim, **1993**
11. A. B. Reitz, ed. "*Inositol Phosphates and Derivatives*", *ACS Symposium Series* 463, American Chemical Society, Washington DC, **1991**.
12. J. L. Meek, F. Davidson and F. W. Hobbs, Jr., *J. Am. Chem. Soc.*, **1988**, *110*, 2317.
13. J. Gigg, R. Gigg, S. Payne and R. Conant, *Carbohydrate Res.*, **1985**, *142*, 132.
14. C. E. McKenna, M. T. Higa, N. H. Cheung and M.-C. McKenna, *Tet. Letters*, **1977**, *1*, 155.
15. C. W. Taylor and A. Richardson, *Pharmac. Ther.* **1991**, *51*, 97-137.
16. B. V. L. Potter, *Nat. Prod. Rep.*, **1990**, 1.
17. G. Powis, *Trends Pharmac.*, **1991**, *12*, 188.
18. Y. Watanabe, T. Ogasawara, S. Ozaki and M. Hirata, *Carbohydr. Res.*, **1994**, *258*, 87.
19. S.-K. Chung and Y. Ryu, *Carbohydr. Res.*, **1994**, *258*, 145.
20. S.-K. Chung, Y.-T. Chang and K.-H. Sohn, *Kor. J. Med. Chem.*, **1994**, *4*, 57.
21. H. W. Lee and Y. Kishi, J. Org. Chem., **1985**, *50*, 4402.
22. S.-K. Chung, Y.-T. Chang and K.-H. Sohn, manuscript in submission.
23. P. J. Cullen, A. P. Dawson and R. F. Irvine, *Biochem. J.*, **1995**, *305*, 139.
24. P. J. Cullen, J. J. Hsuan, O. Truong, A. J. Letcher, T. R. Jackson, A. P. Dawson and R. F. Irvine, *Nature*, **1995**, *376*, 527.
25. P. J. Cullen, S.-K. Chung, Y.-T. Chang, A. P. Dawson and R. F. Irvine, *FEBS Lett.*, **1995**, *358*, 240.
26. I. D. Spiers, C. J. Barker, S.-K. Chung, Y.-T. Chang, S. Freeman, J. M. Gadiner, P. H. Hirst, P. A. Lambert, R. H. Michell, D. R. Poyner, C. H. Schwalbe, A. W. Smith and K. R. H. Solomons, manuscript in submission.
27. J. M. Bond and C. W. Taylor, *Cellular Signalling*, **1991**, *3*, 607
28. R. A. Wilcox, R. A. John Challiss, J. R. Traynor, A. H. Fauq, V. I. Ognayanov, A. P. Kozikowski, and S. R. Nahorski, *J. Biol. Chem.*, **1994**, *269*, 26815.
29. R. A. Wilcox, R. A. John Challiss, G. Baudin, A. Vasella, B. V. L. Potter and S. R. Nahorski, *Biochem. J.*, **1993**, *294*, 191.
30. P. J. Hughes, C. J. Kirk and R. H. Michell, *Biochim. Biophys. Acta*, **1994**, *1223*, 57.

* This paper is dedicated to the memory of Dr. Hogil Kim, the late president of POSTECH.

Industrial bioconversions for chiral molecules, issues and developments

DR STEPHEN C TAYLOR

ZENECA LifeScience Molecules PO Box 2, Belasis Avenue,
BILLINGHAM, Cleveland, TS23 1YN

Abstract

Whilst biological resolution techniques with lipases and esterases have become a well used approach to making single isomer molecules, the direct synthesis of such compounds from non-chiral starting materials such as ketones, using reduction or oxidation biocatalysts, has been more problematical.

This paper considers some of the issues involved and uses a recent example from ZENECA to illustrate how they can be addressed and overcome.

In recent years there has been substantial growth of interest in applying biological catalysts to the production of chemicals by bioconversion processes. The focus for most of this interest has been on developing methods for single isomers of chiral molecules, reflecting both the growing commercial need for such molecules and the lack of cost effective, scaleable chemical methodologies. Although there are needs for these molecules in the agrochemical and food markets, the major sector of relevance is the pharmaceutical industry.

The pharmaceutical industry has been undergoing dramatic change in the 1990s which impacts on the use of bioconversion. At a technical level this has lead to more specific new active molecules where chirality plays a critical role. Furthermore the industry has almost entirely accepted the need for single isomer products rather than racemic ones. The relevance of biotechnology to meeting this need for single isomers is widely recognised and external collaboration is being increasingly utilised by healthcare companies as a way of accessing the particular skills and competencies required to develop and apply large scale bioconversions. However, many of the other changes in the industry are increasing the challenge of using this technology. Drug development timescales are being reduced and phases compressed which means that there is much less time available to develop and scale-up a bioconversion. Downward pressure on costs and a continuing strengthening of the regulatory environment both add to the need, when seeking to apply a bioconversion, for fast and effective selection of appropriate biocatalysts, multipurpose bioconversion assets and flexible downstream processing approaches.

Bioconversions based on ester resolution with lipase and esterase enzymes have become well established with a good supply of "off-the-shelf" stable commercial enzymes available. A skilled, experienced multidisciplinary team can now apply this approach to new pharmaceutical molecules and meet the timescale challenge. However, many of the potentially most valuable bioconversions are based on reduction and oxidation chiral synthesis reactions. In these cases enzymes are typically less stable, have complex cofactor requirements and as a consequence are much less amenable to use as bioconversion catalysts and are certainly difficult to use within the time pressure of a pharmaceutical development programme. ZENECA LifeScience Molecule's approach to this issue has been to focus R&D onto whole microbial-cell bioconversions and to rigorously address the problem of fast micro-organism selection and process scale-up.

The heart of a bioconversion process is the process technology for applying the microbial biocatalyst. ZENECA has established core technology in this area for "redox systems" that is flexible and able to accommodate new biocatalysts essentially based on the use of intact live micro-organisms. However, the front end of the process, where the biocatalyst is selected for a particular substrate, remains a critical area and possibly now the key factor determining development timescale. The importance of this aspect can be illustrated by reference to a recent bioconversion developed and scaled-up by ZENECA for a single isomer hydroxysulfone intermediate required for a new carbonic anhydrase inhibitor (Figure 1). As an optically active alcohol this is representative of the major type of functionality that is currently required by the industry.

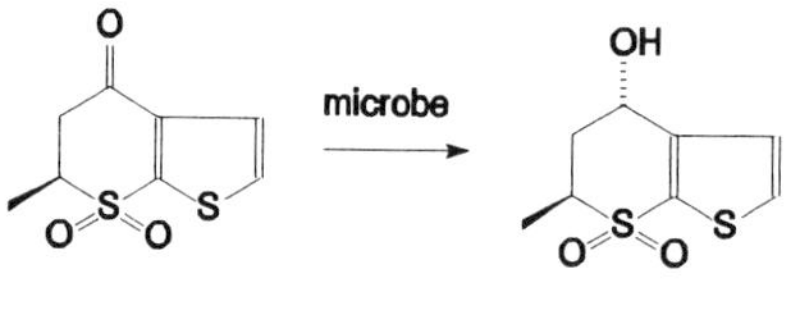

ZENECA'S BIOREDUCTION ROUTE TO HYDROXYSULFONE

FIGURE 1

The target for micro-organism selection was to find and develop a system that would reduce an already optically active ketosulfone interestingly derived by ZENECA from the biological fermentation product, polyhydroxybutyrate, via (R)-3-hydroxybutyrate to the hydroxysulfone with the desired (4S,6S) trans configuration. It was found that a range of yeasts, fungi and bacteria were all able to catalyse the required reduction but with varying selectivity. The apparently best strain in terms of high trans product from a 99% pure starting material, Lactobacillus plantarum, was selected for study (Figure 2).

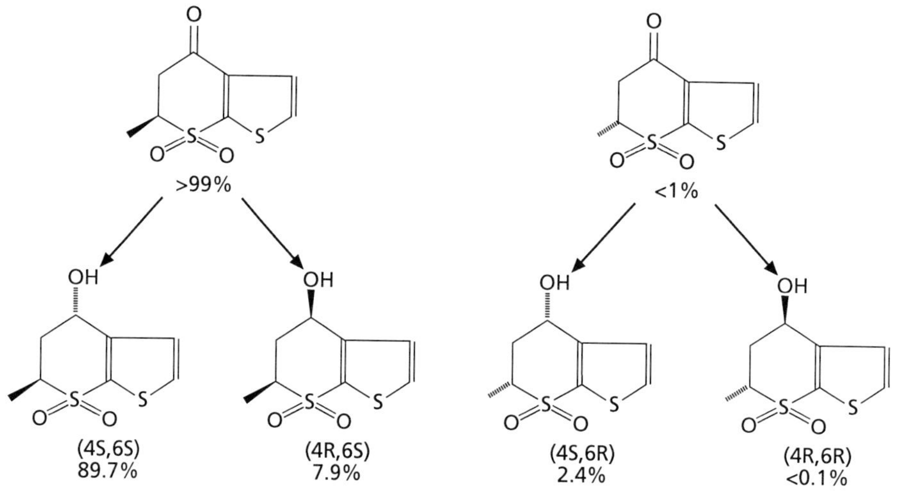

RATIO OF HYDROXYSULFONE DIASTEREOISOMERS OBTAINED FROM A
L. PLANTARUM CATALYSED REACTION. INITIAL pH 7.0.

FIGURE 2

In order to achieve cofactor recycle within the organism, glucose was provided as a co-substrate, the metabolism of which results in a fall in pH of the reaction medium and the production of fairly high optical purity hydroxysulfone. In the absence of glucose there was little reductive activity and no fall in pH. However, the hydroxysulfone that was made was of lower trans : cis ratio than that made in the presence of glucose. This suggested a possible link between pH and stereoselectivity of the reaction. In fact it was determined that (surprisingly) the ketosulfone was extremely pH sensitive and racemised at pH values of above 5.0. Thus in order to obtain high isomer purity product, the reaction would have to be done at a low pH of at least 4.5. Lactobacillus plantarum does not show good activity at this low pH and hence an alternative organism was required. A ZENECA strain of the fungus Neurospora crassa was chosen for further investigation on the basis of its trans : cis ratio of hydroxysulfone products, high conversion, growth characteristics and ability to grow and biotransform at low pH. As a result of this work a process was developed that has now been operated at tonne scale (Figure 3) to give high quality (4S, 6S)-hydroxysulfone.

MK507 HYDROXYSULFONE PROCESS

FERMENTATION

↓

BIOCONVERSION

| CHEMICAL PURITY >99% |
| 4S,6S ISOMER >99% |
| ISOLATED YIELD >80% |

↓

CELL SEPARATION

↓

EXTRACTION & CRYSTALLISATION

FIGURE 3

When the substrate specificity of the dehydrogenase enzyme involved with this reaction was considered, it emerged that it was very limited and no significant activity was found against a range of mono and bicyclic ring structures. In our experience this is not uncommon thus we recognise that a new or different dehydrogenase enzyme is likely to be required for each new substrate. This clearly constrains the ability to establish versatile and relevant bio-reduction systems as commercial, "off-the-shelf" catalysts unlike the situation with ester resolutions for example: hence our ZENECA focus on fast and effective selection of microbial catalysts against defined and specific substrates.

The hydroxysulfone programme illustrates the importance of careful selection of micro-organisms for use in bioconversions and the need to take account of a range of parameters. However, if this is done then the technology can be a powerful way of addressing chirality. Through establishing a library of micro-organisms with increasingly well defined substrate specificities, a fast screening process can be established to identify the appropriate micro-organism/substrate combination. In this way the lack of substrate versatility of many bioreduction systems and thus the need to identify new whole organism catalysts, has minimum effect on the development timescale. It is interesting to compare the timescale of bioconversion development for this hydroxysulfone process to that of the earlier chiral compound, (S-)2-chloropropanoic acid, commercialised by ZENECA in 1989 (Figure 4). A rigorous approach to biocatalyst selection has now reduced by 50% the time taken and when integrated within the overall process development programme ensures that this technology can now fit with the compressed development timescales of the pharmaceutical industry and be the most competitive route option for many molecules.

BIOCONVERSION DEVELOPMENT TIMESCALE

S-2-CPA PROCESS	MONTHS	MK 507 HYDROXYSULFONE PROCESS
0	Start Programme	0
6	Biocatalyst Identified	3
10	100g Scale	5
26	Full Scale / Commercial	14

FIGURE 4

Looking to the future it is clear that more examples will emerge of redox enzymes being successfully applied at commercial scale.   Through pathfinding projects like the hydroxysulfone described here, the bioprocessing issues are being addressed and overcome such that knowledge gained will allow bioconversions for the synthesis of chiral epoxides, sulphoxides and alcohols to become highly cost and time competitive and join ester based resolution as a key technology supporting this industry.   This will be of particular value when linked to the rapid screening and selection of microbial catalysts in the critical first few months of the development of a bioconversion process.

Acknowledgements

The author would like to thank his colleagues in ZENECA BioProducts and ZENECA Fine Chemicals and now the newly formed ZENECA LifeScience Molecules business, whose combined expertise has given success in bioconversion programmes described in this paper and elsewhere.

A new class of an orally active antidiabetic agent, insulin action enhancer: the development of Troglitazone (CS-045)

Takashi Fujita and Takao Yoshioka

Medicinal Chemistry Research Laboratories, Sankyo Co., Ltd.,
No. 2-58, Hiromachi 1-chome, Shinagawa-ku, Tokyo 140, Japan

Abstract: There have been many reports describing relationships between peroxidation and diseases such as diabetes, athereosclerosis and myocardial ischemia (reperfusion) in terms of radical oxidation. Recently, many hindered phenolic compounds known as antioxidants have been studied as potential therapeutic agents especially in the field of inflammation. Our studies on hindered phenols led to the discovery of Troglitazone(CS-045),which is being developed as a hypoglycemic agent with not only hypolipidemic but also lipid peroxide lowering activity. In this symposium we discuss the inhibitory activity of Troglitazone towards peroxidation of LDL in comparison with α-tocopherol and Probucol, and also its quenching activity towards active oxygen species.

Glucose and oxygen are the key substances responsible for sustaining energy in the human living body. Diabetes mellitus complications arise from the body's inability to regulate an abnormally high level of glucose in the plasma. Excess levels of oxygen in the living body can also pose a serious health threat; the so called oxygen toxicity[1] is brought about by active oxygen species such as hydrogen peroxide and oxyl radicals and damages living tissues. These two substances are closely related to aging. Diabetes mellitus is a complex, chronic, progressive disease which eventually impairs the function of kidneys, eyes, and nervous and vascular systems, while active oxygen species are associated with diabetes mellitus and are destructive towards various human tissues as occurring in diabetes mellitus.[2]

Troglitazone has been developed as an oral hypoglycemic agent which enhances the action of insulin in peripheral tissues and liver.[3] Besides its hypoglycemic effects, Troglitazone has an antioxidant activity attributable to its hindered phenol group on the chroman ring. There have been many reports describing relationships between peroxidation and diseases such as diabetes,[4] atherosclerosis[5] and myocardial ischemia (reperfusion) in terms of radical oxidation. Recently, many hindered phenolic compounds functioning as antioxidants have been studied as potential therapeutic agents, especially in the field of inflammation. Examples of reports discussing the relationship between lipid-peroxidation and such angiopathic symptoms are relatively few, except for isolated cases concerning α-tocopherol and Probucol.[6] Our studies on hindered phenols have led to the discovery of Troglitazone(CS-045) as having not only hypoglycemic but also lipid peroxide suppression activities.[7] In this paper we discuss the inhibitory activity of Troglitazone towards peroxidation of LDL and compare its activity with α-tocopherol in terms of atherosclerosis. Furthermore the quenching activity of Troglitazone towards active oxygen species, especially singlet oxygen is also a subject of discussion herein.

Troglitazoe(CS-045)   Probucol

(1) How Troglitazone Was Discovered.

About two decades ago we found that 2,2,6,6-tetramethylpiperidine derivatives had a significant stabilizing effect against photo-degradation of plastics. This work was rudimentary to developing Troglitazone as a therapeutic antioxidant.

Hindered phenols having two alkyl groups at ortho-positions of phenolic hydroxyl group are also used as stabilizers for plastics. It is interesting to note that lipids which are oxidized endogenously may do so nonenzymatically according to the autoxidation mechanism.[8] The resulting lipid peroxides are known to be involved in pathological events such as atherosclerosis, diabetes mellitus, inflammation and several other related ailments. Living organisms, however, utilize natural antioxidants such as α-tocopherol, ascorbic acid, β-carotene, glutathione and superoxide dismutase, thus protecting themselves from peroxidation.

It seemed appropriate therefore to develop such therapeutic agents with antioxidant activity. Probucol was the only drug having antioxidant activity when we started about ten years ago. Primarily we designed hypolipidemic agent (1) having antioxidant activity by considering Gemfibrozil,[9] which was known to be a hypolipidemic, and a hindered phenol group of α-tocopherol. Compound (1) suppressed serum lipid peroxide (s-LPO), triglyceride (TG) and cholesterol (Chol) significantly in alloxan-induced hyperlipoperoxidemic and hyperlipidemic mice. Acetylated vitamin E also inhibited s-LPO but not TG and Chol. Compound 1 decreased TG more effectively than Gemfibrozil in rats. Although compound 1 had potent antioxidant and hypolipidemic activities, it suffered from the drawback of causing an increase in the liver weight of rats.

1   2   3

We considered it appropriate at this stage to design and synthesize benzoxathiole derivatives with a view to targetting them as hypolipidemic agents.[10] Considering Probucol, Clofibrate and α-tocopherol, 5-hydroxy-1,3-benzoxathiole derivatives (2) were designed and synthesized. Although benzoxathioles showed very good peroxidation inhibitory and superoxide quenching activities, their hypolipidenic activities were not so effective as Probucol.
However they inhibited the formation and release of a slow reacting substance of anaphylaxis (SRS-A) and 5-lipoxygenase.

We further designed hypolipidemic and hypoglycemic agents with antioxidant activity. AL-294[11] was known as a hypolipidemic agent and Ciglitazone[12] was also known as a hypoglycemic agent. The moieties of these compounds were connected with antioxidant and lipid peroxide lowering groups to give compounds (3) and Troglitazone. Compound 3 showed a very good hypolipidemic activity but it caused an increase in the liver weight of rats. Troglitazone, showed very potent hypolipidemic and hypoglycemic activities without any adverse effect.

Figure 1 shows the synthetic route to Troglitazone. Trimethylhydroquinone was actylated to acetophenone by a Friedel-Crafts reaction and then in the presence of pyrrolidine acetophenone was condensed with ketone (4) derived from p-nitrophenol to produce chromanone (5). Reduction of carbonyl group of 5 followed by acid-catalyzed dehydration gave chromene (6). Hydrogenation of nitro group and double bond of 6 yielded chromane (7). Diazonium formation of amino group of 7 followed by Meerwein Arylation gave haloester (8) which was reacted with thiourea to form thiazolidine ring (9). Imino group of 9 was hydrolyzed in the presence of acid catalyst to produce Troglitazone.
The biological activities of Troglitazone and its analogues are shown in Table 1. All compounds, even when the phenolic hydroxyl group is acetylated, exhibited good m-LPO inhibitory activity.

Fig. 1 Chemical synthesis of CS-045

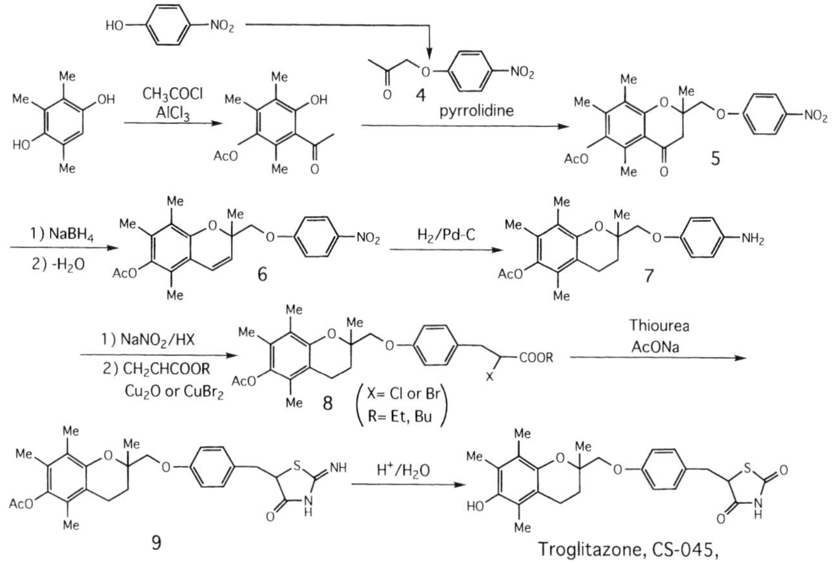

In alloxan-induced hyperlipoperoxidemic and hyperlipidemic mice only Troglitazone decreased s-LPO, TG and Chol significantly. In genetically diabetic KK mice several compounds showed a good hypoglycemic activity. At this juncture Troglitazone was selected as the most promising candidate on the basis of its pharmacological and toxicological profile.

Table 1 Biological activities of the thiazolidine compounds.

	R^1	R^2	R^3	R^4	R^5	n	Z	m-LPO[a] (IC_{50})	KK-mice[b]	ALLOXAN[c] s-LPO	TG	CHOL
CS-045	Me	Me	H	Me	Me	1	O	0.1-0.3	49.0**(38.5*)[d]	54.3*(36.0***)[d]	57.5**(31.9)[d]	26.7***(19.7)[d]
	Me	Me	H	Me	Me	1	NH	< 0.1	-7.5	11.9	16.8	7.2
	Me	Me	Ac	Me	Me	1	O	< 0.1	42.0*(2.6)[d]	6.7	-4.0	4.0
	Me	Me	H	Me	Me	2	O	< 0.1	< 20			
	Me	H	H	t-Bu	H	1	O	< 0.1	51.2*(28.5*)[d]	16.0	-48.6	5.6
	Me	H	H	t-Bu	H	2	O	0.1-0.3	43.6*			
	Me	H	Ac	t-Bu	H	2	NH	0.1-0.3				
	Me	H	H	Me	H	1	O	0.3-1.0	9.5			
	Me	H	Ac	Me	H	1	NH	0.3-1.0	6.3			
	Me	Me	PhCO	Me	Me	1	O	0.1-0.3				
	Me	Me	3-PyCO	Me	Me	1	O	< 0.1	-14.5			
	Me	Me	PrCO	Me	Me	1	O	< 0.1	11.6			
	Me	Me	Ac	MeO	MeO	2	NH	0.3-1.0				
	Me	Me	H	MeO	MeO	2	O	0.3-1.0	12.0			
	H	Me	H	Me	Me	1	O	< 0.1	-5.9			
	Et	Me	H	Me	Me	1	O	0.1-0.3	23.8*	17.6*	14.0	19.5
	i-Bu	Me	H	Me	Me	1	O	0.1-0.3	31.6*	1.5	11.9	12.1

	m-LPO[a] (IC_{50})	KK-mice[b]	s-LPO	TG	CHOL
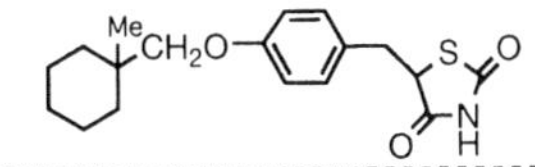 (cyclohexyl structure)	> 1.0	35.7*	18.7	7.9	10.8
AC-Vitamin E	> 1.0		46.1***	27.3	10.7

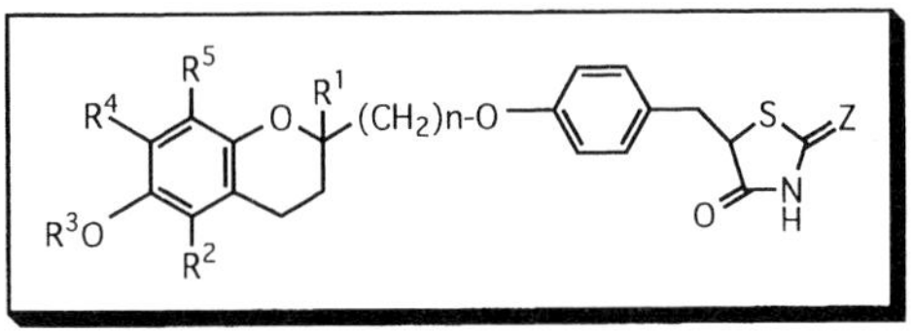

[a] Rat liver microsomal lipid peroxidation, μ g/mL. [b] Hypoglycemic activity in KK-mice, 150mg/kg, po., %decrease. [c] Alloxan induced hyperlipoperoxidemic and hyperlipidemic mice, 100mg/kg, po., %decrease. [d] Dose is 50mg/kg. * p<0.05, ** p<0.02, *** p<0.01.

(2) Antioxidant Activity of Troglitazone from the Atherosclerotic Aspect

Hyperglycemia in diabetic patients is known to cause microangiopathy such as nephropathy, retinopathy and neuropathy. In microangiopathy, glycation[13] and aldose reductase[14] seem to play an important role. On the other hand macroangiopathic events such as ischemic heart disease and celebrovascular disorder also appear frequently in diabetic patients. Atherosclerosis and thrombosis are the predominant causes of these diseases. Active oxygen species and related oxidized products, lipid peroxides, have been singled out as important species in such angiopathic events. Furthermore it is reported that the lipid peroxide level in diabetic patients is higher than that in healthy people.[15] Consequently reduction of the lipid peroxide level and the quenching of active oxygen species are also thought to be effective in suppressing the development of diabetic complications leading to macroangiopathy. In the previous section it is stated that Troglitazone suppressed m-LPO and quenched active oxygen species. Here we would like to show that Troglitazone inhibited the oxidation of low density lipoprotein (LDL) which is reported to be the cause of atherosclerosis.[16] The behavior of Troglitazone in liposomal membrane, a simple model of LDL, studied by electron spin resonance is also shown by comparison with α-tocopherol.

Steinberg and coworkers suggested the mechanism of initiation and formation of atherosclerosis [16] in which oxidatively modified LDL is taken up via scavenger receptors on macrophages and the macrophages are transformed to foam cells. In the course injury endotherial cells and oxidatively modified LDL are responsible for the initiation of atherogenesis. So we found it pertinent to examine the inhibitory effect of Troglitazone in both the oxidation of rabbit LDL induced by Cu^{2+} and foam cell formation in mouse peritoneal macrophages induced by the oxidized LDL. The results are shown in Figure 2. Troglitazone and Probucol inhibited the formation of oxidized LDL at the concentration of 0.5 μM of $CuSO_4$. However, Probucol did not inhibit the oxidation of LDL at 5 μM of $CuSO_4$. Among the antidiabetic thiazolidine compounds only Troglitazone was shown to inhibit the LDL oxidation and Ciglitazone, Pioglitazone and Englitazone failed to inhibit the oxidation of LDL.

Fig. 2   Inhibition of 0.5 μM(A) and 5 μM (B) $CuSO_4$ induced-LDL peroxidation by thiazolidine-2,4-dione derivatives and probucol.

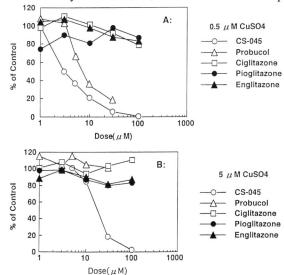

IC50 values of $CuSO_4$ induced-LDL peroxidation

	$CuSO_4$ (μM)	
	0.5	5.0
CS-045	3.0	18.0
Probucol	6.3	>100
Ciglitazone	>100	>100
Pioglitazone	>100	>100
Englitazone	>100	>100

Figure 3 shows ^{14}C-cholesteryl oleate formation in mouse peritoneal macrophages induced by the oxidized LDL obtained in the presence of 30 μM of drug as shown in Fig. 2. Intact LDL did not form ^{14}C-cholesteryl ester and the LDL inhibited by Troglitazone also did not produce it significantly. Probucol suppressed its production slightly. On the other hand the LDL incubated

Fig. 3   ^{14}C-Cholesteryl oleate formation in macrophages.

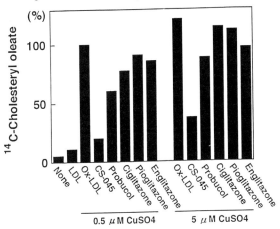

	[14] C-Cholesteryl oleate(%)	
	0.5 μM CuSO4	5 μM CuSO4
None	5.5	-
LDL	8.9	-
Ox-LDL	100	121
CS-045	20.8	37.8
Probucol	60.4	88.1
Ciglitazone	77.1	115
Pioglitazone	90.5	112
Englitazone	86.1	97.0

Drug conc.: 30 μM

Drug conc.: 30 μM

with other thiazolidine compounds formed cholesteryl ester as well as oxidized LDL. These results suggested that Troglitazone may suppress the initiation and progression of atherosclerosis.

We further examined the oxidation of human LDL induced by CuSO$_4$ and the inhibitory effect on it. The inhibitory effect of drugs on LDL peroxidation is shown in Fig. 4.

Fig. 4   Inhibitory effects of drugs against the oxidation of LDL. LDL(200 ug protein/ml) was incubated for 2 hours at 37 C with various concentrations of drugs.

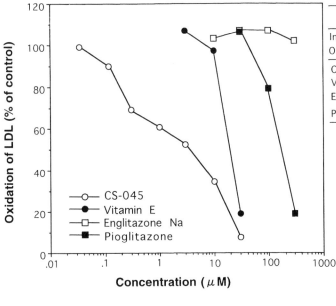

	TBARS (nmol MDA/ml)	IC50 (μM)
Intact-LDL	4.23	
Oxidized-LDL	96.06	
CS-045		2.0
Vitamin E		23.6
Englitazone Na		$\gg$300
Pioglitazone		156.3

Englitazone sodium did not inhibit the oxidation. Pioglitazone HCl was effective at very high concentration. However, Troglitazone showed a very good inhibitory activity. Interestingly Troglitazone is a better antioxidant than α-tocopherol against the oxidation of human LDL. The moiety attributable for this antioxidant activity is considered as being the chromanol. However chromanol moieties in these two compounds are the same. So the problem as to what causes the difference in activities between Troglitazone and α-tocophrol in the oxidation of LDL was indeed a perplexing one. To this end we examined the location of Troglitazone by comparison with α-tocopherol in the liposomal membrane which is a simple model of LDL. Niki and coworkers has clearly demonstrated the location of α-tocopherol in the liposomal membrane by using a spin label technique;[17] stearic acid bearing stable N-oxyl radicals (NS) at either 5, 7, 10, 12 or 16 position and α-tocopherol were incorporated into the liposomal membrane. The decay rate of spin label was measured by electron spin resonance when the liposomal membrane was challenged by the radicals generated in either the interior of the membrane or in the aqueous phase. The sparing efficacy of α-tocopherol was depended on the position of spin label. The nearer the spin label was located to chroman ring of α-tocopherol, the greater the lifetime was preserved. We applied this technique to Troglitazone for the purpose of comparing it with α-tocopherol. The results are shown in Fig. 5 to 6.

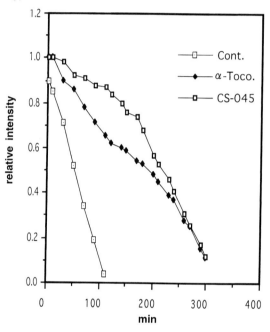

Fig. 5

Consumption of 5NS incorporated into C14:0 PC liposomes during the oxidation initiated with AAPH.
Oxidation reaction was carried out at 37°C in the atmosphere in the absence and presence of α-tocopherol or CS-045.
Concentrations were [5NS]=23μM, [C14:0 PC]=11.6mM, [AAPH]=300mM, and [α-toco.]=[CS-045]=300μM.

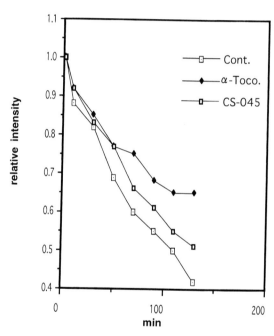

Fig. 6

Consumption of 5NS incorporated into C14:0 PC liposomes during the oxidation initiated with AMVN.
Oxidation reaction was carried out at 50°C in the atmosphere in the absence and presence of α-tocopherol or CS-045.
Concentrations were [5NS]=23μM, [C14:0 PC]=11.6mM, [AMVN]=3mM, and [α-toco.]=[CS-045]=300μM.

Figure 5 shows the time course for consumption of 5-NS, which is a stearic acid spin-labelled at the 5 position and incorporated into the C14:0 PC liposome during oxidation initiated by a water soluble radical initiator AAPH. Oxidation was carried out at 37 C in the atmosphere both in the absence and presence of α-tocopherol and Troglitazone. Radicals generated in aqueous phase are putatively attacked from outside of the membrane. α-Tocopherol and Troglitazone protected N-oxyl (5-NS) and Troglitazone spared more efficiently than α-tocopherol. When 16-NS labeled at 16 position of a stearic acid was used, Troglitazone protected N-oxyl completely at an early stage. The results show that Troglitazone can scavenge radicals more effectively than α-tocopherol; it seems that Troglitazone is located near 16-NS. However, α-tocopherol protected 5-NS more efficiently than Troglitazone as shown in Fig. 6 when a radical was generated in the membrane interior. The same tendency was observed in the case of 16-NS. These results suggest that Troglitazone is not located in the interior of the membrane but nearer to the surface of the membrane than α-tocopherol. One plausible explanation we can offer is that Troglitazone could adhere to the membrane to some extent and/or had greater mobility in the surface region of membrane than α-tocopherol. These results support the results obtained in the inhibitory experiment of Troglitazone on the oxidation of human LDL and thus demonstrate that Troglitazone inhibited the oxidation of LDL more efficiently than α-tocopherol.

Singlet oxygen is also one of active oxygen species produced by neutrophils and macrophages in living body. It was appropriate therefore to examine the quenching of singlet oxygen by Troglitazone. Table 2 shows the singlet oxygen quenching rate constants of Troglitazone and its related compounds. Singlet oxygen was produced photochemically in solution and quenching rate constants were obtained by measuring the decay of the singlet oxygen. From the rate constants Troglitazone has almost an equal quenching activity to that of α-tocopherol. Pentamethylchromanol quenched slightly better than α-tocopherol and Troglitazone. We suppose that pentamethylchromanol has a better mobility than α-tocopherol and Troglitazone. The quenching conferred by benzylthiazolidine compound was not as efficient as that of Troglitazone suggesting that the thiazolidine moiety of Troglitazone did not contribute to the quenching of singlet oxygen. The level of singlet oxygen quenching ability of Troglitazone seemed to match that of α-tocopherol.

Table 2 Quenching of singlet oxygen by CS-045.

Compound	Structure	$(k_r+k_q), M^{-1}s^{-1}$
Troglitazone CS-045		2.14×10^8
ROY-1123		2.91×10^8
α-tocopherol		2.06×10^8
ROY-1350		2.91×10^4

Rate constant of $^1O_2$ quenching by CS-045 and the related compounds.

In conclusion Troglitazone can quench a multitude of active oxygen species including radicals and may contribute to the inhibition and suppression of the initiation and/or progression of macroangiopathy such as atherosclerosis.

References

1. D.B. Balentine, *Pathology of Oxygen Toxicity*, Academic Press, New York,1982 and E.M. Gregory and I. Fridovich, J. Bacteriol.,114,1193-1197(1973)

2. M. Fukuda, H. Ikegami, Y. Kawaguchi, et al, *Diabetes*, 41, 35A (Supplement) and K. Yamada, K, Nonaka, T. Hanafusa, et al, *Diabetes*, 31, 749-753(1982).

3. T. Fujiwara, H. Horikoshi, et al, *Diabetes*, 37, 1549(1988), Y. Iwamoto, T. Kuzuya, et al, *Diabetes Care*, 14, 1083(1991), S.L. Suter, J.M. Olefsky, et al, *Diabetes Care*,15, 193(1992) and H. Horikoshi, T. Fujiwara, et al, *Frontiers in Diabetic Research. Lessons from Animal Diabetes III.* Ed. Shafrir,VI. 8, pp. 320-324(1990) Smith-Gordon,

4. J.V. Hunt, C.C.T. Smith and S.P. Wofll, *Diabetes*, 39, 1420-1424(1990), T. Sakurai, S. Kimura, M. Nakano, et al, *Biochem. Biophys Res. Commun.*, 177, 433-439(1991) and T. Sakurai, S. Kimura, M. Nakano, et al, *Biochem. Biophys. Acta*, 1086, 273-278(1991).

5. J. Glavind, S. Hartmann, J. Clemmesen, K.E. Jessen and H. Dam, *Acta Path. Microbiol. Scand.*, 30, 1(1952), K. Fukuzumi and t. Tanaka, *Oil Chem.*, 10, 659(1961), K. Fukuzumi and Y. Iwata, *Oil Chem.*, 12, 93(1963), K. Fukuzumi, *Oil Chem.*, 14, 119(1965), and K.R. Bruckdorfer, *Current Opinion in Lipidology*, 1, 529-535(1990).

6. K. Hasegawa, K. Iwamoto, Y. Matsushita and M. Yamazaki, *Lipid Peroxide Res.*, 75, 7(1983), F. Tanaka, *Rinsho Kagaku(Jpn J. Clin. Chem.)*, 5, 196(1977), *Chem. Abst.*,87, 199606b(1977) and M.B. Neuworth,R.J. Laufer, J.W. Barnhart, J.A. Sefranka and D.D. McIntosh, *J. Med. Cem.*, 13, 722(1970).

7. T. Yoshioka, T. Fujita, et al, *J. Med. Chem.*, 32, 421(1989).

8. *Autoxidation and Antioxidants* in two volumes, Ed. by W.O. Lundberg, Interscience Publishers, A Division of John Wiley & Sons, New York, London.

9. P.L. Creger, G.W. Moersch, W.A. Neuklis, *Proc. R. Soc. Med.*, 1976, 69, Suppl. 2,3.

10. Y. Aizawa, T. Kanai, et al, *J. Med. Chem.*, 33, 1491-1496(1990).

11. Y. Kawamatsu, H. Asakawa, T. Saraie, et al, *Arzneim.-Forsch.*, 30, 751(1980).

12. T. Sohda, K. Mizuno, E. Imamiya, et al, *Chem. Pharm. Bull.*, 30, 3580(1982).

13. J.V. Hunt, R.T. Dean, et al, *Biochem. J.*, 256, 205-212(1988).

14. D. Dvornik, *Aldose Reductase Inhibition. An Approach to the Prevention of Diabetic Complications*, McGraw-Hill, New York, 1987.

15. Y. Sato, N. Hotta, N. Sakamoto, et al, *Biochem. Med.*, 21, 104-107(1979).

16. M.T. Quinn, S. Parthasarathy, L.G. Fong and D. Steingerg, *Proc. Natl. Acad. Sci. USA*, 84,2995(1987).

17. M. Takahashi, J. Tsuchiya and E. Niki, *J. Am. Chem. Soc.*, 111, 6350-6353(1989).

Kappa agonist phenothiazines: synthesis and S.A.R.

C. Guyon, V. Fardin, C. Garret, B. Plau and G. Taurand

Rhône-Poulenc Rorer S.A. Central Research, Centre de recherches de Vitry-Alfortville 13, quai Jules Guesde, B.P.14, 94403 Vitry-sur-Seine, France.

Abstract: The screening of our library with the aim of finding new kappa receptor ligands allowed the identification of substituted phenothiazines as primary leads. In order to optimize the kappa affinity, various substituents were introduced mainly at positions 2 and 10 on the phenothiazine ring. The simultaneous presence of a 3-(N,N-dialkylamino)-2-propyl chain on the ring nitrogen atom and of a substituent at the position 2 of the phenothiazine ring is required for kappa receptor affinity. RP 60180, (+)-10-[1-methyl-2-(pyrrolidinyl-1-yl)-ethyl]-N-propylphenothiazine-2-carboxamide (IC$_{50}$ = 3 nM), is the representative member of the family. Compounds of this series display agonist activity in an isolated organ assay and were found active in classical tests of analgesia.

INTRODUCTION

Since the concept of multiple opioid receptors was proposed[1], several types of receptors have been identified on the basis of comparative studies with animals. They include: mu (morphine), kappa (ethylketocyclazocine) and sigma (N-allylnormetazocine). Next, tests on isolated organs have disclosed the existence of two other receptors: delta (D-Ala^2-D-Leu^5-enkephalin)[2] and epsilon (β-endorphin). Moreover the existence of receptor subtypes is proposed, particularly for kappa[3] and mu receptors[4].

The interest of kappa agonists was confirmed in the mid 1980s, and many pharmaceutical firms have attempted to find new and potent analgesics devoid of many of the side effects associated with the mu agonists. At that time, the only kappa agonists described in the literature were tifluadom (1), ethylketocyclazocine (2a), bremazocine (2b) and U-50,488 (3)[5]. They display high affinities for the kappa receptor but, except for U-50,488[6], they are not selective for that receptor alone. U-50,488 became, for several research groups, a standard frame for the design of new kappa agonists and, among other things, allowed the discovery of U-62,066 (spiradoline) (4)[7], CI-997 (enadoline) (5)[8] and DUP 747 (6)[9].

a : R = R$_2$ = H, R$_1$ = CH$_3$
b : R = OH, R$_1$ = R$_2$ = CH$_3$

5   6

The first lead compounds were selected through blind screening in the kappa and mu binding tests. RP 21437 **7a** and 23674 **7b** were the two best compounds displaying the highest kappa affinity with an IC_{50} of 250 nM. They belong to the phenothiazine family and are characterized by a 2-(N,N-dimethyl) propylamino chain and a substituent at position 2 on the phenothiazine ring. Several analogues were prepared to disclose structure-activity relationships in this phenothiazine series. The present paper summarizes the results of this approach.

a : NR_1R_2 = NMe_2, R = SO_2—N⟨ ⟩

b : NR_1R_2 = NMe_2, R = CSNHEt

CHEMISTRY

The synthesis of compounds is described in schemes 1 and 2[10].
The phenothiazine ring was coupled with the 2-halogeno-3-dialkylamino-propyl chain according to known methods[11]. This coupling reaction yielded mostly beta isomers which are inactive in opiate binding tests.

Compound **15** was synthesized starting with 2-cyano-phenothiazine **8**. The cyano-ester obtained from **8** by N-alkylation with 2-bromo-propionic acid ethyl ester and sodium hydride, gave the cyano-alcohol **9** upon reduction with sodium borohydride in ethylene glycol. The cyano-mesylate **10** intermediate formed when **9** was treated with methanesulfonyl chloride in the presence of triethylamine, was treated with an excess of secondary amine to yield the cyano-alkylamine **11**. The secondary amine **12** obtained when **10** was treated with primary amine, allowed the introduction of various amine substituents on **11** after N-alkylation with alkyl halides in the presence of sodium carbonate in N,N-dimethylformamide. The thioamide intermediate obtained after the reaction of **11** with hydrogene sulfide in pyridine led to the formation of the N-alkyl-thioamide **13** via a trans-thioamidification with an alkylamine. The latter was transformed into the N-alkyl-amide **15** with mercuric acetate in a water/tetrahydrofurane mixture. The preferred access to the N- alkyl-amide **15** was through the carboxylic acid **14**, which was obtained by hydrolysis of **11** with potassium hydroxyde in ethylene glycol at 180°C and isolated as the hydrochloride. After the transformation of **14** into the carboxylic acid chloride, its coupling with alkylamine gave the N-alkyl-amide **15**. The addition of methyl iodide on **15** led to the ammonium **16**.

The methods used to synthesize amidine derivatives are described in scheme 2. The treatment of the cyano-alkylamine **11** with a hydrochloric acid/methanol mixture gave the corresponding imino-ether which was treated either with primary amine to give the N-alkyl-amidine **17**, or with diamine to give the cyclic amidine **18**. The ammoniums **19** et **20** were obtained by quaternization according to the method described in scheme 1.

The enantiomeric compounds were obtained either by the resolution of **15** using (+) or (-) dibenzoyltartaric acid, or by the resolution of the cyano-alcohol **9** as described in scheme 3. The acid-ester **21** obtained when phtalic anhydride was condensed with **9** and crystallized with (R) or (S) phenylethylamine, gave after saponification the (-)-A and (+)-B forms of **9**.

Scheme 1

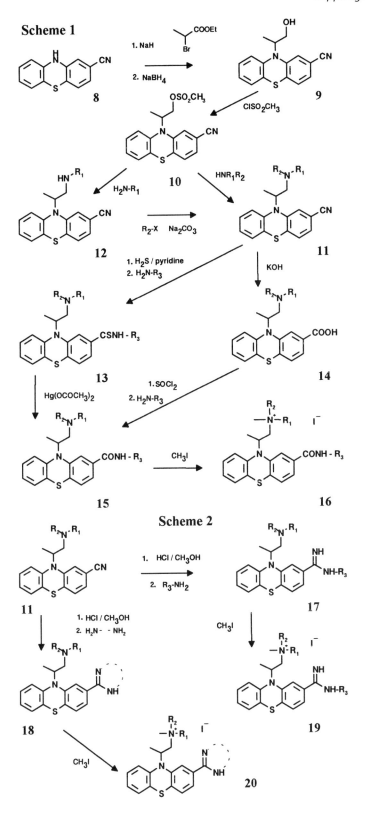

Scheme 2

The enantiomeric compounds respectively from series A and B were synthesized according to the methods described in schemes 1 and 2.

BIOLOGY

The *in vitro* binding affinities of the compounds shown in table 1 were determined by their ability to displace the specific binding of opiate ligands: (^{3}H)-(-)-ethylketocyclazocine (1.2 nM) from kappa receptors in guinea-pig cerebellum homogenates[12] and (^{3}H)-DAMGO (1.2 nM) from mu receptors in guinea-pig brain homogenates after removal cerebellum[13].

In all experiments, morphine was chosen as the standard mu agonist and U-50,488 as the standard kappa agonist.

In vitro, kappa agonist activity was evaluated by the inhibition of electrically-induced contractions in several isolated preparations such as the guinea-pig ileum (mu and kappa receptors)[14] and the rabbit vas deferens (kappa receptors)[15].

In vivo, the antinociceptive effects were evaluated in rodents by using classical nociceptive tests such as the phenylbenzoquinone writing test[16] and the tail-flick test[17], following subcutaneous administration.

Scheme 3

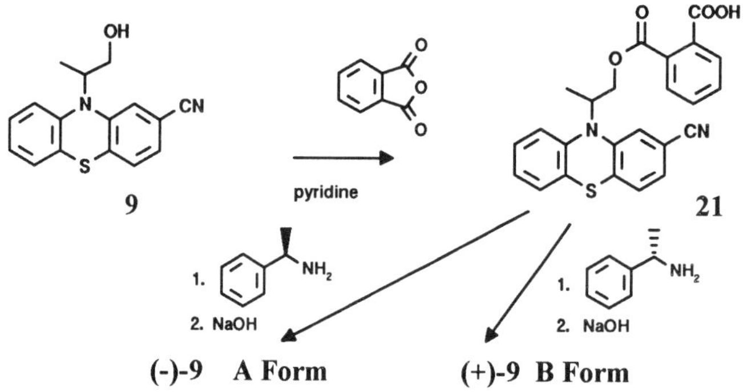

(-)-9   A Form   (+)-9 B Form

RESULTS AND DISCUSSION

Early in our S.A.R. study, it was established on the basis of binding data, that the 3-(N,N-dialkylamino)-2-propyl chain and substitution at the 2-position of the phenothiazine ring, was required for opiate activity. The modifications of the basic moiety and the substitutions have allowed the preparation of either kappa and mu agonists. Compounds 7a and 7b, identified through screening were characterized by submicromolar kappa affinities. A significant increase of potency was observed upon N-alkyl modifications of the 2-thioamide group as in 7b, 22 and 23. Isosteric replacement of this thioamide by an amidine or an amide group yielded a marked gain in kappa affinity as in 24, 25 and 26. The optimal group was the pyrrolidino group 27 and the kappa receptor affinity resides only in the (+)-A form isomer 28, 29. With regards to the modification of the N-alkyl substituent on the amide moiety, the n.propyl substitution appeared to induce the highest kappa/mu selectivity 30, 31. These results led to the selection of RP 60180 28 (IC$_{50}$ = 3 nM, mu/kappa ratio = 70) which has been studied more in depth[18,19]. With the N-benzyl-amide group 32, a marked increase in affinity was observed for both kappa and mu receptors. Ortho aromatic substitutions showed a similar level of affinity (33, 34) whereas a marked decrease was observed with meta and para aromatic substitutions. Although the cyclic amidine derivative, 35, was the most potent kappa agonist, it exhibited a relatively high affinity for the mu receptor (IC$_{50}$ = 0.37 nM, mu/kappa ratio = 17).

7a displayed selectivity for the mu receptor. A dramatic increase in potency for mu receptor was achieved through introduction of N-methyl-N-phenethyl-amino or N-ethyl-N-(3,3-dimethyl-allyl)-amino groups and replacement of sulfonamide by an amidine group 36, 37. As with the kappa receptor, stereospecificity was observed with the mu receptor: the amidine enantiomers, 38 and 39, displayed a mu (A form)/mu (B form) ratio = 100. The additive effects of N-ethyl-N-(3,3-dimethyl-allyl)-amino and amidine groups led to RP 63494 **40**, which possessed a marked affinity for mu opiate sites (IC_{50} = 0.01 nM). Thus, despite its noticeable kappa affinity, this compound displayed an excellent selectivity (kappa/mu ratio = 1100) when compared to morphine (IC_{50} = 6 nM, kappa / mu ratio = 75).

After quaternization of RP 60180 **28** and RP 62622 **39** with methyl iodide, we obtained the compounds **41** (RP 61059) and **43** (RP 63219) which possess, as their parent compounds, preferential affinities for kappa and mu sites, respectively. With other alkyl halides a marked decrease of kappa and mu affinities was observed.

TABLE 1. Modifications of the amine moiety and substitution at the 2-position of phenothiazine **7**. Opioid receptor affinities (IC_{50} nM).

No	NR_1R_2	R	Form	Kappa	Mu
U-50,488				12	900
Morphine				450	6
7a	NMe_2	SO_2-N	A/B	210	17
7b	"	CSNHEt	A/B	280	280
22	"	CSNH-n.Pr	A/B	135	380
23	"	CSNH-n.Bu	A/B	229	1660
24	NEt_2	CSNH-n.Pr	A/B	40	225
25	"	C(=NH)NH-n.Pr	A/B	26	121
26	"	CONH-n.Pr	A/B	22	294
27	Pyrrolidine	"	A/B	9	235
28 RP 60180	"	"	A	3	215
29	Pyrrolidine	CONH-n.Pr	B	550	326
30	"	CONH-n.Bu	A	3	88
31	"	CONH-i.Bu	A	1.4	55
32	"	$CONHCH_2$-Ph	A	1.5	3.5
33	"	$CONHCH_2$-oF-Ph	A	1	22
34	"	$CONHCH_2$-oNO_2-Ph	A	4	55
35	"		A	0.37	6.3
36		$CSNH_2$	A/B	2600	7

37		"	A/B	1650	6
38	Pyrrolidine	C(=NH)NH$_2$	A	53	9
39	"	"	B	30	0.09
40 RP 63494		"	B	11	0.01
41 RP 61059		CONH-n.Pr	A	40	240
42	"	CONHCH$_2$-Ph	A	16	18
43 RP 63219	"	C(=NH)NH$_2$	B	71	0,07

TABLE 2. Kappa agonist activity of RP 60180 in isolated organs.

	Rabbit vas deferens (IC$_{50}$, nM)	Guinea-pig ileum (IC$_{50}$, nM)
RP 60180	15	1.5
U-50,488	23	4.4
Morphine	In.10000	70

RP 60180 dose dependently inhibits electrically-induced contractions in several isolated preparations, these effects are naloxone reversible. Thus, as U-50,488, RP 60180 is a potent kappa agonist *in vitro*.

TABLE 3. Antinociceptive effects of RP60180 in rodents (ED$_{50}$, mg/kg, s.c.).

	RP 60180	U-50,488	Morphine
PBQ test (mouse)	0.08	0.08	0.04
Hot plate (mouse)	>30	>40	7.5
Tail-flick (mouse)	0.8	1.1	2.4
Tail-flick (rat)	>20	>20	2.5
Paw-pressure (rat)	1	3.1	0.3

In vivo, RP 60180 has potent antinoceptive activities at doses similar or superior to that of morphine. Using chemical (PBQ) and thermal (tail-flick) nociceptive stimuli in mice, RP60180 has a similar profile to U-50,488 and morphine. However, in contrast with morphine, the efficacy of kappa agonists against thermal stimuli depends on the species used. These antinoceptive effects are reversed by naloxone. In addition, in cats, even at high doses, RP 60180 does not induce the typical behavioural morphine-like syndrome.

CONCLUSIONS

In this paper, we present new pharmacological activities in the phenothiazine series. The N-alkylation of the phenothiazine ring with 3-(N,N-dialkylamino)-2-propyl chain led, after resolution of the chiral center, to two distinct opiate series: a kappa agonist family (A form) with RP 60180 as representative compound, and a mu agonist family (B form) with RP 63494.

RP 60180 is a potent kappa agonist with an original chemical structure among this pharmacological class. On the basis of its pharmacological properties, it has been chosen for clinical studies .

ACKNOWLEDGMENTS

We thank MM. F.Chenot, J.M.Duchesne, G.Duran, M.Folcke, G.Galea, A.Juin, G.Lamblin, H.N'Guyen, C.Michel, C.Planiol and M.Tillier for contributing to the synthesis of the compounds; and A.Carruette, O.Flamand, A.Jolly for biological evaluations. We thank M. Vuilhorgne and coworkers for extensive NMR investigations and interpretation of data. We are greatful to M.C.Dubroeucq for valuable comments.

REFERENCES

1. a, Martin, W.R., Eades, C.G., Thompson, J.A., Huppler, R.E. and Gilbert, P.E. J.Pharmacol. Exp. Ther. 1976, 197: 517-532.
   b, Gilbert, P.E., Martin, W.R. *Ibid.* 1976, 198: 66.
2. Lord, J.A.H., Waterfield, A.A., Hughes, J. and Kosterlitz, H.W. Nature 1977, 267: 495-499.
3. Wollermann, M., Benyhe and S., Simon, J., Life Sci. 1993, 52: 599-611.
4. Froimowitz, M., Pick,C.G. and Pasternak, G.W. J. Med. Chem. 1992, 35: 1521-1525.
5. Horwell, D.C. Drugs of the future 1988, 12: 1061-1071.
6. Szmuszkowicz, J. and Von Voigtlander, P.F. J. Med. Chem. 1982, 25: 1126-1127.
7. Von Voiglander, P.F. and Lewis, R.A. J. Pharmacol. Exp. Ther. 1988, 246: 259-262.
8. Hunter, J.C., Leighton, G.E., Meecham, K.G. et al. Brit. J. Pharm. 1990, 101: 183-189.
9. Rajagopalan, P., Scribner, R.M., Pennev, P., Schmidt, W.K., Tam, S.W. and Steinffels, G.F. Bioorg. Med. Chem. Lett. 1992, 2: 715-720.
10. Rhône-Poulenc E.P. 346238, 1989; E.P. 346239, 1989; E.P. 346240, 1989; E.P. 419360, 1991.
11. a, Rhône-Poulenc D.E. 824944, 1945; D.E. 841750, 1949.
    b, Charpentier, P. C.R. Acad. Sci. (France) 1947, 225: 306-308.
12. Robson, L.E. Foote, R.W., Maurer, R. and Kosterlitz, H.W. Neuroscience 1984, 12: 621-627.
13. Handa, B.K., Lane, A.C., Lord, J.A., Morgan, B.A., Rance, M.J. and Smith, C.F.C. Eur. J. Pharmacol. 1981, 70: 531-540.
14. Sheeham, M.J., Hayes, A.G. and Tyers, M.B. J. Pharmacol. 1966, 129: 19-24.
15. Oka, T., Hegishi, K., Suda, M., Matsumiya, T., Inazu, T. and Ueki, M. Eur. J. Pharmacol. 1980, 73: 235-236.
16. Blumberg, H., Wolf and P.S., Dayton, H.B. Proc. Soc. Exp. Biol. Med. 1965, 118: 763.
17. D'Amour, F.E. and Smith, D.L. J. Pharmacol. Exp. Ther. 1941, 72: 74.
19. Fardin, V., Plau, B., Carruette, A., Guyon, C., Bardon, T., Taurand, G., Laduron, P.M. and Garret, C. Pain 1990, Suppl. 5, S 192.
19. Fardin, V., Bardon, T., Taurand, G., Reibaud, M., Guillet, M.C., Guyon, C., Carruette, A., Plau, B., Laduron, P.M. and Garret, C. Fundam. and Clini. Pharmacol. 1990, 4: 561.

A new class of diacidic nonpeptide angiotensin II receptor antagonists

Pharmaceutical Research Laboratories I, Takeda Chemical Industries, Ltd.,
17-85 Jusohonmachi 2-chome, Yodogawa-ku, Osaka 532, Japan.

Blockade of the action of angiotensin II (AII) has long been a target for development of novel antihypertensive agents. We recently discovered a novel class of potent nonpeptide AII receptor antagonists, benzimidazole-7-carboxylic acids (e.g., CV-11974). TCV-116, the prodrug of CV-11974, showed highly potent AII antagonistic and antihypertensive activities at oral administration. Structure-activity relationship (SAR) studies revealed that the adjacent arrangement of a lipophilic substituent, a tetrazolylbiphenyl moiety and a carboxyl group was the important structural requirement for potent AII antagonistic activity. Our efforts to find a new acidic bioisostere as a tetrazole replacement, resulted in the discovery of TAK-536 having 5-oxo-1,2,4-oxadiazole ring, which showed both potent AII antagonistic and antihypertensive activity and good oral bioavailability comparable to that of TCV-116.

I. Introduction

The renin-angiotensin system (RAS) plays an important role in the regulation of blood pressure and electrolyte homeostasis. Angiotensin II (AII) is the biologically active component of the RAS and is responsible for most of the peripheral effects of this system. There are three commonly described classes of effective inhibitors of the RAS : renin inhibitors, angiotensin converting enzyme (ACE) inhibitors and AII receptor antagonists. In recent years, renin inhibitors with high specificity and affinity for human renin have been reported, but they have yet to be marketed. ACE inhibitors such as captopril, enalapril and others are very effective for the treatment of most types of hypertension and congestive heart failure. However, their lack of specificity is a major reason for exploring alternative therapy. Some of the adverse effects of ACE inhibitors such as dry cough and angioedema have been attributed to the multisubstrate action of ACE. AII receptor antagonists may specifically block the RAS independently of the source of AII at the receptor site. Saralasin was the first specific peptide antagonist of AII administered to humans, and it was found to reduce blood pressure in hypertensive patients with high renin levels. However, long-term antihypertensive treatment was not possible because these peptide antagonists have low oral bioavailability and very short half-lives.

Our efforts during the last two decades have focused on finding another way to interfere with the RAS. A variety of benzylimidazole-5-acetic acid derivatives were prepared from our original starting material, 2-amino-3,3-dichloroacrylonitrile (ADAN)[1], and were found to have diuretic and antihypertensive activities. During our studies to determine the mode of the pharmacological actions, we found that benzylimidazole acetic acid CV-2198 (1) has

AII receptor blocking activity. To clarify SAR of benzylimidazole acetic acids for AII antagonism, continued research on the chemical modification of CV-2198 (1) culminated in the discovery of CV-2961 (2) having AII antagonist activity 100-fold stronger than CV-2198 [2]. These benzylimidazoleacetic acids were the first nonpeptide AII receptor antagonists, and this finding prompted many researchers to enhance the potency of the prototype. It led to the discovery of the novel series of molecules, DuP 753 [3] and PD-123177 [4], from independent research groups, Du Pont and Park-Davis groups, respectively. DuP 753 (losartan) was an orally active AT1-selective antagonist and has been recently marketed. A second AII antagonist, PD-123177, was an AT2-selective antagonist and did not show significant antihypertensive activity.

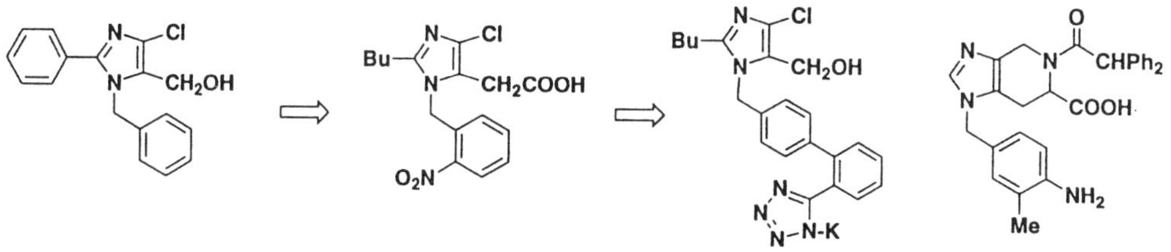

Fig. 1

II. Synthesis and AII receptor antagonist activities of benzimidazole-7-carboxylic acid derivatives

Our strategy of developing more potent AII antagonists was to take advantage of the established structure-activity relationships (SAR) of the benzylimidazoleacetic acids and to incorporate a biphenyl tetrazole moiety which is critical for its AII binding properties as described by the Du Pont group. We designed heterocyclic compounds with key structural features to exhibit potent AII antagonism : a butyl side chain and a biphenyltetrazole moiety. A series of azoles and fused azoles such as pyrrole, pyrazole, triazole, benzimidazole and imidazopyridine were prepared and evaluated for AII antagonism. The affinity of these compounds for the AII receptor was comparable to that of DuP 753, but their *in vivo* AII antagonist activities were less potent. We presumed that this reduction in *in vivo* activity could be ascribed to the absence of the acetic acid moiety which was found to be essential for AII antagonist activity of the benzylimidazoleacetic acids.

A. CV-11194

We then turned our attention to designing fused heterocycles possessing a carboxyl group, particularly benzimidazolecarboxylic acids, by connecting the methylene group of acetic acid group with the 4-position of imidazole (Fig. 2). A series of 2-butylbenzimidazole derivatives (9) bearing a biphenylmethyl moiety was prepared by reductive cyclization of valeroanilides (8) as shown in Scheme I and evaluated for angiotensin II (AII) receptor antagonistic activity (*in vitro* and *in vivo*). Binding affinity was determined using bovine adrenal cortical membrane. Substitution at the 4-,5-, or 6-position reduced the affinity relative to that of the unsubstituted compound (9e, R^1=H). However, most of the compounds with a substituent at the 7-position showed high binding affinity comparable to that of DuP 753 (losartan). In functional studies, a carboxyl group was found to be very important for potent antagonistic activity. Comparison of 2-butylbenzimidazole-7-, 6-, 5-, and 4-carboxylic acids (9a-d) in an AII-induced rabbit aortic ring contraction assay clearly demonstrated the importance of the substitutional position of the carboxyl group. The optimum substituent at the 7-position of the benzimidazole ring was found to be a carboxyl group.

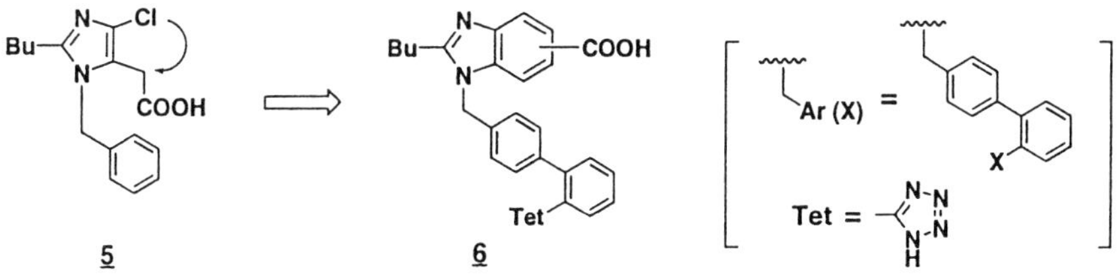

Fig. 2

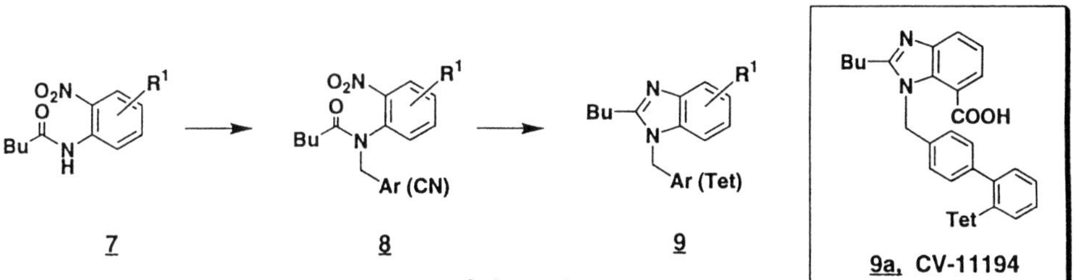

Scheme I

The representative compound, 2-butyl-1-[[2'-(1*H*-tetrazol-5-yl)biphenyl-4-yl]methyl]-1*H*-benzimidazole-7-carboxylic acid, CV-11194 (9a), inhibited the specific binding of [^{125}I]AII to bovine adrenal cortical membrane and the AII-induced contraction of rabbit aortic strips with an IC$_{50}$values of 5.5×10^{-7}M and 1.6×10^{-10}M, respectively, while the compound had no effect on the contraction induced by norepinephrine or KCl. Orally administered CV-11194 at doses of 0.3-10mg/kg dose-dependently inhibited the AII-induced pressor response in rats and dogs. CV-11194 at 1mg/kg, p.o. reduced blood pressure in spontaneously hypertensive rats (SHR). CV-11194 was a novel and highly potent AT1 receptor antagonist and showed long-acting antihypertensive activity [5].

B. CV-11974

Thus, we discovered a novel and orally active nonpeptide AII receptor antagonists, CV-11194, which is more potent and longer acting than DuP 753. Next, we tried to clarify extensive SAR of benzimidazole-7-carboxylic acids having various substituents at the 2-position of the benzimidazole ring. A series of 2-substituted-1-[(biphenyl-4-yl)methyl]-1*H*-benzimidazole-7-carboxylic acids (14,15,16,17) was prepared from the key intermediates (11, 12,13) as shown in Scheme II and evaluated for AII receptor antagonistic activities.

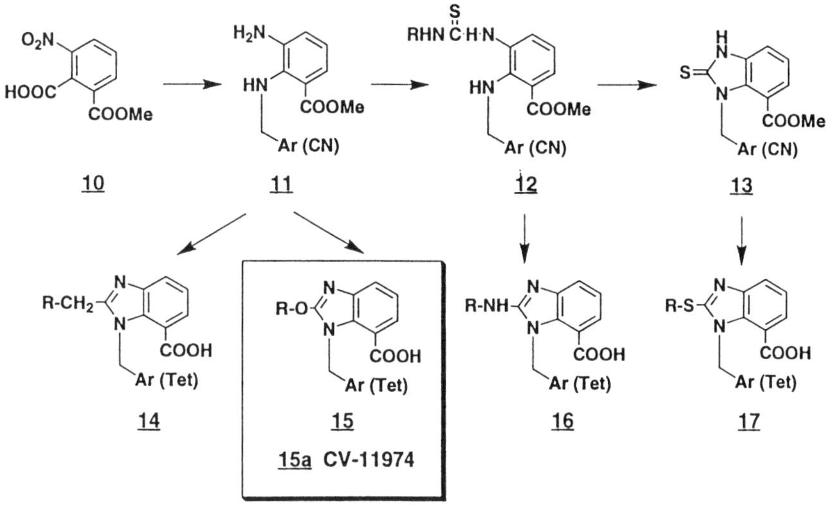

Scheme II

Some of the benzimidazoles showed high affinity for the AII receptor (IC_{50} value, 10^{-6}-10^{-7}M) and inhibited the AII-induced pressor response at 1 or 3 mg/kg, p.o., and the inhibitory effects were more potent than those of CV-11194 and DuP 753. The SAR studies on the binding affinity and the inhibition of AII-induced pressor response suggested that straight chains of a certain length (e.g., EtO, EtS, Pr) were the best as substituents at the 2-position and that their steric factors, lipophilicity, and electronic effects affected the potency of the AII antagonistic action. Three pharmacophoric elements were particularly important for potent and orally active AII antagonistic activity and a long-acting hypotensive effect [6]. Comparison of 2-ethoxybenzimidazole-7-, 6-, 5- and 4-carboxylic acids (15a-d) in an *in vitro* assay clearly demonstrated the importance of the presence of the carboxyl group at 7-position as shown in TABLE 1. The representative compound, 2-ethoxy-1-[[2'-(1H-tetrazol-5-yl)biphenyl-4-yl]methyl]-1H-benzimidazole-7-caboxylic acid, CV-11974 (15a), inhibited the specific binding of [^{125}I]AII to rabbit aortic membrane with an IC_{50} value of 2.86×10^{-8}M. The AII-induced contraction of rabbit aortic strips was potently antagonized by CV-11974 in a noncompetitive manner (pD_2'=9.97); it had no effects on the contraction induced by norepinephrine, KCl, serotonin, $PGF_{2\alpha}$ or endothelin. CV-11974 given intravenously inhibited the AII-induced pressor response in rats with ID_{50} values of 0.033mg/kg. CV-11974 at 0.1-1mg/kg, i.v. reduced blood pressure dose-dependently in spontaneously hypertensive rats (SHR)[7]. These findings demonstrated that CV-11974 is a highly potent and insurmountable AT1-selective antagonist.

TABLE 1. AII Antagonist Activities of 2-Etoxybenzimidazole-7-carboxylic Acid Derivatives

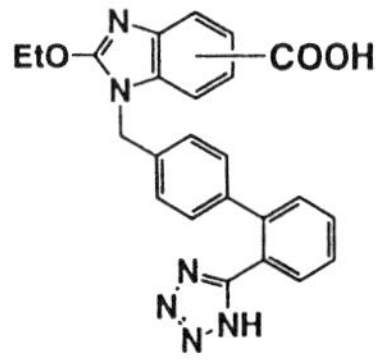

Compound		Receptor Binding (IC50, µM)	AII-induced contraction (IC50, nM)
15a	7-COOH (CV-11974)	0.078	0.2
15b	6-COOH	0.93	1.9
15c	5-COOH	12.6	190
15d	4-COOH	44.7	130

C. TCV-116

Although CV-11194 (9a) and CV-11974 (15a) are very potent AII receptor antagonists, they were found to be poorly absorbed after oral administration. It can be assumed that this is due to the highly polar character of these AII antagonists possessing two acidic groups, a carboxyl group and a tetrazole ring, and that a transient masking of these groups could improve oral absorbtion. To improve the oral bioavailability (BA) of CV-11194 and CV-11974, chemical modification of those compounds to yield prodrugs has been examined. Various ester prodrugs were prepared for this study by the synthetic method outlined in Scheme III.

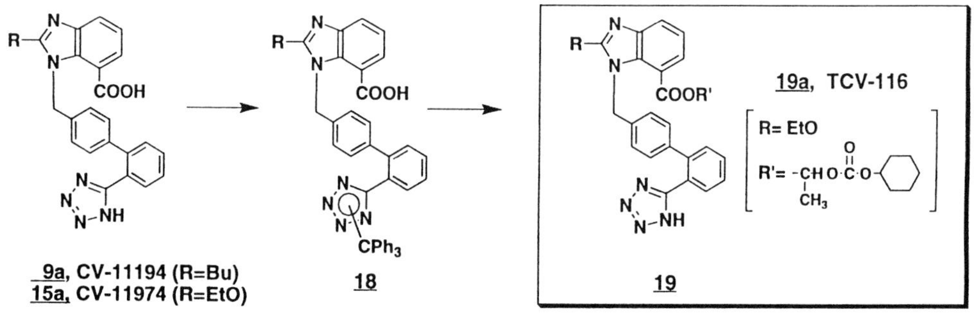

9a, CV-11194 (R=Bu)
15a, CV-11974 (R=EtO)

18

19

19a, TCV-116

R= EtO

Scheme III

After selective tritylation of the tetrazole rings in **9a** and **15a**, treatment of N-tritylated benzimidazole-7-carboxylic acids (18) with a variety of alkyl halides, followed by deprotection with hydrochloric acid, afforded esters of **9a** and **15a**. Although simple alkyl esters did not improve the BA, double esters were found to be effective prodrugs. Their inhibitory effect on the AII-induced pressor response and oral BA in rats were examined [8]. Pivaloyloxymethyl and 1-(cyclohexyloxycarbonyloxy)ethyl esters of **9a** and **15a** showed marked increases in oral BA which significantly potentiated the inhibitory effect of the parent compounds (9a and 15a) on the AII-induced pressor response as shown in TABLE 2. Among them, 1-(cyclohexyloxycarbonyloxy)ethyl 2-ethoxy-1-[[2'-(1*H*-tetrazol-5-yl)biphenyl-4-yl]methyl]-1*H*-benzimidazole-7-carboxylate, TCV-116 (19a) was selected for further evaluation [9].

TABLE 2. Bioavailability and Inhibitory Effect on AII-induced Pressor Responce

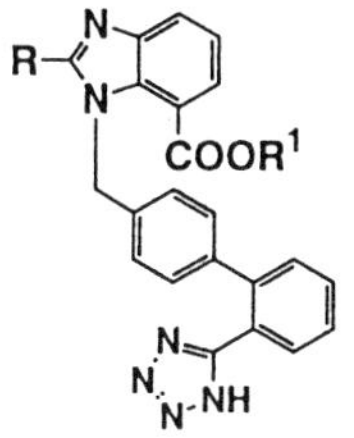

	R	R^1	dose (mg/kg, po)	% inhibition, time(hr) 1	3	7	24	BA (%)
	Bu	H (CV-11194)	1	33	49	53	42	5.7
		-CH$_2$OCO-t-Bu	1	90	95	86	50	52.8
		-CH(Me)OCOO-◯	1	85	81	79	47	60.8
	EtO	H (CV-11974)	0.1	24	67	85	43	5.0
		-CH$_2$OCO-t-Bu	0.1	77	86	86	43	34.9
		-CH(Me)OCOO-◯	0.1	76	87	75	36	33.4

In SHR, TCV-116 (0.1mg/kg, p.o.) had a sustained antihypertensive effect that lasted for more than 10hr and the dose that reduced the blood pressure by an average of 25mmHg for 24hr (ED$_{25}$) was 0.68mg/kg. Repeated p.o. administration of TCV-116 (1mg/kg) to SHR once daily for 2 weeks reduced the blood pressure by 30 to 50 mmHg over 24hr without any heart rate change [10]. Thus,we successfully performed chemical modification of novel AII receptor antagonists, (9a and 15a), to improve oral absorption. On the basis of its profile, TCV-116 has been selected for clinical evaluation as an antihypertensive agent and is the first example of applying the double ester prodrug technique to an orally effective AII antagonists. We believe that this double ester prodrug technique is synthetically easy for good overall yield from diacidic parent compounds and can be used with other diacidic nonpeptide AII antagonists to increase their oral absorption.

D. Other heterocyclic carboxylic acid derivatives

The SAR studies on various benzimidazole derivatives related to CV-11974 revealed that the adjacent arrangement of three pharmacophoric elements, a tetrazolylbiphenylmethyl moiety, a lipophilic substituent and a carboxyl group, was the common and important structural requirement for potent AII antagonistic activity. The presence of the carboxyl group at the 7-position was also very important for *in vivo* and insurmountable AII antagonism.

To examine extensive SAR, a variety of heterocyclic compounds, imidazo[4,5-c]pyridine (20), thieno[3,4-d]imidazole (21) and imidazo[1,2-b]pyrazole (22) derivatives, having the key structural features mentioned above were prepared and evaluated for *in vitro* and *in vivo* AII antagonistic activities.

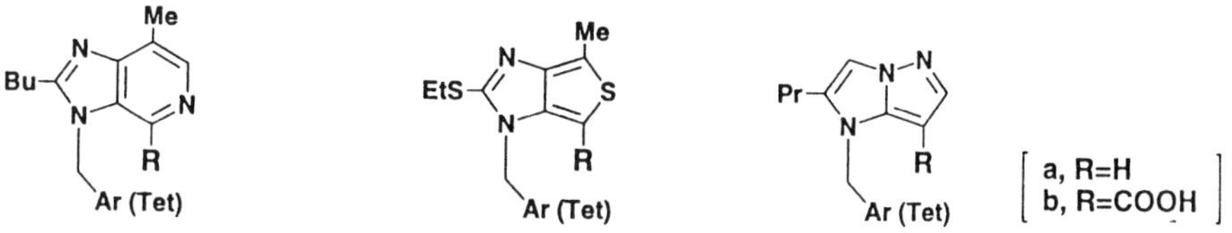

20a,b   **21a,b**   **22a,b**   [a, R=H b, R=COOH]

Fig. 3

In the functional assay using rabbit aorta, the carboxylic acids (20b, 21b, 22b) were 10- to 100-fold more potent than their parent compounds (20a, 21a, 22a) regardless of the chemical structures of the heterocyclic moieties. The carboxylic acid derivatives (20b, 21b, 22b) showed highly potent and insurmountable AII receptor antagonism, whereas the parent compounds showed less potent and surmountable antagonism. The *in vivo* AII antagonistic activities of these compounds were evaluated for inhibition of AII-induced pressor response in conscious normotensive rats. The carboxylic acids (20b, 21b, 22b) were more effective than their parent compounds (20a, 21a, 22a). This result was consistent with the one in *in vitro* functional assay.

This study demonstrated that the adjacent arrangement of three pharmacophoric elements mentioned above not only in benzimidazole but also other heterocycles, was the key structural feature for the potent AII antagonistic activity in *in vitro* and *in vivo*. In all cases, the presence of the carboxyl group is crucial for high potency, insurmountable antagonism and long duration of action[11].

III. A new acidic bioisostere, 5-oxo-1,2,4-oxadiazole and TAK-536

As shown in Fig.4, our efforts have been mainly directed to modification of heterocycles (part A) of the compound (23) having the biphenyl tetrazole moiety. To find a new class of AII antagonist, we turned our attention to chemical modification of the biphenyl tetrazole moiety (part B). The heterocycle (A) was fixed by the benzimidazole-7-carboxylic acid moiety, one of the most potent heterocycles possessing a biphenyl tetrazole moiety, and a series of 2-ethoxy (or 2-butyl) benzimidazole-7-carboxylic acids was prepared and evaluated for AII antagonist activity, in which one benzene ring (X and/or Y) of biphenyl group was replaced by heterocycles (e.g., pyridine, pyrrole, quinone and pyrimidine). However, these compounds were found to have the same or weaker activity than the biphenylmethyl derivatives. Therefore, we searched for a novel tetrazole replacement by other five or six-membered acidic heterocycles.

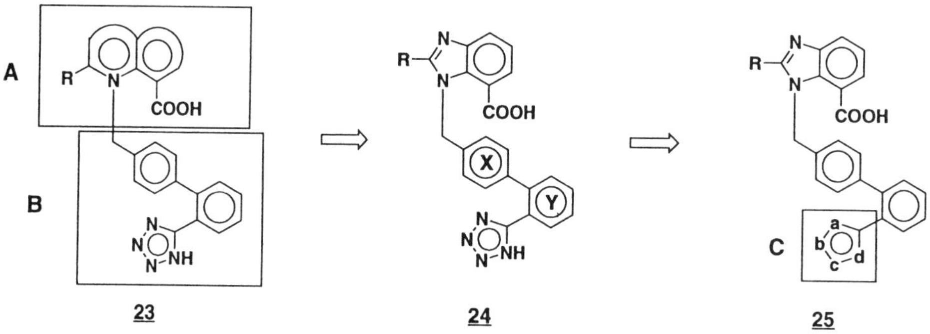

Fig. 4

As shown in Fig. 5, various benzimidazole-7-carboxylic acids possessing five- or six-membered heterocycles as a tetrazol replacement were synthesized and evaluated for AII receptor antagonistic activity. Some of them showed AII antagonistic activity comparable to that of the tetrazole derivative. Herein we focus on a 5-oxo-1,2,4-oxadiazole ring, which is a novel class of the tetrazole replacement, and report the synthesis, AII antagonistic activities and SAR of 2-substituted benzimidazole-7-carboxylic acids possessing the 5-oxo-1,2,4-oxadiazole ring as a novel acidic bioisostere.

A series of 2-substituted benzimidazole-7-carboxylic acids (29) having the 5-oxo-1,2,4-oxadiazole ring was synthesized and evaluated for AII antagonistic activities. As shown in Scheme IV, 5-oxo-1,2,4-oxadiazole derivatives (29) were prepared starting from the nitriles (26). The amidoximes (27), which were prepared from nitriles (26) with hydroxylamine, were reached with chloroformate in the presence of triethylamine to afford acylamidoximes (28). Cyclization of **28** by heating or treatment with 1,8-diazabicyclo[5.4.0]undec-7-ene provided the 5-oxo-1,2,4-oxadiazolederivatives, followed by hydrolysis to give the corresponding 2-substituted benzimidazole-7-carboxylic acids (29). The SAR of the benzimidazole derivatives possessing a 5-oxo-1,2,4-oxadiazole ring was similar to that of the tetrazole counterparts. 2-Ethoxy-1-[[2'-(5-oxo-2,5-dihydro-1,2,4-oxadiazol-3-yl)biphenyl-4-yl]methyl]-1*H*-benzimidazole-7-carboxylic acid, TAK-536 (29a) was a new highly potent AT1-selective receptor antagonist selected for further pharmacological studies.

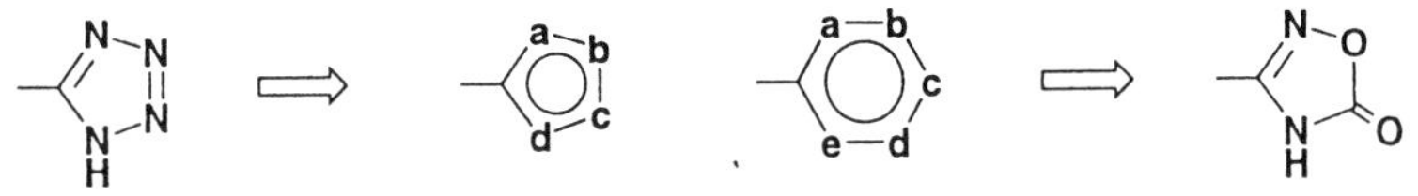

$$\text{a - e : -NH-, -NH=, -O-, -S-, -CO-, -SO-}$$

Fig. 5

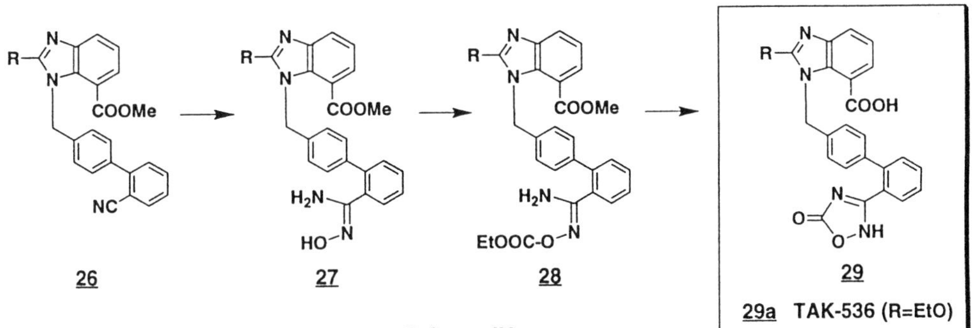

Scheme IV

In isolated aorta helical strips, TAK-536 inhibited AII-induced contraction as potently as the tetrazole derivative (CV-11974) in a non-competitive manner and the pD_2' values of TAK-536 and CV-11974 were 9.93 and 9.97, respectively. In conscious normotensive rats, TAK-536 at more than 0.01mg/kg,p.o. inhibited AII-induced pressor response dose-dependently ($ID_{50} = 0.06$mg/kg, p.o.) as strong as that of TCV-116. In SHR, TAK-536 at more than 0.1mg/kg,p.o. reduced blood pressure dose-dependently ($ED_{25} = 0.40$mg/kg, p.o.) and the antihypertensive effect of TAK-536 was comparable to that of TCV-116 ($ED_{25} = 0.68$mg/kg, p.o.) as shown in Fig. 6.

Fig. 6 Antihypertensive Effects of TAK-536 and TCV-116 in SHR

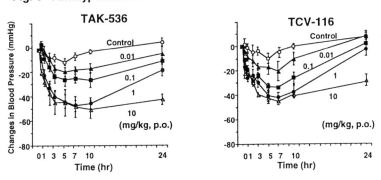

These findings suggested that the BA of TAK-536 was improved in comparison with that of CV-11974. Actually, the BA of TAK-536 in rats was found to be 20%. This indicated that further chemical modification of TAK-536 was unnecessary for improvement of its oral BA. This difference of BA may be explained by their physicochemical properties, the observed partition coefficients (log P) between 1-octanol and water and the pKa values. TAK-536 (log P = 0.90) was found to be more liphophilic than CV-11974 (log P = 0.32) and the 5-oxo-1,2,4-

oxadiazole ring (pKa = 6.1) showed somewhat higher pKa value in comparison to that of the tetrazole ring (pKa = 5.3). The higher lipophilicity and pKa value of TAK-536 may result in the improvement of the BA. TAK-536 is currently undergoing clinical development as an antihypertensive agent [12].

IV. Conclusion

A series of benzimidazole-7-carboxylic acids were prepared and evaluated for in vitro and in vivo AII antagonist activities. The SAR studies demonstrated that the adjacent arrangement of the tree pharmacophores, a tetrazolylbiphenyl methyl moiety, a lipophilic side chain and a carboxyl group was the important structural requirement for potent AII antagonist activities. Among them, 2-ethoxybenzimidazole-7-carboxylic acid, CV-11974, is a most potent selective AT1 receptor antagonist, but its oral bioavailability (BA) is not sufficient for development. The double ester prodrug approaches improved the oral BA, and the prodrug of CV-11974, TCV-116, shows oral activities and potent long-acting antihypertensive effects in animal models. TAK-536 bearing a novel acidic bioisostere, oxadiazole ring as tetrazole replacement shows AII receptor antagonist activity and antihypertensive effect comparable to those of TCV-116 at oral administration. TCV-116 and TAK-536 are under clinical evaluation.

References

1) K.Matsumura, M. Kuritani, H.Shimazu, N. Hashimoto, *Chem. Pharm. Bull.*, **24**, 960 (1976).

2) Y. Furukawa, T. Naka, S. Kishimoto, M. Tomimoto, Y. Mastushita, A. Miyake, K. Itoh, K. Nishikawa, *J. Takeda Res. Lab.*, **50**, 56 (1991).

3) D. J. Carini, J. V. Duncia, P. E. Aldrich, A. T. Chiu, A. L. Johnson, M. E. Pierce, W. A. Price, J. B. Santella, G. J. III, Wells, R. R. Wexler, P. C. Wong, S. -E. Yoo, and P. B. M. W. M. Timmermans, *J. Med. Chem.*, **34**, 2525 (1991).

4) C. J. Blankley, J. C. Hodges, S. R. Klutchko, R. J. Himmelsbach, A. Chucholowski, C. J. Connolly, S. T. Neergard, M. S. van Niseuwenhze, A. Sebastian, J. Quin, A. D. Essenburg, D. M. Cohen, *J. Med. Chem.*, **34**, 3248 (1991).

5) K. Kubo, Y. Inada, Y. Kohara, Y. Sugiura, M. Ojima, K. Itoh, Y. Furukawa, K. Nishikawa and T. Naka, *J. Med.Chem.*, **36**, 1772 (1993).

6) K. Kubo, Y. Kohara, E. Imamiya, Y. Sugiura, Y. Inada, Y. Furukawa, K. Nishikawa and T. Naka, *J. Med. Chem.*, **36**, 2182 (1993).

7) M. Noda, Y. Shibouta, Y. Inada, M. Ojima, T. Wada, T. Sanada, K. Kubo, Y. Kohara, T. Naka and K. Nishikawa, *Biochem. Pharm.*, **46**, 311 (1993).

8) Y. Yoshimura, N. Tada, K. Kubo, M. Miyamoto, Y. Inada, K. Nishikawa and T. Naka, *Int. J. Pharm.*, **103**, 1 (1994).

9) K. Kubo, Y. Kohara, Y. Yoshimura, Y. Inada, Y. Shibouta, Y. Furukawa, T. Kato, K. Nishikawa and T. Naka, *J. Med. Chem.*, **36**, 2343 (1993).

10) Y. Inada, T. Wada, Y. Shibouta, M. Ojima, T. Sanada, K. Ohtsuki, K. Ito, K. Kubo, Y. Kohara, T. Naka and K. Nishikawa, *J. Pharm. Ex. Ther.*, **268**, 1540 (1993).

11) N. Cho, K. Kubo, S. Furuya, Y. Sugiura, T. Yasuma, Y. Kojiara, M. Ojima, Y. Inada, K. Nishikawa and T. Naka, *Bioorganic Med. Chem. Lett.*, **4**, 35 (1994).

12) Y. Kohara, E. Imamiya, K. Kubo, T. Wada, Y. Inada and T. Naka, *Bioorganic Med. Chem. Lett.*, in press (1995).

Synthetic studies using sulfate radicals generated from peroxysulfates

Yong Hae Kim[*], Jae Chul Jung, and Hyun Chul Choi

Department of Chemistry, Korea Advanced Institute of Science and Technology, 373-1, Kusong Dong, Yusong Gu, Taejon 305-701, Korea

Abstract : Tetra-n-butylammonium peroxysulfate **1** ((Bu$_4$N)$_2$S$_2$O$_8$) has been successfully prepared and turned out to be a useful tetra-n-butylammonium sulfate radical **2** (Bu$_4$NSO$_4$·) source.   In contrast to metal peroxydisulfates such as potassium or sodium peroxydisulfate, **1** is miscible in most of the organic solvents of acetone, benzene, methylene chloride, ether and acetonitrile. Thus, **1** has been found to be useful for the various interesting organic synthesis in organic solvents without accompaning the side products derived from hydroxyl or perhydroxy radicals which are gernerally generated in aqueous solvents with metal peroxydisulfates. Primary, secandary, and tertiary alcohols can be converted into the corresponding 2-tetrahydrofuranyl- or 2-tetrahydropyranyl ether in excellent yields in the presence of **1** and tetrahydrofuran or tetrahydropyran respectively.   β-Masked formylated products were obtained by 1,4-addition of electron deficient olefins with **1** in anhydrous 1,3-dioxolane. While various nitroolefins reacted with **1** in anhydrous 1,3-dioxolane to give the corresponding α-masked formylated ketones.   These new methodologies can be applied to synthesize the biologically active compounds such as aristeromycin and carbohydrates.

Oxidation reactions with peroxysulfates are widely encountered in research   and industrial laboratories.   Peroxymonosulfuric acid (Caro's   acid) (1) peroxymonosulfates (2,3), and metal peroxysulfides are stable and commercially available.   The known inorganic peroxydisulfates of K$_2$S$_2$O$_8$ or Na$_2$S$_2$O$_8$ and ammonium peroxydisulfate are insoluble in organic solvents but soluble in water and show a strong oxidizing abilities in aqueous media.   Thermal and photochemical decomposition of S$_2$O$_8^{2-}$ provides the radical anion ·OSO$_3^-$ (4, 5) which is an effective electron transfer agent (4).   It has been demonstrated that the alkoxy radical is generated from a sulfate radical anion and alcohols in aqueous medium by one electron transfer from the alcohols to **2**, and that the alkoxy radical fragments further to the alkyl radicals (6,7). In gernaral, these metal peroxydisulfates are used in aqueous media and accompanied with the various side products.   In order to obtain the organic solvents soluble peroxydisulfate, tetra-n- butylammonium peroxydisulfate **1** was synthesized by the reaction of tetra-n-butaylammonium sulfonic acid with potassium peroxydisulfate (**1** white solid, mp. 119 ^{0}C decomp.) (8).

$$2 \text{ n-Bu}_4\text{NHSO}_4 + \text{K}_2\text{S}_2\text{O}_8 \longrightarrow \text{n-Bu}_4\text{N}^+ \text{O}-\overset{\overset{O}{\|}}{\underset{\underset{O}{\|}}{S}}-\text{O}-\text{O}-\overset{\overset{O}{\|}}{\underset{\underset{O}{\|}}{S}}-\text{O}^- {}^+\text{NBu}_4\text{-n}$$

1

In contrast to metal peroxydisulfate (K$_2$S$_2$O$_8$ or Na$_2$S$_2$O$_8$) or ammonium peroxydisulfate ((NH$_4$)$_2$S$_2$O$_8$), **1** is soluble in aprotic solvents such as chloroform, dichloromethane, acetonitrile, dichloroethane and acetone.   Thus, it turned out to be an excellent source for the formation of sulfate ion radical (·OSO$_3$) under anhydrous conditions (9).   The new oxidizing peroxydisulfate **1** oxidizes various primary and secondary alcohols such as benzylic and allylic alcohols and also alkyl and aryl secondary alcohols to the corresponding aldehydes and ketones, respectively, in almost quantitative yields in aprotic solvents under anhydrous conditions (9).

$$n\text{-}Bu_4N^+ \; \overset{O}{\underset{O}{\overset{\|}{O\text{-}S}}}\text{-}O\text{-}O\text{-}\overset{O}{\underset{O}{\overset{\|}{S}}}\text{-}O^{-} \; {}^+NBu_4\text{-}n \; \rightleftharpoons \; 2 \; n\text{-}Bu_4N^+ \; \overset{O}{\underset{O}{\overset{\|}{O\text{-}S}}}\text{-}O \cdot$$

1   **2**

$$\overset{R_1}{\underset{R_2}{>}}CHOH \; + \; 1 \; \xrightarrow{\; \underset{\text{or } Me_2CO}{C_6H_6, \; CH_2Cl_2} \;} \; \overset{R_1}{\underset{R_2}{>}}C{=}O$$

3   **4**

Under these conditions, primary or secondary alcohols are oxidized to aldehydes or ketones (87 %-99 %) in the presence of alkenes and pyridyl functionalities without reacting with these groups. Under anhydrous conditions, side-reactions occurring by hydroxyl radicals and an aqueous medium can be avoided and hence a thermodynamically stable allylic or benzylic carbon radical may be involved in this reation.

$$R_1R_2CHOH \; \xrightarrow{\; {}^-OSO_3\cdot \quad SO_4^{2-} \;} \; R_1R_2\overset{\cdot}{C}OH \; \xrightarrow{\; {}^-OSO_3\cdot \quad SO_4^{2-} \;} \; R_1R_2C{=}O$$

Both tetrahydrofuranyl ethers and tetrahydropyranyl ethers are utilized for the protection of alcohols. However, it has been a problem to prepare tetrahydrofuranyl ethers by tetrahydrofuranylation of alcohols in mineral acid conditions because the ethers are sensitive and hydrolized under acidic conditions (10). Recently, it has been found that various alcohols containing functional moieties are readily tetrahydrofuranylated in excellent yields by the reaction of tetra-n-butylammonoum disulfate with tetrahydrofuran as a solvent (9). Some of the results are given in TABLE 1.

$$n\text{-}Bu_4N^+ \; \overset{O}{\underset{O}{\overset{\|}{O\text{-}S}}}\text{-}O\text{-}O\text{-}\overset{O}{\underset{O}{\overset{\|}{S}}}\text{-}O^{-} \; {}^+NBu_4\text{-}n \; \rightleftharpoons \; 2 \; n\text{-}Bu_4N^+ \; \overset{O}{\underset{O}{\overset{\|}{O\text{-}S}}}\text{-}O \cdot$$

1   **2**

This reaction is carried out without using an acid catalyst under mild conditions. The tetrahydrofuranylation of alcohols appears to involve the tetrahydrofuranyl radical **5**, followed by attack by the alcohol on the oxonium ion intermediate **6**, which is formed by one electron transfer. The formation of **5** was verified by the isolation and identification of its dimer **8**.

TABLE 1.   Protection of alcohols with (n-Bu$_4$N)$_2$S$_2$O$_8$ in THF

$$\text{ROH} \quad + \quad \text{(TBA)}_2\text{S}_2\text{O}_8 \quad \xrightarrow{\text{THF}} \quad \text{RO-}\!\!\left\langle\!\!\begin{array}{c}\\O\end{array}\!\!\right.$$

Run	Alcohol (0.7 mmol)	(TBA)$_2$S$_2$O$_8$ (mmol)	Time (h)	Product	Yield[a] (%)
1	CH$_3$—⟨ ⟩—CH$_2$OH	1.4	4	CH$_3$—⟨ ⟩—CH$_2$O-⟨O⟩	97
2	⟨ ⟩—OH	1.4	3.5	⟨ ⟩—O-⟨O⟩	91
3	(structure) OH	1.4	7.5	(structure) O-⟨O⟩	95
4	(structure) OH	1.4	4	(structure) O-⟨O⟩	88
5	(structure) OH	0.77	1.5	(structure) O-⟨O⟩	81
6	(structure with S) CH$_2$OH	1.4	3.5	(structure with S) CH$_2$O-⟨O⟩	85
7	(structure) OH	0.77	0.67	(structure) O-⟨O⟩	94
8	(structure) OH	0.77	1	(structure) O-⟨O⟩	87[b]
9	(structure) OH	1.4	6	(structure) O-⟨O⟩	96

a. Isolated yields. b. GC yields.

The same reaction of **1** with alcohols in tetrahydropyran resulted in exellent yields of tetrahydropyranylated ethers **9** (11).

ROH + **1** → RO—(dioxolanyl) **9**
85 ~ 95 %

In order to trap the tetrahydrofuranyl radical **5**, various electron-deficient olefins were reacted with **1** in anhydrous 1,3-dioxolane to give the corresponding dioxolanylated products at the β-position of the olefins.   These new reactions are equivalent to the β-masked formylation by 1,4-addition. Cyclic α,β-unsaturated ketones, esteric alkenes, and sulfonyl alkenes reacted with **1** in 1,3-dioxolane to afford the corresponding 1,3-dioxanyl adducts in excellent yields (12).   The results obtained are summarized in TABLE 2.

$$n\text{-Bu}_4N^{+-}O\text{—}\overset{O}{\underset{O}{\overset{\|}{S}}}\text{—}O\cdot + \underset{O\_O}{} \longrightarrow n\text{-Bu}_4N^{+-}O\text{—}\overset{O}{\underset{O}{\overset{\|}{S}}}\text{—OH} + \underset{O\_O \cdot}{} \quad \mathbf{10}$$

$$\underset{O\_O \cdot}{} + \underset{R_2}{\overset{R_1}{=}}EWG \longrightarrow \overset{O}{\underset{R_1 \; R_2 \; H}{}}\text{—}\overset{EWG}{\underset{}{C:}}$$

While, 1-nitro-cyclohexane was reacted **1** in anhydrous 1,3-dioxolane at 20 °C under argon atmosphere, α-masked formylated cyclohexanone **13** was obtained.

12 + 2 eq. **1** →[20 °C, Ar / 86 %] **13**

Thus, it is possible to introduce the formyl moiety at either α- or β-position of cyclohexanone (Run 5 in TABLE 3).

+ **1** →[90%] **11**

TABLE 2. Masked formylation to the electron-deficient olefin

Run	Olefin	Time(h)	Product	Yield(%)[a]
1	CH$_3$O$_2$C—CH=CH—CO$_2$CH$_3$	1.5		98
2	CH$_3$O$_2$C—CH=CH—CO$_2$CH$_3$	1.5		98
3		18		97
4		54		73
5		44		90
6	CH$_3$CH:CHCO$_2$CH$_3$	15	CH₃ CH-CHCO$_2$CH$_3$	98
7		3		99
8		20		71

a. Isolated yields.

It is interesting that α,β-unsaturated carbonyl compounds resulted in the β-masked formylated ketones **11** but that nitroolefins gave the α-masked formylated ketones **13**. Various nitroolefins were converted to the α-masked formylated ketones. The results obtained are summarized in TABLE 3 (13).

TABLE 3. Preparation of α–masked formylating ketone.

$$\underset{12}{\overset{R_2 \quad NO_2}{\underset{R_3 \quad R_1}{>=<}}} + TBA_2S_2O_8 \xrightarrow[\text{(1:4), -40 °C}]{O\smile O/ CH_2Cl_2} \underset{13}{\overset{R_2 \quad O}{\underset{R_3}{>}}\!\!-\!R_1}$$

Run	Substrate	Time(h)	products	Yield(%)[a]
1	(nitroolefin, long chain)	23	(dioxolane product)	83
2	(nitroolefin)	36	(dioxolane product)	85
3[b]	(nitroolefin)	46	(dioxolane product)	82
4	Ph∼NO₂	25	Ph...NO₂	75
5[b]	(nitrocyclohexene)	25	(dioxolane product)	86
6	(nitrocyclopentene)	35	(dioxolane product)	78

a: Isolated yield. b: Solvent : 1,3-dioxolane only.

The mechanism of denitration and addition of 1,3-dioxolanyl moiety is interesting. The reaction appears to be initiated by formation of 1,3-dioxolane-2-yl radical **10** which attacks the β-carbon of nitroolefins **12** to form **14**. A coupling of **14** with sulfate radical forms the product **13**. The oxygen atom of the α-formylated product is considered to be originated from the oxygens of **1**. The possibility of cleavage of C=N bond species by moisture or water was excluded by $H_2^{18}O$ experiment.

During the reaction and after the complete reaction, $H_2^{18}O$ was added to the reaction mixture and then product was isolated. The ^{18}O incorporated product could not be detected by GC-Mass spectroscopy. Chiral butyrolactones have shown considerable potential as synthetic intermediates in asymmetric synthesis of the carbohydrates or acyclic molecules (14). In order to introduce the formyl function at C-3 position of butenolides, several butenolides have been prepared. The butenolides reacted with 1,3-dioxolane and **1** to give the corresponding β-formylated products in excellent yields (TABLE 4)(15).

TABLE 4. β–Masked formylation of α,β–unsaturated lactones with 1 and 1,3- dioxane

Run	Substrate	Time(min)	Yield[a] (%)
1		26	90[b]
2		20	92
3		18	94
4		25	87
5		15	96

a. Isolated yields. b. Determined by GC.

(S)-5-(t-butyl-diphenylsiloxymethyl)-2-(5H)-furanone **15** was synthesized and treated with **1** and 1,3-dioxolane to appord one diastreoisomer **16** in high stereoselectivity by *trans* addition (15).

15 → **16**   96 %, ~100 % de

The diasteroselective addition has been applied to enantiospecific synthesis of (-) aristereomycin **17**.

82 % yield
~99 % de

overall yield : 50 %

17
aristeromycin

ACNOWLEDGEMENT   The authors gratefully acknowledge support of this work by Center or Biomolecules and Korea Advanced Institute of Science and Technology.

REFERENCES

1.   Nishhara, A. and Kubota. *J. Org. Chem.* **1968**, *33*, 2525.
2.   Trost, B. M. and Curran, D. P. *Tetrahedron Lett.*, **1981**, *22*, 1287.
3.   Trost, B. M. and Blaslau, R. *J. Org. Chem.* **1988**, *53*, 532.
4.   Latimer, M. *Oxidation States of the Elements and their Potentiols in Aqueous Solution;* Prentice-Nall; New York, 1952, p. 78.
5.   House, D. A. *Chem. Rev.* **1962**, *62*, 185 and related reference cited therein.
6.   Ledwith, A.; Russel, P. J.; Sutcliffe, L. H. *J. Chem. Soc. Perkin Trans.* **1973**, *2*, 63.
7.   Caronna, T.; Citterio, A.; Grossi, L.; Minisci, F.; Ogawa, K. *Tetrahedron* **1976**, *32*, 2741.
8.   Jung, J. C., Choi, H. C., and Kim, Y. H. *Tetrahedron Lett.* **1993**, *34*, 3581.
9.   Kim Y. H. and Jung, J. C. unpublished data.
10.   Caronna, T., Citterio, A., Grossi, L., Minisci, F., and Ogawa, K. *Tetrahedron* **1976**, *23*, 2741.
11.   Choi, H. C., Jung, J. C., Cho, K. I., and Kim, Y. H. *Heteroatom Chem..* in press.
12.   Kim, Y. H., and Jung, J. C. unpublished data.
13.   Kim, Y. H., and Jung, J. C. unpublished data.
14.   (a). Hanessian, S. and Sahoo, S. P. *Tetrahedron Lett.* **1985**, *26*, 5627
      (b). *ibid*, **1985**, *26*, 5631.
15.   Kim, Y. H., Choi, H. C., and Jung, J. C. unpublished data.

Catalytic asymmetric carbon–carbon bond-forming reaction utilizing rare earth metal complexes

Masakatsu Shibasaki*, Hiroaki Sasai and Takayoshi Arai

Faculty of Pharmaceutical Sciences, University of Tokyo
Hongo, Bunkyo-ku, Tokyo 113, Japan

Abstract: Novel optically active rare earth complexes have made possible a catalytic asymmetric nitroaldol reaction for the first time.   Structural elucidation reveals that the complexes consist of one rare earth metal, three lithium atoms, and three BINOL units.   Applications of the catalytic asymmetric nitroaldol reaction to syntheses of several β-blockers and *erythro*-AHPA have been also achieved. Although the lithium containing rare earth-BINOL complexes are not effective for Michael reactions, the optically active lanthanum-sodium-BINOL complex, prepared from La(O-*i*-Pr)$_3$, (*R*)-BINOL (3 mol equiv), and NaO-*t*-Bu (3 mol equiv), has been found to be quite effective as an asymmetric catalyst for various Michael reactions to give adducts in up to 92% ee.   In addition, the optically active lanthanum-potassium-BINOL complex has been found to be useful for catalytic asymmetric hydrophosphonylation to imines.

Recently we have found that rare earth alkoxides exhibit basic character, which can be utilized in aldol, cyanosilylation and nitroaldol reactions. (1)   Furthermore, we have succeeded in preparing the optically active rare earth-BINOL complexes for the first time and demonstrating that they are useful basic catalysts for asymmetric nitroaldol reactions. (1,2)   Typical results utilizing La-BINOL complex are shown in Scheme 1.   These are the first examples of a catalytic asymmetric nitroaldol reaction.

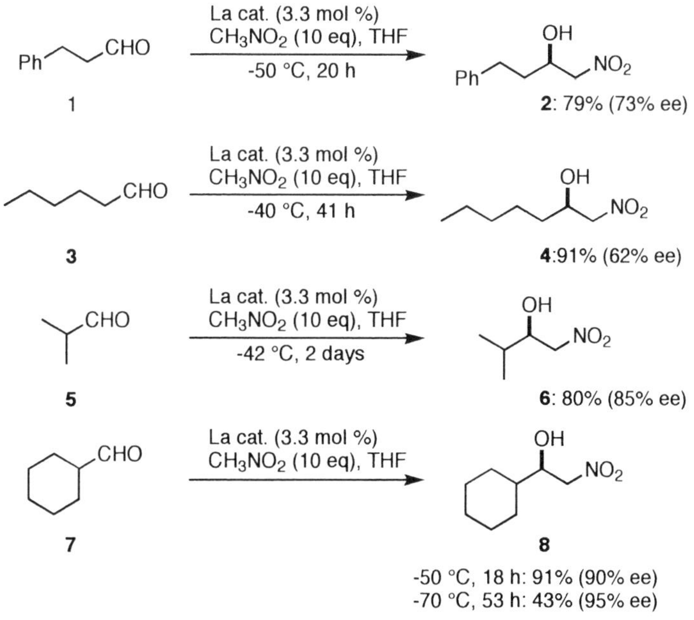

Scheme 1. La-BINOL complex catalyzed asymmetric nitroaldol reactions.

The La-BINOL complex and other rare earth-BINOL complexes are readily prepared starting from rare earth (La, Pr, Nd, Sm, Eu, Gd, Tb, Yb, and Y) trichlorides, di-Li salts of (S)-BINOL (1-2 mol equiv), NaOH (1 mol equiv), and H_2O (10 mol equiv). (3) The rare earth complexes are self assembled in THF and the solution can be directly used as an asymmetric catalyst. The addition of H_2O and NaOH is essential to obtain nitroaldols in high optical purity. The role of H_2O and NaOH are considered to be as shown in Fig. 1.

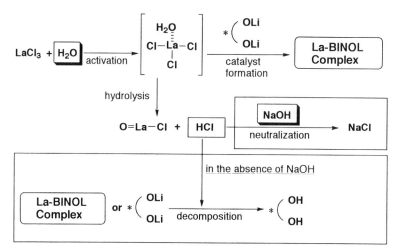

Fig. 1.   Schematic view of the role of H_2O and NaOH in the optically active La complex formation.

Rare earth metals are generally regarded as a group of seventeen elements with similar properties, especially with respect to their chemical reactivity.   However in the case of the above mentioned nitroaldol reaction, we have observed pronounced differences both in the reactivity and in the enantioselectivity among the various rare earth metals used. The unique relationship between ionic radius of rare earth metals and the optical purities of the nitroaldols is depicted in Fig. 2. For example, when benzaldehyde and nitromethane were used as a starting material, the Eu-(S)-BINOL complex gave (R)-9 in 72% ee (91% yield) in contrast to 37% ee (81% yield) in the case of the La-BINOL complex (-40 °C, 40 h). (3)

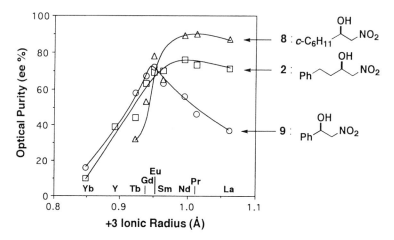

Fig. 2.   Effect of the ionic radii of rare earth metals on the optical purities of nitroaldol derivatives.

The relatively simple ^{13}C and ^{1}H NMR peak patterns of the La-BINOL catalyst suggest (4,5) either a simple structure for the catalyst or the magnetic equivalence of the BINOL moiety in case the catalyst has an oligomeric structure. With the exception of the La catalyst complex, an NMR study of all the other rare earth complexes provided little information on catalyst structure due to the paramagnetism of the rare earth elements.

By conventional EI- and FAB-MS methods, we could obtain only obscure spectra with complex fragment peaks. In contrast, the laser desorption/ionization time-of-flight mass spectrometry (LDI-TOF MS) (6,7) proved to be quite a powerful tool in the analysis of the structure of rare earth BINOL complexes. By LDI-TOF MS method anionic and cationic species could be detected independently according to the measurement mode. In both anionic and cationic mode, tris-binaphthoxy mixed-metal complex of rare earth and Li came up as a candidate for the framework of the catalyst. Representative spectra in anionic mode are shown in Fig. 3. Although the LDI-TOF MS has a mass accuracy of about ±0.1%, (6) the proposed framework was strongly supported by the similarity of the mass spectra of the various rare earth complexes; since rare earth elements have their own atomic weight and isotope abundance distribution. The other BINOL-rare earth complexes also showed a peak pattern similar to La complex, suggesting that these rare earth complexes have fundamentally similar structures. Surprisingly no Cl atom containing fragment was detected.

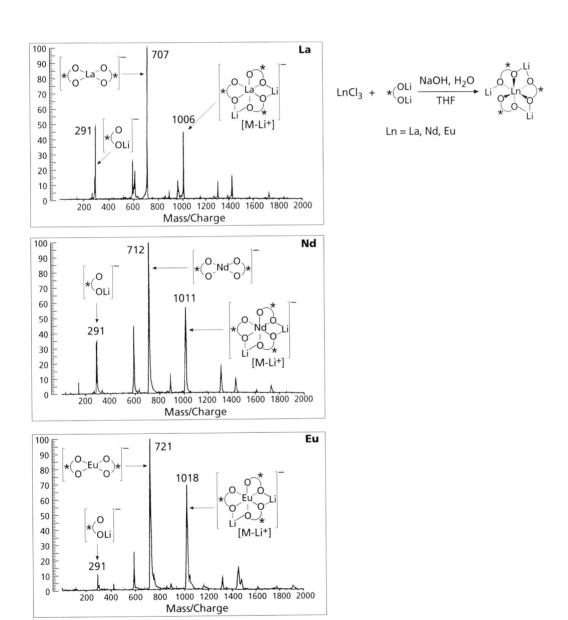

Fig. 3. LDI-TOFMS spectra of La, Nd, and Eu BINOL complex in anion mode.

Although several attempts were made to obtain a X-ray grade crystal of a rare earth complex, these were largely unsuccessful. One reason for the low crystalline character of these rare earth complexes might be contamination due to soluble LiCl in the catalyst solution. Therefore we turned our attention to preparing the crystals under Li free conditions. Starting from rare earth (La, Pr, Nd, Sm, Eu, Gd, Tb, Dy, Er, Yb, and Y) trichlorides, di-Na salt of (*S*)-BINOL, NaO-*t*-Bu, and H₂O, (8) we were pleased to find that several of the rare earth complexes could be crystallized from THF. As shown in Fig. 4, X-ray crystallographic analyses have revealed that La, Eu, Pr, and Nd containing crystals have almost the same structure except for the distance of the atoms around the center rare earth metal. (9) For example the bond length between the rare earth metals and oxygens of BINOL were Eu-O: 2.286 and 2.312; Nd-O: 2.338 and 2.363; Pr-O: 2.365 and 2.386; and La-O: 2.423 and 2.425 Å. These results suggest that small changes in the structure of catalyst (*ca.* 0.1 Å in bond length) cause a drastic change in the optical purity of nitroaldols produced. (3, 4)

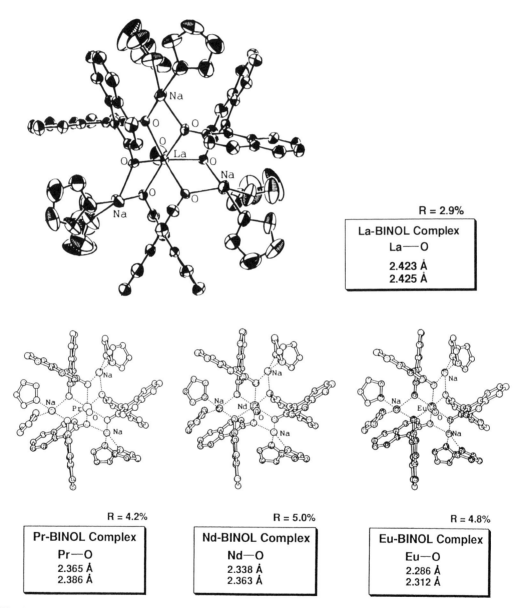

R = 2.9%

| La-BINOL Complex |
| La—O |
| 2.423 Å |
| 2.425 Å |

R = 4.2%

| Pr-BINOL Complex |
| Pr—O |
| 2.365 Å |
| 2.386 Å |

R = 5.0%

| Nd-BINOL Complex |
| Nd—O |
| 2.338 Å |
| 2.363 Å |

R = 4.8%

| Eu-BINOL Complex |
| Eu—O |
| 2.286 Å |
| 2.312 Å |

Fig. 4. Structure of rare earth-sodium-BINOL complexes.

All the above mentioned crystals were basic and acted as catalysts for the nitroaldol reaction. However, the nitroaldols thus obtained were mostly racemic mixtures. For example in the case of reaction of 3-phenylpropanal (**1**) with nitromethane at -40 °C, the results were as follows; Eu: 72% yield, 3% ee (*S*) configuration; Nd: 83%, 7% ee (*R*); Pr: 80%, 9% ee (*R*)). On the other hand, only after they were stirred for 3 days with 3 equivalents of LiCl in THF, did the solutions prove to be efficient asymmetric catalysts (Eu: 84%, 70% ee; Nd: 71%, 68% ee; Pr: 84%, 73% ee). Again by applying the LDI-TOF MS method, all Li free crystals showed similar mass pattern (cf. Fig. 3) except that they contained Na instead of Li. Moreover, complete exchange of Na with Li, on addition of LiCl to the catalyst, was also observed with LDI-TOF MS.

The disclosed structure of the rare earth complexes led us to prepare the La-Li-BINOL complex (LLB) in a rationally designed procedure. (4) Namely, to a stirred suspension of mono-Li salt of (*S*)-BINOL, was added a THF solution of La(O-*i*-Pr)$_3$ (0.33 molar equiv) at room temperature. (10) This catalyst solution also gave almost the same LDI-TOF mass spectra as depicted in Fig. 3, and also showed excellent catalytic activity in the asymmetric nitroaldol reaction.

Nitroaldols are readily converted into β-amino alcohols and/or α-hydroxy carbonyl compounds. As an effective application of the LLB catalyzed nitroaldol reaction, convenient syntheses of three kinds of optically active β-blockers are presented in Scheme 2. (3, 11, 12)

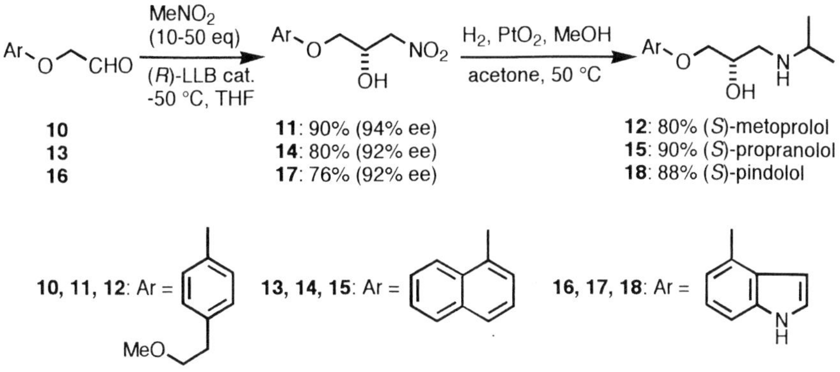
MeNO2
(10-50 eq)
(R)-LLB cat.
-50 °C, THF
H2, PtO2, MeOH
acetone, 50 °C
10
13
16
11: 90% (94% ee)
14: 80% (92% ee)
17: 76% (92% ee)
12: 80% (S)-metoprolol
15: 90% (S)-propranolol
18: 88% (S)-pindolol
10, 11, 12: Ar =
13, 14, 15: Ar =
16, 17, 18: Ar =

Scheme 2. Catalytic asymmetric syntheses of β-blockers using (*R*)-LLB as a catalyst.

Furthermore, diastereoselective catalytic asymmetric nitroaldol reaction of optically active α-amino-aldehydes with nitromethane proceed in a highly diastereoselective manner. A typical adduct, (2*R*,3*S*)-3-phthaloylamino-2-hydroxy-1-nitro-4-phenylbutane was conveniently converted to (2*S*,3*S*)-3-amino-2-hydroxy-4-phenylbutanoic acid (**21**: *erythro*-AHPA; phenylnorstatine), a component of the HIV protease inhibitor KNI-272 (Fig. 5). (13) As shown in Table 1, commonly used bases such as LDA and NaO-*t*-Bu gave the nitroaldol **20** with lower diastereo- and enantioselectivity.

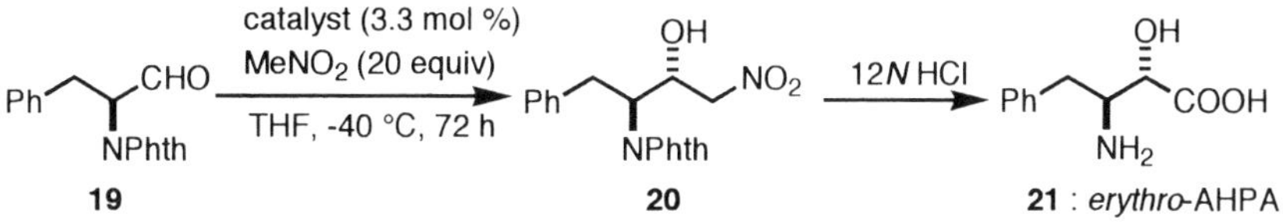
catalyst (3.3 mol %)
MeNO2 (20 equiv)
THF, -40 °C, 72 h
12N HCl
19
20
21 : erythro-AHPA

TABLE 1

entry	catalyst	yield of **20** (%)	*erythro* (% ee) : *threo*
1	(*R*)-LLB	92	99 (96) : 1
2	(*S*)-LLB	96	74 (90) : 26
3	La(O-*i*-Pr)$_3$	52	89 (95) : 11
4	NaO-*t*-Bu	95	62 (65) : 38
5	LDA (1 equiv)	80	74 (49) : 26

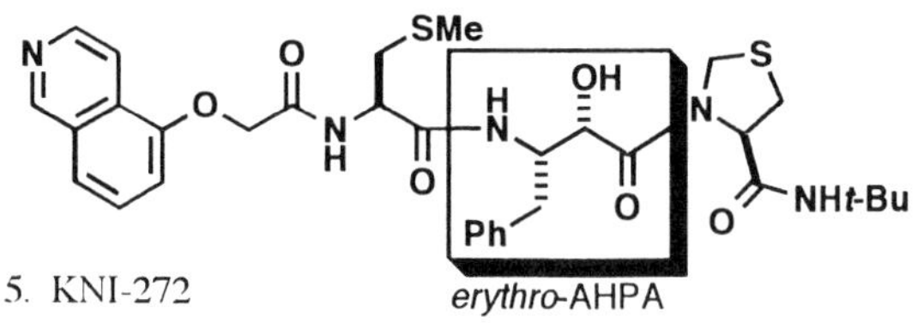
erythro-AHPA

Fig. 5. KNI-272

In the case of a catalytic asymmetric Michael reaction which is another important carbon-carbon bond-forming reaction, LLB and other rare earth-Li-BINOL complexes have been found to be quite ineffective. For example, treatment of cyclopentenone (**31**) with dibenzyl methylmalonate (**25**) in the presence of 3 mol % of LLB gave the Michael adduct **32** of only 13% ee in 26% yield.

We were pleased to find that the La-Na-BINOL complex (LSB) was quite effective in catalytic asymmetric Michael reaction of various enones with malonates to give Michael adducts in up to 92 % ee in almost quantitative yield. (14) Typical results are summarized in Table 2. Quite recently we had succeeded in developing the alkali metal free lanthanum BINOL complex as an effective catalyst for asymmetric Michael reactions. (15) Compared to the alkali metal free lanthanum BINOL catalyst, LSB catalyzed Michael reactions were found to proceed smoothly and with high enantioselectivities even at room temperature. As shown in entry 12, LSB catalyzed the reaction of trans chalcone (**33**) with **27** to give **34** in 77% ee (93% yield). The previous method using the alkali metal free lanthanum BINOL complex gave **34** only in 7% ee.

TABLE 2. Catalytic asymmetric Michael reactions promoted by (R)-LnMB (10 mol %)

22; n = 2   **23**; R^1 = Bn, R^2 = H   **24**; n = 2, R^1 = Bn, R^2 = H
31; n = 1   **25**; R^1 = Bn, R^2 = Me   **26**; n = 2, R^1 = Bn, R^2 = Me
                 27; R^1 = Me, R^2 = H   **28**; n = 2, R^1 = Me, R^2 = H
                 29; R^1 = Et, R^2 = H   **30**; n = 2, R^1 = Et, R^2 = H
                                                 32; n = 1, R^1 = Bn, R^2 = Me

entry	enone	Michael donor	product	cat.	solvent	temp. (°C)	time (h)	yield (%)	ee (%)
1	22	23	24	LSB	THF	0	24	97	88
2	22	23	24	LSB	THF	rt	12	98	85
3	22	23	24	LSB	toluene	rt	12	96	82
4	22	23	24	LLB	THF	rt	12	78	2
5	22	23	24	LPB	THF	rt	12	99	48
6	22	25	26	LSB	THF	0	24	91	92
7	22	25	26	LSB	THF	rt	12	96	90
8	22	27	28	LSB	THF	rt	12	98	83
9	22	29	30	LSB	THF	rt	12	97	81
10	31	25	32	LSB	THF	-40	36	89	72
11	33	27	34	LSB	THF	-50	36	62	0
12	33	27	34	LSB	toluene	-50	24	93	77
13	33	27	34	PrSB	toluene	-50	24	96	56
14	33	27	34	GdSB	toluene	-50	24	54	6

α-Aminophosphonic acids **37** are interesting compounds in the design of enzyme inhibitors. The concept of mimicking tetrahedral transition states of enzyme-mediated peptide bond hydrolysis led in the past years to the successful design and synthesis of phosphonamide containing peptides as a promising new class of protainase inhibitors. It is not surprising that the absolute configuration of the α-carbon strongly influences the biological properties of **37**. We have found that the optically active La-K-BINOL (LPB) complex is useful for the catalytic asymmetric synthesis of β-aminophosphonates. (16) The result is summarized in TABLE 3.

TABLE 3. Catalytic asymmetric hydrophosphonylation

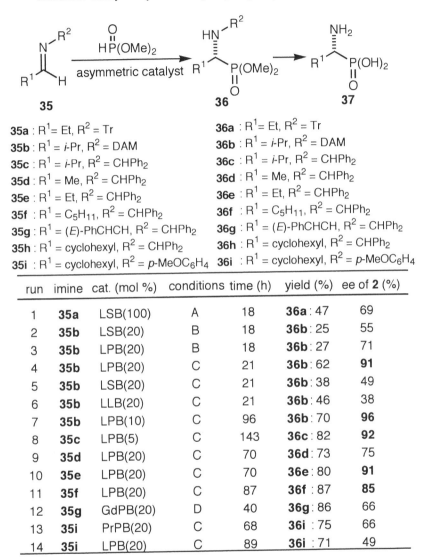

35a : R^1= Et, R^2 = Tr   **36a** : R^1= Et, R^2 = Tr
35b : R^1 = *i*-Pr, R^2 = DAM   **36b** : R^1 = *i*-Pr, R^2 = DAM
35c : R^1 = *i*-Pr, R^2 = CHPh$_2$ **36c** : R^1 = *i*-Pr, R^2 = CHPh$_2$
35d : R^1 = Me, R^2 = CHPh$_2$ **36d** : R^1 = Me, R^2 = CHPh$_2$
35e : R^1 = Et, R^2 = CHPh$_2$ **36e** : R^1 = Et, R^2 = CHPh$_2$
35f : R^1 = C$_5$H$_{11}$, R^2 = CHPh$_2$ **36f** : R^1 = C$_5$H$_{11}$, R^2 = CHPh$_2$
35g : R^1 = (*E*)-PhCHCH, R^2 = CHPh$_2$ **36g** : R^1 = (*E*)-PhCHCH, R^2 = CHPh$_2$
35h : R^1 = cyclohexyl, R^2 = CHPh$_2$ **36h** : R^1 = cyclohexyl, R^2 = CHPh$_2$
35i : R^1 = cyclohexyl, R^2 = *p*-MeOC$_6$H$_4$ **36i** : R^1 = cyclohexyl, R^2 = *p*-MeOC$_6$H$_4$

run	imine	cat. (mol %)	conditions	time (h)	yield (%)	ee of **2** (%)
1	**35a**	LSB(100)	A	18	**36a** : 47	69
2	**35b**	LSB(20)	B	18	**36b** : 25	55
3	**35b**	LPB(20)	B	18	**36b** : 27	71
4	**35b**	LPB(20)	C	21	**36b** : 62	**91**
5	**35b**	LSB(20)	C	21	**36b** : 38	49
6	**35b**	LLB(20)	C	21	**36b** : 46	38
7	**35b**	LPB(10)	C	96	**36b** : 70	**96**
8	**35c**	LPB(5)	C	143	**36c** : 82	**92**
9	**35d**	LPB(20)	C	70	**36d** : 73	75
10	**35e**	LPB(20)	C	70	**36e** : 80	**91**
11	**35f**	LPB(20)	C	87	**36f** : 87	**85**
12	**35g**	GdPB(20)	D	40	**36g** : 86	66
13	**35i**	PrPB(20)	C	68	**36i** : 75	66
14	**35i**	LPB(20)	C	89	**36i** : 71	49

All reactions were performed in the presence of 5 equiv of dimethyl phosphite except run 7 and 8 (1.5 equiv). Condition A ; 60 °C in THF. Condition B ; Room temperature in THF. Condition C ; Room temperature in toluene - THF (7/1). Condition D ; 50 °C in toluene - THF (7/1).

In conclusion three types of asymmetric La-BINOL complexes are now available; namely LLB which is quite effective in catalytic asymmetric nitroaldol reactions, and LSB is very effective in catalytic asymmetric Michael reactions. In addition, LPB is useful for catalytic asymmetric hydrophosphonylation to imines. Very interestingly, the three complexes complement each other in their ability to catalyze asymmetric nitroaldol, asymmetric Michael and asymmetric hydrophosphonylation reactions. Mechanistic studies are in progress.

Acknowledgment. We are greatly indebted to our talented coworkers, Dr. Takeyuki Suzuki, Mr. Shigeru Arai, Mr. Noriie Itoh, and Ms. Yoshie Kirio. We are also grateful to Dr. Tadamasa Date and Mr. Kimio Okamura of Tanabe Seiyaku Co. Ltd., for their assistance with X-ray crystallographic analysis, and Mr. Koichi Tanaka of Shimadzu Corporation for his assistance with LDI-TOF MS measurement. This study was financially supported by a Grant-in-Aid for Scientific Research on Priority Areas from the Ministry of Education, Science and Culture, Japan.

References and notes

1. H. Sasai, T. Suzuki, S. Arai, T. Arai, and M. Shibasaki, *J. Am. Chem. Soc.*, **114**, 4418 (1992).
2. H. Sasai, T. Suzuki, N. Itoh, and M. Shibasaki, *Tetrahedron Lett.*, **34**, 851 (1993).
3. H. Sasai, T. Suzuki, N. Itoh, S. Arai, and M. Shibasaki, *ibid.*, **34**, 2657 (1993).
4. H. Sasai, T. Suzuki, N. Itoh, K. Tanaka, T. Date, K. Okamura, M. Shibasaki, *J. Am. Chem. Soc.*, **115**, 10372 (1993).
5. M. Shibasaki, and H. Sasai, *J. Synth. Org. Chem. Jpn.*, **51**, 972 (1993).
6. Performed by Shimadzu/Kratos KOMPACT MALDI III apparatus.
7. F. Hillenkamp, and M. Karas, *Methods in ENZYMOLOGY, volume 193 ; Mass Spectrometry* , J. A. McCloskey, Eds.; p. 280, Academic Press, Inc., San Diego (1990).
8. The molar ratio was 1:2:1:11, respectively. See reference 3.
9. Crystal data for the rare earth BINOL complexes: $EuNa_3C_{60}H_{36}O_6\bullet6THF\bullet H_2O$, space group $P6_3$, $a = 15.279$ (2) Å, $c = 18.454$ (8) Å; $\alpha = 90.00$ (0)°, $\beta = 90.00$ (0)°, $\gamma = 120.00$ (0)°, $Z = 2$, The structure was solved by direct methods and refined to $R(F) = 0.048$, $R_w(F) = 0.056$. $NdNa_3C_{60}H_{36}O_6\bullet6THF\bullet H_2O$, space group $P6_3$, $a = 15.295$ (1) Å, $c = 18.459$ (2) Å; $\alpha = 90.00$ (0)°, $\beta = 90.00$ (0)°, $\gamma = 120.00$ (0)°, $Z = 2$, The structure was solved by direct methods and refined to $R(F) = 0.050$, $R_w(F) = 0.049$. $PrNa_3C_{60}H_{36}O_6\bullet6THF\bullet H_2O$, space group $P6_3$, $a = 15.309$ (2) Å, $b = 15.309$ (2) Å, $c = 18.452$ (8) Å; $\alpha = 90.00$ (0)°, $\beta = 90.00$ (0)°, $\gamma = 120.00$ (0)°, $Z = 2$, The structure was solved by direct methods and refined to $R(F) = 0.042$, $R_w(F) = 0.048$. $LaNa_3C_{60}H_{36}O_6\bullet6THF\bullet H_2O$, space group $P6_3$, $a = 15.355$ (1) Å, $c = 18.388$ (0) Å; $\alpha = 90.00$ (0)°, $\beta = 90.00$ (0)°, $\gamma = 120.00$ (0)°, $Z = 2$, The structure was solved by direct methods and refined to $R(F) = 0.029$, $R_w(F) = 0.030$.
10. Purchased from Soekawa Chemical Co., Ltd., Tokyo, Japan.
11. H. Sasai, N. Itoh, T. Suzuki, and M. Shibasaki, *Tetrahedron Lett.*, **34**, 855 (1993).
12. H. Sasai, Y. M. A. Yamada, T. Suzuki, and M. Shibasaki, *Tetrahedron*, **50**, 12313 (1994).
13. H. Sasai, W.-S. Kim, T. Suzuki, M. Mitsuda, J. Hasegawa, T. Ohashi, and M. Shibasaki, *Tetrahedron Lett.*, **35**, 6123 (1994).
14. H. Sasai, T. Arai, Y. Satow, K. N. Houk, and M. Shibasaki, *J. Am. Chem. Soc.*, **117**, 6194 (1995).
15. H. Sasai, T. Arai, and M. Shibasaki, *J. Am. Chem. Soc.*, **116**, 1571 (1994).
16. H. Sasai, S. Arai, Y. Tahara, and M. Shibasaki, *J. Org. Chem.*, accepted for publication.

Design and development of practical asymmetric syntheses of drug candidates

Ichiro Shinkai

Merck Research Laboratories, PO Box 2000, Rahway, New Jersey, U.S.A.

Abstract

   Recent progress in asymmetric syntheses has made a significant impact on strategies for drug development in the pharmaceutical industry. Under the recent guidelines by the Food and Drug Administration (FDA), many drug firms are no longer considering the development of racemic forms of chiral drugs. This lecture presents a few examples of highly practical stereoselective syntheses of drug candidates, shown below, discovered and developed at Merck Research Laboratories.

L-738,372   L-743,726

1. L-738,372[1]

   Ongoing worldwide development of a new non-nucleosidal reverse transcriptase inhibitor, L-738,372 required a practical stereoselective synthesis in order to pursue animal safety studies and clinical trials. Stereocontrolled additions of carbon nucleophiles to aldehydes and aldimines have been developed with chiral auxiliaries on the electrophiles or nucleophiles and noncovalent bound chiral additives [2]. Homogeneous THF solutions of acetylides and alkoxides were more selective than the suspensions obtained with toluene.

The chiral additive method appears to be most advantageous since it avoids the auxiliary attachment and removal steps. For practical reasons, the search focused on readily available amino alcohols.

Ephedrine derivatives showed some selectivity, the most promising results were obtained with the cinchona alkaloids. Quinine and dihydroquinine both gave the required S-isomer while quinidine gave the other R-isomer with similar stereoselectivity. We chose quinine for further development due to its availability and cost.

The protecting group on the distal nitrogen also played a key role in stereoselectivity of the addition. There is a substantial electronic influence as well as the effect of steric bulk at this remote position. The selectivity (98% ee) provided by the 9-anthrylmethyl group made it the protecting group of the choice for the synthesis of L-738,372. The scope of the reaction regarding imine variation remains to be explored.

2.   L-743,726[3]

We have developed a highly enantioselective synthesis of L-743,726 based on a lithium cyclopropylacetylide addition to trifluoromethyl ketone intermediate in the presence of an ephedrine alkoxide (96-98 % ee).

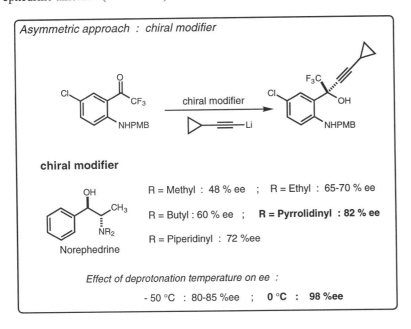

Since the enantiomeric excess of the product alcohol using lithiated cinchona alkaloids was only in the moderate range of 50 - 60 %, we turned towards ephedrine alkaloids as another class of readily accessible amino alcohols. It was reported[4] that the ephedrine nitrogen substituents can play an important role in the % ee of the product during the asymmetric alkylzinc addition to aldehydes in the presence of amino alcohols. The best ligand we found was 1-phenyl-2-(1-pyrrolidinyl)propan-1-ol. Upon optimizing the reaction conditions with this amino alcohol, we noticed that lithiations performed at 0 °C and cooled to - 50 °C prior to adding the ketone gave desired addition product in high stereoselectivity (96-98 %ee). The optical purity of the product was enhanced to 99.5 % ee by crystallization of the product in 80 % yield. Thus, we have developed a highly enantioselective acetylide addition based on lithiated N-pyrrolidinyl norephedrine. We are currently studying the unusual temperature effect on the stereoselectivity of this reaction.

Acknowledgement

I would like to thank my colleagues in the Process Research group at Merck, especially Drs. and Messrs Mark Hoffman, Nobu Yasuda, and Ann DeCamp for their enantioselective acetylide addition to cyclic N-acyl ketimines; Andrew Thompson, Edward Corley, and Martha Huntington for their work on use of an ephedrine alkoxide to mediate enantioselective addition of an acetylide to a prochiral ketone; in addition, Dr. Edward J.J. Grabowski, Dr. Andrew S. Thompson and Dr. Paul J. Reider for their scientific and managerial leadership in guiding these projects.

References

1) Hoffman, M.A; Yasuda, N.; DeCamp, A.E.; Grabowski, E.J.J. *J. Org. Chem.* **1995**, 60, 1590.

2) (a) Miao, C.K.; Sorcek, R.; Johnes, P.-J. *Tetrahedron Lett.* **1993**, *34*, 2259. (b) Poliniaszek, R. P.;Dillard, L. W. *Tetrahedron Lett.* **1990**, *31*, 797. (c) Denmark, S. E.; Weber, T.; Piotrowski, D. W. *J. Am. Chem. Soc.* **1987**, *109*, 2224. (d) Enders, D.; Schubert, H.; Nubling, C. *Angew. Chem. Int. Ed. Engl.* **1986**, *25*, 1109. (e) Takahashi, H.; Inagaki, H.; *Chem. Pharm. Bull.* **1982**, *30*, 972. (f) Ojima, I.; Inaba, S.; *Tetrahedron Chemistry Lett. Lett.* **1980**, *21*, 2081. (g) Takahashi, H.; Tomita, K.; Otomasu, H. *J. Chem. Soc., Chem. Commun.* **1979**, 668. (h) Murahashi, S.-I; Sun, J.; Tsuda, T. *Tetrahedron Lett.* **1993**, *34*, 2645. (i) Ojima, I; Habus, I. *Tetrahedron Lett.* **1990**, *31*, 4289. (j) Tomioka, K.; Inoue, I.; Shindo, M.; Koga, K. *Tetrahedron Lett.* **1991**, *32*, 3095. (k) Itsuno, S.; Yanaka, H.; Hachisuka, C.; Ito, K. *J. Chem. Soc. Perkin Trans. I* **1991**, 1341. (l) Soai, K.; Hatanaka, T.; Miyazawa, J. *J. Chem. Soc., Chem. Commun.* **1992**, 1097. (m) Denmark, S. E.; Nakajima, N.; Nicaise, O. J.-C. *J. Am. Chem. Soc.* **1994**, *116*, 8797.

3) Thompson, A. S.; Corley, E. G.; Huntington, M. F.; Grabowski, E. J. J. Unpublished.

4) (a) Mukaiyama, T.; Suzuki, K.; Soai, K.; Sato, T. *Chemistry Lett.* **1979**, 447. (b) Soai, K.; Yokoyama, S.; Hayasaka, T. *J. Org. Chem.* **1991**, *56*, 4264. (c) Niwa, S.; Soai, K. *J. Chem. Soc., Perkin Trans I*, **1990**, 993. (d) Mukaiyama, T.; Suzuki, K. *Chemistry Lett.* **1980,** 255.

Search for biologically active compounds from tropical plants

Kurt Hostettmann, Olivier Potterat and Jean-Luc Wolfender

Institut de Pharmacognosie et Phytochimie, Université de Lausanne, B.E.P.
CH-1015 Lausanne-Dorigny, Switzerland.

Abstract: Chemical and biological screening approaches are discussed for the discovery of new interesting plant constituents. HPLC coupled to diode array detection (LC-UV) and mass spectrometry (LC-MS) is in particular presented as a highly efficient method for the rapid identification and targeted isolation of new promising molecules. This technique has been used to investigate Gentianaceae and Guttiferae species. The structures of several xanthones could be fully determined on-line. Considerable interest has recently focused on these polyphenols since they are selective and reversible inhibitors of monoamine oxidase (MAO). LC-MS also enabled selective isolation of new types of secoiridoid glucosides.

Higher plants represent a tremendous source of new lead compounds with a variety of applications. Recent examples of development of drugs of plant origin include the new antimalarial agent artemisinin from *Artemisia annua* and the anticancer drug taxol from some yew species. Among the approximately 250,000 known plant species, only a small percentage have been phytochemically investigated, and the fraction submitted to biological or pharmacological screening is even smaller. Moreover plants contain hundreds or thousands of metabolites, only a narrow spectrum of which is touched upon during a phytochemical investigation. The plant kingdom thus represents a largely unexplored reservoir of valuable compounds to be discovered. Obviously, not every one of the 250'000 different species can be submitted to a complete panel of pharmacological and biological assays. The selection of the plants to be studied is therefore a crucial factor for the ultimate success of the investigation. Besides random collection of plant material, targeted collection based on consideration of chemotaxonomic relationships and the exploitation of ethnobotanical information is currently carried out. Ethnopharmacological screens taking into account the empirical knowledge of traditional medicine appear to be the most likely to yield pharmacologically active compounds. Field observations may be particularly useful in screening programs aimed at antimicrobial or insect deterrent/insecticidal activities.

Despite the emergence of powerful techniques, getting pure, pharmacologically active constituents from a plant remains a long and tedious process. This interdisciplinary task requires collaboration of botanists, chemists, pharmacologists and toxicologists. Numerous stages have to be carried out, including proper botanical identification of plant material; biological and chemical screening of crude extracts; isolation of active constituents by several consecutive chromatographic steps; elucidation of chemical structure, and detailed pharmacological and toxicological characterization of pure constituents.

BIOLOGICAL ASSAYS

With the possibilities of modern spectroscopy - NMR, MS and X-ray diffraction, structure elucidation has become rather straightforward. The major challenge today is the discovery of novel leads with promising activities. Isolation of active principles from plant extracts is still largely empirical and requires experience and patience. Crucial is the availability of suitable bioassays. In the screening phase, detection of compounds with the desired biological activity in complex extracts depends on the reliability and sensitivity of the test systems used. Further, bioassays are essential to monitor the required effects throughout the isolation process, in a procedure known as 'activity-guided fractionation'. Thus, all fractions are tested and those continuing to exhibit activity are carried out through further isolation and purification until pure active principles are obtained. The discovery of promising plant extracts and the subsequent activity-guided isolation put specific requirements on the bioassays to be used. Bioassays must be simple, inexpensive and rapid in order to be compatible with a large number of samples which include extracts from the screening phase and all fractions obtained during the isolation procedure. They must be sensitive enough to detect active principles which are generally present only in small concentrations in crude extracts. Their selectivity should be such that the number of false positives is reasonably small. When developing or selecting a test system, the choice of a suitable target is of particular importance. Targets for bioassays can be lower organisms, such as microorganisms, insects, crustaceans and molluscs; isolated subcellular systems, such as enzymes, receptors and organelles; cultured cells of human or animal origin; and whole animals. With a deeper knowledge about the fundamentals of diseases and the molecular basis of living processes, mechanism-based assays using subcellular systems have become increasingly popular. Because of their selectivity and sensitivity, combined with good reproducibility and high sample throughput, this type of assay is given preference for large screening programs. However, such assays do not detect compounds with unknown mechanisms of action and unspecific interactions lead to false positive results (1).

CHEMICAL SCREENING APPROACH

The number of available targets for biological screening is limited. Moreover bioassays are quite often not reliably predictive for clinical efficiency. As further drawback, the bioassay-guided fractionation strategy can lead to the frequent reisolation of known metabolites. The chemical screening of crude plant extracts thus constitutes an efficient complementary approach allowing localization and targeted isolation of new types of constituents with potential activities. This procedure also enables recognition of known metabolites at the earliest stage of separation, thus avoiding a time and money consuming isolation of common compounds. Once novelty or utility of a given constituent is established, it is selectively isolated and can be submitted to a broad panel of biological and pharmacological assays. To be successful, such an approach requires hyphenated techniques able to provide at the same time efficient separation of the metabolites, selectivity and sensitivity of detection, and valuable structural information on-line. Mass spectrometry provides a type of 'universal' LC detection and gives important structural information on-line (molecular weight and fragments). By coupling HPLC with both mass spectrometry (LC-MS) and UV diode array detection (LC-UV), a large amount of data can be obtained about the constituents of a plant extract before starting any isolation work. Indeed UV spectra give useful complementary information (type of chromophore or pattern of substitution) to those obtained with LC-MS. While LC-UV as already been widely used for the analysis of crude plant extract, LC-MS has been introduced more recently and is still not widespread in natural product chemistry. The coupling of LC with MS is indeed not as simple and straightforward as with UV. Several factors have to be considered, in particular the amount of column effluent that has to be introduced in the MS vacuum system, the composition of the eluent, and the type of compounds that must be analyzed. To cope with these problems, a considerable number of interfaces have been built up, each of them having its own characteristics and range of application. Thermospray (TSP) (3) and continuous flow FAB (CF-FAB) (4) have been used in our laboratories in order to ionize a wide range of plant metabolites with various polarities and of various molecular weights.

The use of LC-MS combined with LC-UV has been applied for the investigation of tropical Gentianaceae and Guttiferae species. Considerable interest has recently focused on both plant families because of the presence in several species of xanthones. Xanthones are potent inhibitors of monoamine oxidase (MAO), with some selectivity towards MAO type A (5). Monoamine oxidase plays a key role in the regulation of certain physiological amines in the human body. It causes a deactivation by deamination of neurotransmittors such as catecholamins and serotonin, and of endogenous amines such as tyramine. Inhibitors of MAO, in particular of the type A isoenzyme, have a potential as antidepressive drugs since they increase both noradrenaline and serotonin levels in the brain. Xanthones occur only in a limited number of plant families. Xanthone-O-glycosides are found in two families only (Gentianaceae and Polygalaceae) while C-glycosides and aglycones are more widely distributed among angiosperms, ferns and fungi.

CHIRONIA KREBSII (GENTIANACEAE)

Chironia krebsii Griseb. is an endemic species of Malawi growing in marshy areas of the high plateaux. LC-UV analysis of the methanolic extract of *C. krebsii* revealed the presence of numerous xanthones, and enabled complete attribution of the peaks of the chromatogram to xanthones (**1-17**) and secoiridoids (**18-19**), respectively. In the UV spectra of xanthones, the number of bands and the absorption maxima gave preliminary information about the type of oxygenation pattern encountered.

In order to obtain information on the molecular weight, the substituents and the sugar sequence (glycosides) of the xanthones, a TSP LC-MS analysis was carried out. TSP LC-MS was performed in the positive ion mode with the filament off mode and by using NH_4OAc as buffer. The source was at 280°C and the vaporizer at 100°C. Under these conditions, the total ion current trace showed ionization of all peaks detected in the UV trace. The TSP-MS spectra of the xanthone aglycones recorded on line from the HPLC chromatogram exhibited $[M+H]^+$ pseudomolecular ions as main peak. The number of hydroxyl and methoxyl substituents could be calculated after subtracting the molecular weight of the xanthone nucleus (MW 196) from that of the aglycone. The TSP-MS spectra of xanthone glycosides usually showed weak pseudomolecular ions corresponding to $[M+H]^+$ and $[M+Na]^+$ adducts and a main fragment due to the

	R^1	R^2
1	H	CH_3
2	H	H
3	Glc	H
4	Prim	CH_3
5	Prim	H
6	H	Prim

	R^1	R^2	R^3	R^4
7	H	H	H	H
8	H	CH_3	H	H
9	H	CH_3	CH_3	CH_3
10	Prim	CH_3	H	H

	R^1	R^2	R^3
11	H	H	H
12	H	H	CH_3
13	H	CH_3	CH_3
14	Prim	CH_3	CH_3

	R^1	R^2	R^3
15	H	CH_3	H
16	H	H	H

Secoiridoids:

18 Swertiamarin
19 Sweroside

17

aglycone moiety $[A+H]^+$. In the case of diglycosides, a weak intermediate fragment due to the cleavage of a first sugar unit was observed. The TSP-MS spectra of the xanthone diglycosides found in *C. krebsii* presented a first loss of 132 amu corresponding to a pentosyl residue, followed by a loss of 162 amu (hexosyl residue) leading to the aglycone ion. These losses were attributed to a primeverosyl moiety, a disaccharide often encountered in the Gentianaceae family. The loss of 162 amu in the spectra of the monoglycosidic xanthones was attributable to the cleavage of a glucosyl moiety.

In most cases, additional data about the positions of free hydroxyl groups were required to permit a full structural determination on-line. Such information was obtained from LC-UV analyses after postcolumn addition of UV shift reagents (6) (7). A strong bathochromic shift observed upon addition of $AlCl_3$ in acidic mobile phase was characteristic of free hydroxyl group *peri* to the carbonyl function. Complexation with boric acid enabled detection of ortho-dihydroxyl subsistent. A NaOAc induced shift indicated the presence of hydroxyl groups at positions 3 or 6. By combining data provided by post-column derivatization with those already obtained by LC-UV and LC-MS analyses, a full on-line structure determination of most xanthones from *C. krebsii* was possible (7). The structural data obtained on-line for the three xanthone aglycones 7-9 are summarized in Fig. 1. Xanthones 1-17 were isolated in order to test their IMAO properties (8). Several of them showed potent, selective and reversible inhibition of MAO A. Further characterization of their IMAO properties is in progress.

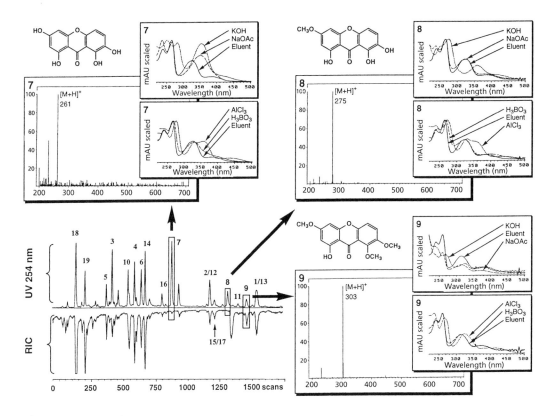

Fig. 1 Summary of spectral data obtained on-line for xanthones **7-9**.HPLC: RP-18 Novapak column (4µm, 150 x 3.9 mm i.d.), CH₃CN-H₂O gradient (0.05% TFA): 5:95 -> 65:35 in 50 min (1 ml/min).

LISIANTHIUS SEEMANII (GENTIANACEAE)

Lisianthius seemanii (Griseb.) Kuntze is an herbaceous plant growing in tropical America. The LC-UV analysis of the methanolic extract of this species did not show the presence of polyphenols. The UV spectra of all the peaks recorded on the UV trace exhibited only one maximum around 240 nm, characteristic of the α,β-unsaturated ketone function found in secoiridoid glycosides, the bitter principles of Gentianaceae. TSP LC-MS analysis allowed identification of swertiamarin, gentiopicroside and sweroside, and detected a dimeric secoiridoid, lisianthioside (**20**), as main component of the methanolic extract. The TSP-MS spectrum of **20** recorded with ammonium acetate as buffer, exhibited an intense pseudomolecular ion at m/z 734 and an important fragment at m/z 359 (Fig. 2). This information was however insufficient for the unambiguous attribution of the ion at m/z 734 to $[M+NH_4]^+$ or $[M+NH_4-H_2O]^+$ since no $[M+H]^+$ ion was present. Thus, a second analysis of the extract was carried out with diaminoethane as buffer. The spectrum recorded with diaminoethane exhibited an intense pseudomolecular ion at m/z 777 ($[M+61]^+$), confirming the molecular weight of **20** to be 716 amu. The fragment ion at m/z 359 corresponded to a protonated entity with half the mass of the parent molecule. These two complementary TSP LC-MS analyses confirmed that lisianthioside was a dimer of sweroside. The full structural determination of this novel type of secoiridoid was achieved after isolation of the pure substance from *Lisianthius jefensis* (9).

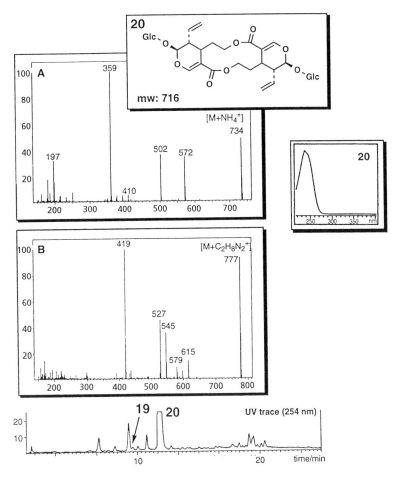

Fig. 2 TSP LC-MS analyses of **20** in a root methanolic extract of *L. seemanii*.
TSP: positive ion mode; A) NH₄OAc buffer, B) diaminoethane buffer. HPLC: RP-18 Novapak column (4µm, 150 x 3.9 mm i.d.), CH₃CN-H₂O gradient (0.05% TFA): 5:95 -> 7:3 in 50 min

GENTIANA RHODANTHA (GENTIANACEAE)

Gentiana rhodantha Fr. is a small herb growing in south-western China which is used as a stomachic remedy. The LC-UV analysis of *G. rhodantha* showed the presence of only one predominant xanthone (**21**) and several secoiridoids (Fig. 3). Compound **21** was rapidly identified as the widespread xanthone C-glycoside mangiferin. The TSP-MS spectrum of **21** exhibited a pseudomolecular ion at m/z 423, characteristic fragments for C-glycosides at [M+H-90]+ and [M+H-120]+ and a weak aglycone ion at m/z 261 (xanthone with 4 OH). The computer fitting of the UV spectrum of **21** with our in-house UV spectral library definitively confirmed identification. The secoiridoids swertiamarin (MW 374), kingiside (MW 404), epi-kingiside (MW 404) and sweroside (MW 358) were easily identified from their TSP LC-MS spectra recorded on-line and not further investigated. Two slower running peaks (**22** and **23**) also exhibited the characteristic UV spectra of secoiridioids. TSP LC-MS analyses of **22** and **23** gave in each case a spectrum identical to that obtained for sweroside. These compounds however were less polar than common secoiridoids. In order to obtain complementary information on these metabolites, a second LC-MS analysis with CF-FAB was carried out, using the same HPLC conditions. The total ion current recorded in the negative ion mode showed a strong MS response for both compounds. In the CF-FAB spectra of **22** and **23**, very intense [M-H]⁻ pseudomolecular ions were detected at m/z 913 and 1629, respectively. In addition, fragments were observed corresponding to the loss of 'sweroside like' units (-358 amu). Following the LC-MS screening results, a targeted isolation of **22** and **23** was undertaken. Full structure determination was achieved with the help of 1D and 2D NMR experiments, and by means of chemical degradation (10). Compound **22** consisted of two secoiridoid moieties linked together with a monoterpene unit through ester functions. Compound **23** was found to contain two more secoiridoid units (359 amu) attached to the carboxylic groups of **22**. Both compounds are natural products of a new type.

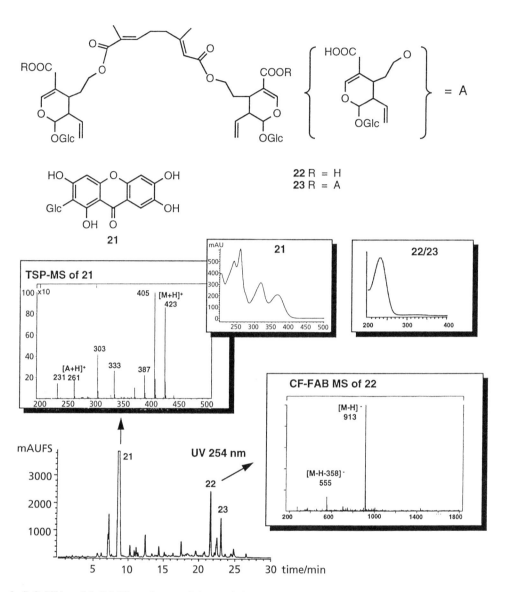

Fig. 3 LC-UV and LC-MS analyses of the enriched BuOH fraction of a root methanolic extract of *G. rhodantha*. TSP: positive ion mode; CF-FAB: negative ion mode, glycerol. HPLC: RP-18 Novapak column (4µm, 150 x 3.9 mm i.d.), CH$_3$CN-H$_2$O gradient (0.05% TFA): 5:95 -> 1:1 in 30 min (0,9 ml/min).

HYPERICUM BRASILIENSE (GUTTIFERAE)

Hypericum brasiliense Choisy is a herbaceous plant of south and south east Brazil. The dichloromethane extract of the roots exhibited IMAO properties, as well as antifungal activity against the plant pathogenic fungus *Cladosporium cucumerinum* in a bioautographic assay (11). LC-UV analyses revealed the presence of several xanthones. Fungicidal testing and LC-UV analyses carried out throughout the fractionation procedure allowed isolation of the active principles. The xanthones **24-26** and a new γ-pyrone derivative (**27**) displayed potent IMAO properties, with MAO A *vs* MAO B selectivities. Moreover, they seemed to act in a reversible and time-independent manner. Compound **24** exhibited an IC_{50} of 0.73 µM towards MAO A and was by far the most active. Compounds **24-27** were also found to be responsible for the antifungal activity of the extract (12).

24 R = H
25 R = CH$_3$

26

27

CONCLUSION

The few examples presented here demonstrate the enormous potential of LC-UV combined with LC-MS for the rapid screening of plant constituents and the targeted isolation of promising molecules. With the development of powerful hyphenated techniques, the chemical screening has become an highly efficient approach which is fully complementary to biological screening and bioassay-activity fractionation. However the great deal of information obtained on plant constituents will be useful for the discovery of new potent drugs only if suitable pharmacological models exist. Effective collaboration between botanists, phytochemists and pharmacologists is more than ever required.

REFERENCES

1. M. Hamburger and K. Hostettmann. *Phytochemistry* **30**, 3864 (1991).
2. J.L. Wolfender, M. Maillard and K. Hostettmann *Phytochem. Anal.* **5**, 153 (1994).
3. O. Suzuki, Y. Katsumata, M. Oya, V.M. Chari, B. Vermes, H. Wagner and K. Hostetmann. *Planta Med.* **42**, 17 (1981).
4. C.R. Blakley and M.L. Vestal. *Anal. Chem* **55**, 750 (1983).
5. R.M. Caprioli, F. Tan and J.S. Cotrell. *Anal. Chem.* **58**, 2949 (1986).
6. K. Hostettmann, B. Domon, D. Schaufelberger and M. Hostettmann. *J. Chromatogr.* **283**, 137 (1984).
7. J.L. Wolfender and K. Hostettmann. *J. Chromatogr.***647**, 191 (1993).
8. J.L. Wolfender, M. Hamburger, J.D. Msonthi and K. Hostettmann. *Phytochemistry* **30**, 3625 (1991).
9. M. Hamburger, M. Hostettmann, H. Stoeckli-Evans, P.N. Solis, M.P. Gupta and K. Hostettmann. *Helv. Chim. Acta* **73**, 1845 (1990).
10. W.G. Ma, N. Fuzzati, J.L. Wolfender and K. Hostettmann. *Helv. Chim. Acta* **77**, (1994).
11. A.L. Homans and A. Fuchs. *J. Chromatogr.* **51**, 327 (1970).
12. L. Rocha, A. Marston, M.A.C. Kaplan, H. Stoeckli-Evans, U. Thull, B. Testa and K. Hostettmann. *Phytochemistry* **36**, 1381 (1994).

Marine pharmaceuticals:
receptor/ligand interactions and cell signalling

Ronald J Quinn

Queensland Pharmaceutical Research Institute, Griffith University, Brisbane, 4111, Australia

Abstract: Marine organisms have not produced significant numbers of traditional medicines. However the large number of toxic marine organisms has resulted in the discovery of many bioactive constituents. The causative agent of diarrhetic shellfish poisoning is okadaic acid. The binding of the okadaic acid class compounds to their receptors, the catalytic subunits of protein phosphatases 1 (PP1) and 2A (PP2A), causes an inhibition of phosphatase activity resulting in modulation of cellular events controlled by phosphorylation/dephosphorylation mechanisms. A general biochemical pathway of tumor promotion, the okadaic acid pathway, which is mediated through inhibition of PP1 and PP2A has been shown to occur.

We now report the activity of synthetic cyclic peptides and cantharidin analogues against PP2A. Some of the cyclic peptides which lack the Adda side chain had inhibitory activity against PP2A. Several cantharidin analogues had inhibitory activity against PP2A. Consideration of the other cyclic peptide members of the okadaic acid class compounds has led us to suggest a new proposal for face selectivity which simplifies the existing pharmacophore model. We propose a face selective conserved acid pharmacophore model for the interaction of the okadaic acid class compounds with protein phosphatases 1 and 2A.

Traditional medicines have developed because of beneficial properties observed over long periods of use. In traditional use, observation of deleterious effects has often resulted in the cessation of use. Marine organisms have not produced significant numbers of traditional medicines but they have produced a large number of toxic compounds. These compounds have become important tools to probe biological control systems, an understanding of which may result in the longer term in useful therapeutic intervention.

In various parts of the world, normally edible shellfish may become toxic as a result of feeding on toxic organisms, mostly dinoflagellates, and thus accumulating their toxins. Consumption of contaminated shellfish can result in poisoning characterised by gastrointesinal illness of varying severity. The dominating symptoms are diarrhoea, nausea, vomiting and abdominal pain. This poisoning is known as diarrhetic shellfish poisoning (DSP), named after its predominant human symptom. The first known incidences of gastrointestinal illness associated with consumption of mussels exposed to dinoflagellates can be traced back to the Netherlands during the 1960's and 1970's. The presence of *Prorocentrum* species of dinoflagellate in the estuaries of the Waddensea in 1961 and 1976, and in the Easterscheldt in 1971 reportedly coincided with gastrointestinal illness which afflicted several people after consumption of boiled mussels. The intestines of the poisonous mussels were infested with the dinoflagellates *Prorocentrum redfieldii*, *Prorocentrum micans* and *Dinophysis acuminata*. DSP is not fatal to humans but does have a high morbidity rate and worldwide occurrence.

Okadaic acid (1) was found to be the causative agent for diahrettic shellfish poisoning. It was first isolated from the sponges *Halichondria okadai and H. melanodocia*.(1) Dinophysisotoxin-1 (35(S)-methylokadaic acid) was isolated from the hepatopancreas of the mussel *Mytilus edulis* infested with *Dinophysis fortii* in 1982 by Murata. Dinophysisotoxin-1 along with okadaic acid was later isolated from the dinoflagellate *Prorocentrum lima* in 1982 by Murakami et al. Okadaic acid was later isolated from toxic mussel *Mytilus edulis* found in Japan in 1985 by Yasumoto et al and found to be produced by the dinoflagellates, *Dinophysis acuminata*, *Dinophysis fortii* and *Prorocentrum concavum*. (2)

Okadaic acid is a regulator of mitosis in the cell cycle and specifically inhibits protein phosphatases 1 and 2A resulting in the accumulation of phosphorylated protein in cells. (3-5)

As well as okadaic acid there are other natural products which inhibit protein phosphatases 1 and 2A. (6) Collectively, these protein phosphatase inhibitors are referred to as the okadaic acid class compounds. (7) The binding of the okadaic acid class compounds to their receptors, the catalytic subunits of protein phosphatases 1 and 2A, causes an inhibition of phosphatase activity resulting in modulation of cellular events controlled by phosphorylation/dephosphorylation mechanisms. (3,8-16) These compounds all bind to the same site of the catalytic subunits of the protein phosphatases 1 and 2A as they all inhibit specific [^{3}H]okadaic acid binding. A general biochemical pathway of tumour promotion, the okadaic acid pathway, which is mediated through inhibition of protein phosphatases 1 and 2A has been shown to occur. (17) These natural-product ligands are very useful tools to probe the cellular processes they modulate.

Molecular modelling has allowed common regions of okadaic acid, calyculin A and microcystin-LR to be recognised and a pharmacophore model developed. (18)

Okadaic acid is a polyether compound with a molecular backbone spanning six tetrahydropyran rings and one tetrahydrofuran ring, terminating in a carboxylic acid. The microcystins are potent hepatotoxic cyclic heptapeptides produced by certain species of cyanobacteria. (19) These peptides effect hepatocytes resulting in haemorrhagic necrosis and death by haemorrhagic shock. The microcystins are specific, high affinity inhibitors of the protein phosphatases 1 and 2A. (8) The molecular recognition between the microcystins (such as microcystin-LR (<u>2</u>)) and protein phosphatases 1 and 2A, resulting in inhibition of phosphatase activity and hyperphosphorylation of proteins, begins a signal transduction process within the cytoplasm of the cell leading to propagation of the signal to the nucleus and regulation of gene expression. (20-21) Calyculin A is a novel spiro ketal compound bearing phosphoric acid, oxazole, nitrile and amide functionalities. Calyculin A (<u>3</u>) from the marine sponge *Discodermia calyx* collected in the Gulf of Sagami. (22)   It was reported to inhibit the development of *Asterina pectinifera* embryos and was cytotoxic against L1210. Calyculin A was found to inhibit the catalytic subunit of PP1, with IC$_{50}$=1.6nM, and the catalytic subunit of PP2A, with IC$_{50}$=0.5nM. Calyculin A inhibited PP1 and PP2A with similar potency unlike okadaic acid which inhibited the catalytic subunit of PP2A, (IC$_{50}$=1nM), to a much greater extent than the catalytic subunit of PP1, (IC$_{50}$= 330nM). (14)

<u>1</u>

<u>2</u>

<u>3</u>

Our conserved acid binding domain pharmacophore model, consists of a central core, containing one conserved acidic group and two potential hydrogen bonding sites, and a non-polar side chain. (18) The essential features of this pharmacophore model are as shown.

A new member of okadaic acid class compounds has been recently reported. The dried body of the Chinese blister beetle was first used by the people of China as a traditional medicine over 2000 years ago. Cantharidin (exo,exo-2,3-dimethyl-7-oxobicyclo[2.2.1]heptane-2,3-dicarboxylic acid anhydride) <u>4</u> is a naturally occurring toxin present in over 1500 different species of the Chinese blister beetle (*Mylabris phalerata* or *M. cichorii*). (23-25) The name, Chinese 'blister' beetle, was derived from the vesicating properties of the beetle, the application of the pulverised bodies or even simple contact with the insect can produce blistering of the skin. The toxin was isolated from the Spanish fly, *Cantharis vesicatoria* (23), otherwise known as the European blister beetle, by Robiquet in 1812 (26) The Europeans used the toxin in the treatment of warts, as an abortifacient and even as an aphrodisiac. (24, 25, 27) By the early 1900's cantharidin was determined too toxic for use as an internal medicine. 0.1mg of the toxin applied to the skin can cause violent superficial irritation resulting in blistering as well as irritating the kidneys. (27) It is now known that oral ingestion leads to severe irritation and ulceration of the gastrointestinal and urinary tract epithelial linings and that intraperitineal injection causes severe congestion and edema of the liver with an LD_{50} of 1mg/kg. (27-30) Livestock toxicosis due to the consumption of feed containing the blister beetles continues to present a problem for ranchers. (31) Ingestion of only a few beetles provides enough cantharidin to be lethal to an animal as large as a horse. (32-34) Several human poisonings have recently been reported as a result of ingestion of the toxin for its supposed aphrodisiac qualities. (35) Cantharidin, mixed with bay rum, has been used in the commericial preparation of hair restorers. (36)

Cantharidin was found to bind, with high affinity, to a specific cantharidin-binding protein (CBP) isolated from mouse liver. (37) The strong binding affinity of cantharidin to this protein supports its known toxic effects in mammals. Examination of the peptide sequence revealed high homology with PP2A AC type. PP2A AC type contains only two subunits, A (60kDa) and C (36kDa). The CBP was also shown to have significant phosphorylase a phosphatase activity. Okadaic acid inhibited the specific [^{3}H]cantharidin binding to CBP identifying the protein as either a type 1 or 2A protein phosphatase.

Subsequently, the inhibitory activity of cantharidin was compared with okadaic acid on the activity of the purified catalytic subunits of PP1, PP2A and PP2B. (38) The results showed that like okadaic acid (IC_{50} 0.4nM for PP2A, C type), cantharidin inhibited PP2A at a lower concentration (IC_{50} 0.16µM) than PP1 (IC_{50} 1.7µM), and that PP2B is inhibited only at much higher concentrations (IC_{50} >1000µM). This established that cantharidin was a member of the okadaic acid class compounds. (38, 39) Cantharidin is particularly interesting as it was the first compound of the group to be isolated yet the last to be classified as a protein phosphatase inhibitor. It is also the smallest, least flexible compound within the group, and there is limited structure activity data available.

We synthesised thirteen cantharidin analogues. The compounds were tested for their inhibitory activity against partially purified PP2A (heterotrimeric complex ABC type) isolated from mouse brain. The heterotrimeric complex was chosen to allow correlation with data previously reported for the okadaic acid class compounds. The IC_{50} value of cantharidin against PP2A (ABC type) was determined to be 8µM. The thirteen cantharidin analogues were screened for the inhibition of PP2A at a concentration of 1mM. Three showed greater than 80% inhibition of PP2A, two showed approximately 50% inhibition and the remaining showed less than 40% inhibition.

4	**5**	**6**	**7**
IC_{50} 8 µM	40 µM	70 µM	300 µM

Compounds 5, 6 and 7 were the most potent of the thirteen cantharidin analogues. The three compounds contained a 7-oxo moiety and lacked the bridgehead methyls. Compounds 5 and 6 contained an anhydride system, like cantharidin. Cantharidin was still the most potent compound. Comparison of these four compounds indicated that the presence of the bridgehead methyls was not essential for activity and that the presence of the double bond, although increasing the rigidity of the ring system, did not significantly affect activity. Our study showed that removal of the bridgehead methyls resulted in a decrease in potency. The inhibition of PP2A (ABC type) by cantharidin (4) was decreased 9-fold in the 'bisnor' compound 5.

5	**8**	**9**	**10**
IC_{50} 40 µM	48% at 1 mM	24% at 1 mM	17% at 1 µM

The 7-oxo moiety is required for inhibition of PP2A. Comparing compounds 5 and 8, there is a decrease in inhibition as a result of the conversion of the 7-oxo moiety to a 7-methylene. Removal of the bridge of 5 and 8, with consequent stereochemical changes, to give 9 resulted in a further loss of activity. Ring opening of the anhydride of 5 to give 7 resulted in a 7.5-fold decrease in activity. Similar results were observed for the compounds 8 and 10.

It is challenging to arrive at an understanding of the mode of binding of cantharidin to PP1 and PP2A considering the smaller size relative to the other known members of the okadaic acid class compounds. Preliminary modelling of cantharidin in to the conserved acid region has produced a number of possible orientations. Further work is required to clarify the impact of cantharidin on the pharmacophore model and to determine its usefulness as a template to derive other synthetic analogues with greater potency.

The solution structure of microcystin-LR and nodularin-V (motuporin) has recently been determined by NMR spectroscopy. (40) This will allow refinement of our pharmacophore model based on solution structures rather than calculated minima. As well as providing the solution structures a superimposition of microcystin-LR and nodularin-V was presented. These molecules were superimposed via the backbone segment Masp-Arg-Adda of microcystin-LR (11) and Masp-Val-Adda of nodularin-V (12). (40)

11

12

The microcystins have the general structure cyclo(D-Ala-L-X-D-*erythro*-β-methylisoAsp-L-Y-Adda-D-isoGlu-N-methyldehydroAla) where X and Y are variable L-amino acids. X can be L-Leu, L-Ala, L-Arg, L-Hty, L-Met, L-Met(O), L-Phe, L-Trp, L-Tyr while Y can be L-Arg, L-Aba, L-Ala, L-Har, L-Hph, L-Leu, L-Met, L-Phe, L-Tyr, L-Val. The nodularins lack the variable amino acid X and the D-Ala while the N-methyldehydroAla is replaced by N-methyldehydroaminobutyric acid. Y can be L-Arg or L-Val.

A more rigorous comparison of the microcystins and nodularins reveals that the 'variant' amino residues X and Y are not the only changes that can be tolerated whilst showing no loss in activity. D-Ala can be replaced by D-Ser in the microcystins, D-Ala is not present in the nodularins, suggesting that it is not essential for activity. The D-*erythro*-β-methyl-aspartic acid residue, can tolerate the loss of the methyl group with little effect on the biological activity. Mdha, can be replaced by L-Ala, Dha, Mser or L-Ser in the microcystins and is *N*-methyl dehydroaminobutyric acid in the nodularins. Modification of the methoxy of Adda can be tolerated. Microcystin is unable to accommodate any residue other than D-isoGlu. This position is non-variant. Esterification of the D-isoGlu residue results in total loss of activity. (41)

We have published experimental evidence demonstrating that the L-Arg to L-Ala change in microcystin-LR / microcystin-LA has no effect on potency both as inhibitors of phosphatase activity of PP1 and PP2A and [^{3}H] okadaic acid binding to PP1 and PP2A. (42)

This information coupled with the pharmacophoric pattern of the other members of the okadaic acid class compounds suggests that only the hydrophobic side chain, Adda, and the conserved acid, D-isoGlu, are essential for binding. The potential hydrogen bonding sites are situated within the Adda and D-isoGlu residues. The enzymes appear not to recognise the other amino acid residues within the molecules.

For microcystin-LR and nodularin-V, we propose that a superimposition that corresponds to binding to PP1 and PP2A should involve Adda-isoGlu as in 13 rather than Masp-Arg-Adda as in 11. This superimposition makes very little difference to the relative positions of both the isoGlu and isoMasp. Superimposition as proposed via Adda-isoGlu still results in Mdha (the residue involved in covalent linkage of microcystin-LR to PP1c/PP2Ac) occupying a different region to Mdhb in nodularin-V providing an equally plausible rationale for why microcystin-LR binds covalently to PP1c/PP2Ac whilst nodularin-V does not. In terms of enzyme recognition, however, it is an important to discriminate Adda-isoGlu.

13

We are undertaking synthesis in order to provide experimental evidence for the pharmacophore model and report some preliminary biochemical activity of our precursor molecules. We have chosen as a starting point the microcystins and established a target, based on the structures of the known microcystins as indicated above.

14   15   16

Cyclisation of a linear peptide was achieved at site 1 but was unsuccessful at site 2. Four linear peptides were cyclised on resin at site 1. This work provided cyclic peptides suitable for elaboration with Adda like hydrophobic side chains.

Compound 14 showed 45 % inhibition at 1mM and 15 exhibited 42% inhibition at 1mM. A synthetic peptide 16 cyclised via the α-carboxylic acid groups to investigate the effect alteration of the size of the cyclic core has on phosphatase activity proved the most potent inhibitor. The IC_{50} value for 16 was determined to be 0.5 mM. Unexpectedly, the results showed that these compounds which are Adda deficient precursors modelled on the microcystins, had some inhibitory activity against PP2A. This supports the concept that not only the Adda is essential, but that the cyclic system plays an important role in the interaction of the microcystins with the catalytic subunit of PP2A.

The four compounds, microcystin-LA, nodularin-V (motuporin), okadaic acid and calyculin A, are the most appropriate for development of a pharmacophore model. Alignment of microcystin-LA, nodularin-V, okadaic acid and calyculin A using the Conserved Acid Binding Domain model shows that the common features are present on a face of the molecules. The superimposition of microcystin-LA and nodularin-V highlights the similarity of the 'recognised' face, and the huge difference of the remaining regions   The protein phosphatases 1 and 2A only interact with one face of the molecules and that face must contain both Adda and D-isoGlu residues. The concept of face selectivity simplifies the existing pharmacophore model. The biological data confirms that the face tolerates very little change whereas the rest of the molecules can be altered and still retain activity. We propose a face selective conserved acid pharmacophore model for the interaction of the okadaic acid class compounds with protein phosphatases 1 and 2A. (41)

Acknowledgments: I thank my collaborators in this research. Cherie Taylor and Dr. Adam McCluskey at the Queensland Pharmaceutical Research Institute, Griffith University; Dr . Hirota Fujiki and Dr. Masami Suganuma at the Saitama Cancer Center Research Institute; Dr. Paul Alewood, Center for Drug Design and

Development, University of Queensland. We thank the Australian Research Council and the Monbusho International Scientific Research Program from the Ministry of Education, Science and Culture, Japan for support of this research. We acknowledge the award of an Australian Postgraduate Award to CT.

References

1. K. Tachibana *et al J. Am. Chem. Soc.* **103**, 2469 (1981).
2. R. W. Dickey *et al Toxicon* **28**, 371 (1990).
3. M. Suganuma *et al FEBS LETTERS* **250**, 615 (1989).
4. K. Yamashita *et al EMBO J.* **9**, 4331 (1990).
5. M. Felix *et al EMBO Journal* **9**, 675 (1990).
6. M. Suganuma *et al Toxicon* **30**, 873 (1992).
7. H. Fujiki *et al* In *Toxic Microcystis*, CRC Press (in press).
8. S. Yoshizawa *et al Cancer Res. Clin. Oncol.* **116**, 609 (1990).
9. S. Shenolikar and A. C. Nairn In *Advances in Secondary Messenger and Phosphoprotein Research*, p.1, Raven Press, New York (1991).
10. J. Hescheler *et al Pflügers Arch. ges Physiol.* **412**, 248 (1988).
11. M. Suganuma *et al Proc. Natl Acad. Sci. USA.* **85**, 1768 (1988).
12. T. Sassa *et al Biochem. Biophys. Res. Commun.* **159**, 939 (1989).
13. C. Bialojan and A. Takai *Biochem. J.* **256**, 283 (1988).
14. H. Ishihara *et al Biochem. Biophys. Res. Comm.* **159**, 871 (1989).
15. R. E. Honkanen *et al The Journal of Biological Chemistry* **265**, 19401 (1990).
16. M. Suganuma *et al Cancer Res* **50**, 3521 (1990).
17. H. Fujiki *et al* In *Relevance of Animal Studies to the Evaluation of Human Cancer Risk*, p.337, John Wiley & Sons, New York (1992).
18. R. J. Quinn *et al BioMed. Chem. Letters* **3**, 1029 (1993).
19. K. L. Rinehart *et al J. Am. Chem. Soc.* **110**, 8557 (1988).
20. H. Fujiki *et al Mol. Carcinogenesis* **2**, 184 (1989).
21. S.-J. Kim *et al Cell Regulation* **1**, 269 (1990).
22. Y. Kato *et al J. Am. Chem. Soc.* **108**, 2780 (1986).
23. G.-S. Wang *J. Ethnoparmacol.* **26**, 147 (1989).
24. G. W. K. Cavill and D. V. Clark In *Naturally Occuring Insecticides*, Marcel Dekker, New York, (1971).
25. W. W. Oaks *et al A.M.A. Arch. Int. Med.* **105**, 106 (1960).
26. M. T. James and R. F. Harwood *Herm's Medical Entomology*, 6th ed., p.403, Macmillan Company, London (1969).
27. M. T. Goldfarb *et al Dermatologic Clinics* **9**, 287 (1991).
28. M. Matsuzawa *et al J. Agric. Food Chem.* **35**, 823 (1987).
29. F. K. Bagatell *et al Toxicol. Appl. Pharmacol.* **15**, 249 (1969).
30. L. C. Nickolls and D. Tear *Br. Med. J.* **2**, 1384 (1954).
31. D. G. Schmitz *J. Vet. Intern. Med.* **3**, 208 (1989).
32. V. R. Beasley *et al J. Amer. Vet. Med. Assoc.* **182**, 283 (1983).
33. A. C. Ray *et al Amer. J. Vet. Res.* **41**, 932 (1980).
34. T. R. Schoeb and R. J. Panciera *Vet. Pathol.* **16**, 18 (1979).
35. A. Polettini *et al Forensic. Sci. Int.* **56(I)**, 37 (1992).
36. J. L. Cloudsley In *Insects and History*, p.204, Weidenfeld & Nicolson, London, (1976).
37. Y.-M. Li and J. E. Casida *Proc. Natl. Acad. Sci. USA* **89**, 11867 (1992).
38. R. E. Honkanen *FEBS Lett.* **330**, 283 (1993).
39. H. Fujiki and M. Suganuma *Adv. Canc. Res.* **61**, 143 (1993).
40. J. R. Bagu *et al* Nature *Struct.Biol.* **2**, 114 (1995).
41. R. J. Quinn *et al* In *Toxic Microcystis*, p.235, CRC Press (in press).
42. R. Nishiwaki-Matsushima *et al BioMed. Chem. Letters* **2**, 673 (1992).

Bioactive constituents
of Chinese traditional medicine

Da-yuan ZHU.

Shanghai Institute of Materia Medica, Chinese Academy of Sciences.
294 Tai-yuan Road, Shanghai 200031, China

ABSTRACT Chinese traditional medicine is great and tremendous treasure, an accumulation of experience over thousand of years. China has very rich resources of medical herbs. This paper introduce the studies on the bioactive constituents of Chinese traditional and folk medicines. Among them, some bioactive constituents have been developed and used in clinic as new drugs and other have been studied chemically and pharmacologically in more detail. They may have potential clinical use and now still are in studying, probably will be developed as new drugs and used in clinic in the future.

In Chinese ancient literatures, there were several famous medicinal classics. Early in the late Ming Dynasty in the year of 1590, the prominent medical scientist Li Shi-zhen, who devoted his lifetime (38 years) to summarize all pharmacological knowledge accumulated in China at that time and combined with his own practical experiences, composed the most famous and comprehensive work of historical significance in medical science, and wrote "Ben-Cao-Gang-Mu" ("Compendium of Materia Medica") in which he collected 1,892 kinds of medicines with 11,096 prescriptions. He described the morphology, collecting season, part of plant used, properties and therapeutic effect of every plant in detail. This is a classical pharmacopoeia of China, and is a required course and necessary reference in the study of Chinese medicine.

Our government now attaches great importance to systematize and develope the legacy of traditional Chinese medicine. Several great books have been published, such as "Zhong Yao Zhi" ("Monograph of Chinese Medicines and Drugs"); "Zhong Yao Da Ci Dian" ("Big Dictionary of Chinese Medicines and Drugs") in which including 5,767 traditional Chinese medicines and Chinese pharmacopoeia, 1990 edition, including 784 Chinese traditional medicines.

Recently, national specialists that are relative to Chinese traditional medicine organized by the Chinese State Council carried out comprehensive investigation to the resource of Chinese traditional medicine. They announced the 12,807 kinds of Chinese traditional medicine of which 11,146 kinds of them are plant medicine, 1,581 kinds of them are animal medicine and 80 kinds of them are mineral medicine.

On the basis of the above data and experience thousands of Chinese scientists have been researching in the fields of chemistry, pharmacology and clinical evaluation and gained elegant achievements.

In the following, I will introduce the studies of the bioactive constituents of Chinese traditional medicine. First, some bioactive constituents from Chinese traditional medicine have been developed into new medicines.

As all know, antimalaria active principle-artemisinin was isolated from *Artemisia Annua L.* which has been used as a cure for fever and malaria for many centuries. It is a sesquiterpene lactone containing a peroxide, and unlike most of other antimalarials, it lacks a nitrogen ring system. Artemisinin has been used successfully in over ten thousands malaria patients in China, with both chloroquin-sensitive and chloroquine resistant strains of *Plasmodium falciparum* and *vivax*. Although it is fast-acting, effective against malaria, low toxicity, the problems encountered with recrudescence, (28 days recrudescence rate 45.8%), poor solubility in water or oil as well as absorption via oral was not well, led to study of metabolism transformation and its modification [1].

The metabolism of artemisinin was studied in human. The results showed that medicine via oral was excreted mostly as original type. Other metabolites had no anti-malarial activity. The common characteristic of metabolites was loss of peroxide group [2,3]. The peroxide group in artemisinin is essential for activity. When artemisinin is reduced with sodium borohydride, the lactone moiety of the molecule is converted into a lactal, dihydroartemisinin, in which the peroxide is preserved. The lactal is an even stronger antimalarial. The hydroxy group of the molecule provides the way for synthesizing various derivatives. Three types of derivatives of dihydroartemisinin have been studied intensively. Over 300 derivatives of dihydroartemisinin were prepared and screened. Among them, artemether showed higher antimalarial activity in animal tests compared with artemisinin and other derivatives, artemether is superior in many aspects, such as higher antimalarial efficiency against rodent malaria, better solubility in oil, chemical stability and simple preparative procedure. Therefore, comprehensive studies have been made on its pharmacodynamics, toxicology, drug metabolism and clinical trials. Clinically, it shows an excellent therapeutic effect against falciparum malaria, including the chloroquine or multi-resistant malaria and severe cerebral malaria. The schizonticide action of artemether is much more potent than quinine and other antimalarials currently used. 28-days recrudescence rate was 6.7%. [4]

Huperzine A (Hup-A) is a new alkaloid isolated from *Huperzia serrata* which is used for the treatment of contusion, strain, hematuria and swelling in Chinese folk medicine [5] . In vitro test Hup-A exhibited marked anticholinesterase activity. The inhibitory effect of Hup-A is about 3 times more potent than that of physostigmine with AChE but less than that of physostigmine when tested with serum. Hup-A exhibited a higher selectivity towards AChE. It was demonstrates that Hup-A belongs to the mixed and reversible Che inhibitor. Nowadays, Hup-A is most potent Ache inhibitor compared to other natural ChE inhibitors under investigation. [6]

A comprehensive analysis of the effect of Hup-A on the central cholinergic system in rat demonstrated that Hup-A, when administered either i.m. or i.p., induced a long-lasting inhibition of AChE activity in brain(up to 360 min) and an increase of the Ach levels up to 40% at 60 min. There was considerable regional variation in the degree of Ach elevation after Hup-A injection with maximal values seen in fronal (125%) and parietal (105%) cortex and less increase (22-65%) in other brain regions [7]. In behavioral studies with Hup-A has been shown to improve mice and rats performance on Y-maze, to protect young and aged mice against NaNO$_2$, scopolamine, cycloheximide and electroconvulsive shock-induced disruption of passive avoidance responses [8]. Phase I clinical studies of safety, tolerance and pharmacokinetics of oral Hup-A were conducted in 22 young healthy volunteers. No significant side effects were observed at doses of 0.18-0.54 mg. [9]

The clinical trials with Hup-A have been reported since 1986. Cheng et al. was the first to report the clinical effects of Hup-A on 128 cases of myasthenia gravies, 99% of the patients showed controlled or improved clinical manifestations of the disease. The duration of action of Hup-A was longer than neostigmine. Parasympathominetic side effects, with the exception of nausea, were minimal when compared with those caused by neostigmine. [10]

In a comparative study with hydergine, Hup-A appeared to improve memory for 1-4 hr after its injection in a sample of 100 aged individuals suffering from memory impairment. Memory function was assessed according to the method of Buschke and Fuld. Side effects were rarely noted. The therapeutic effects of Hup-A were studied by random, match and double-blind method on 56 patients of multi-infarct dementia or senile dementia and 104 patients of senile and presenile simple memory disorders. Each group was divided into two subgroups, in which one was treated with saline and the other with Hup-A (50 μg b.i.d/4 weeks, and 30 μg b.i.d./2 weeks). The curative effects were evaluated by Wechsler memory scale. Hup-A significantly improved the memory quotient of the patients in both treated groups (P< 0.01) with minimal observed side effects. Only a few patients exhibited slight dizziness. [11]

Recently, Phase II clinical trail with Hup-A has been conducted in patients of Alzheimer's disease (AD) in China by using multiple-centers and double-blind method 103 cases of patients who all met AD criteria of DSM-III R were selected for the study. About 59% of patients treated with Hup-A exhibited significant improvements on memory, cognitive, activity of daily living with rare side effects observed. [12]

In order to search for new agents which might possess high activity, longer action duration and less toxicity, we led to its chemical modification 25 derivatives of Hup-A were prepared by our group. Preliminaly test, some compounds exhibited higher selectivity towards AChE and lower toxicity than Hup-A. [13]

Chuanxing (*Ligusticum chuanxing*) is famous "activate blood and relieve stasis" drug in traditional Chinese medicine, tetramethylpyrazine (TTMP) is the main active principle. It had spasmolytic action on isolated rabbit aorta, improved acute rabbits myocardium ischemia, increased coronary artery flow, prolonged the mice survival time under hypoxia, inhibited platelet aggregation. This compound has been proved to possess good curative effect in treating coronary heart disease, ischemia cerebrovascular disease and angiopathy in clinic. [14]

The new pharmacological actions were found by further investigation. TTMP could increase renal blood flow and diuresis. Now it was tried out to treat renoprival disease and renal function failure in clinic. [15]

It has strong effects of scavenging cyctoxic oxygen free radicals (O_2^-, LOO^-, OH^-) using spin trapping technic and chemiluminescence methods [16]. It reduced the experimental pulmonary metastasis of B16-melanoma. The antimetastatic action of TTMP may be related not only to lowering the effect on the level of TXA_2, thereby decreasing the TXA_2/PGI_2 ratio, but also to promoting the effect the NK cell activity [17].

New-breviscapin (NB) is the soluble sodium and calcium salts of 4'-OH- scutellarin-7-0-glucuronide, extracted from Chinese herb *Erigeron breviscapus*, which has been used in the treatment of occlusive vascular diseases. It was demonstrated that NB increased cerebral blood flow of experimental animal, reduced resistance of cerebral vascular, improved cerebral-circulation increased permeability of blood cerebral barrier, increased nutritional myocardium blood flow, improved immunological function of antibody and macrophages and inhibited platelet aggregation induced by ADP in vitro and against thrombus formation in vivo. It inhibited platelet TXB_2 production without alteration of hydroxyeicosatetraenoic acid. It also inhibited 6-ketoprostaglandin $F_{1\alpha}$ prodution by endothelial cell. For leukocytes, it did not affect TXB_2 production. However it potentiated the effect of calcimycin in stimulating leukotriene B_4 [18]. Cultured confluent human umbilical vein endothelial cells were incubated with NB. The releases of tissue-type plasminogen activator, and epoprostend from endothelial cells were stimulated by NB but no significant effect on plasminogen activator inhibitor activity was seen. NB induced a production on thrombodulin (TM) with the cells, an expression of TM on the surface of the cells, and a release of TM from the cells. Those data provide a new evidence that NB is a stimulant to fibrinolysis and anticoagulation of endothelial cells [19]. NB have been used for treatment of coronary heart disease, paralysis and rheumatagia in clinic.

The roots of *Pueraria lobate Willd* have been used to treat for headache, hypertension in Chinese traditional medicine. Puerarin isolated from this plant dilated cononary artery, decreased BP, HR and PRA in SHR by blocking the β-adrenergic receptor [20]. Pueraria antagonized the cardiac arrhythmia by chlorofom-epinephrine in rabbits and ventricular extrasystole and ventricular tachycardia induced by ouabaine in guinea pigs [21]. Puerarin had no mutagenesis or teratogenesis action [22]. It has been used for treatment of coronary heart disease such as angina pectoris and myocardial infarction.

The roots of Aconitum are well-known as Chinese traditional medicine. There are 170 Aconitum species, which are found throughout China located in the provinces of the southwestern and northwestern China. 44 Aconitum species have been recorded to be used medicinally for various ailments. Chinese Aconitum has been used frequently as a component of numerous prescriptions which could be used for the treatment of cold dampness evil, rheumatoid arthritis, pain and swelling induced by trauma and fracture. Up to date, about 20 Aconitum species have been subjected to scientific testing and experimentation. Over 170 alkaloids, were isolated from Chinese Aconitum plant. Some alkaloids, such as 3-acetylaconitine (AAC), bulleyaconitine (BUL) and lappaconitine (LA) completed preclinical and clinical trials, and have now been introduced into clinic for the treatment of several kinds of chronic pain, rheumatoid arthritis, and so on. [23,24,25]

Oleanolic acid (OA) is a triterpenoid compound that exists widely in medical herbs, such as *Ligustrum lucidum Ait*, *Tinospora sagittata (Oliv. Gaynep)*. Pretreatment with OA dramatically diminished CC14-, bromobenzene-, acetaminophen-, phalloidin, and cadmium-induced liver injury, and decreased the hepatotoxicity of D-galactosamine plus endotoxin, thioacetamide, furosemide, and colchicine. OA has no effect on the toxicity of dinothylni-trosamine, α-amanitin, chloroform, and ally alcohol [26]. OA has extensive inhibitory activities on type I, II and III allergic reactions [27,28]. It has been used for treatment of acute icterohepatitis.

In the following description, some herbs have been studied chemical and pharmacologically in detail, and now are still being studied.

Bolbostemma paniculatum (Maxi.) Faraquet has been used as a folk medicine for centuries in China. Tubeimoside A isolated from this plant is a triterpenoid saponin with an inter saccharide chain bridged by disrotalic acid to a unique macrocyclic structure. It showed the inhibitory effects on inflammatory mouse ear edema and on ^{3}H-incorporation into phospholipids of cultured cells induced by 1α-O-tetrdecanoyl-phorbol-13-acetate. It exhibited potent antitumor activity and antitumor promoting activity on skin tumor formation induced by phorbol ester tumor promoter in 7,12-dimethylbenz (α) anthracen-treated mice by topical application. It had an inhibitory action on the infection of HIV-1 isolates and would be a promising candidate for treatment of AIDS [29,30,31,32,33,34].

The roots of *Polyalthia nemoralis* A.DC were used for treatment of antimalaria and hepatitis. Antimalaria active zincpolyanemine has been obtained from this plant [35]. In the course of continuing search for antimalaria constituent one new natural product, capric bis (pyridine-N-oxide-2-thiolate) was isolated. Those metallic compounds possessed potent antimalarial, antimycotic and antiseptis activities [36].

In the past few years, some species of the genus cepharotaxus have attracted considerable attention as sources of antitumor alkaloid. Several antitumor alkaloid such as harringtonine, homoharringtonine, have been isolated in large scale for clinical use in China. Recently, two new alkaloids, neoharringtonine and anhydroharringtonine with significant antileukemic activity were isolated from *Cephalotaxus fortunci Hook f.* [37]. Hainanensine, an antitumor alkaloid isolated from *C hainanensis* is poorly soluble in water or oil, led to effects to its chemical modification. Among them HH07 A inhibited the growth of L1210 and HL-60 cells in vitro. It exerted inhibitory effect on the ascitic of L1210 and S-180 in mice [38].

Hernandezine isolated from *Thalictram glandulossimum* was found to be effective for mouse-bearing P388 leukemia, S-180 ascites and C26 colon cancer. Preliminary results showed that hernandezine blocked cell-cycle transfer from G1 to S phase, and its cytocidal might be cell cycle specific [39].

Momorcharaside A and B, new saponines were isolated from the seeds of *Momordica charantia*. Momorcharaside A exhibited obvious inhibition of DNA and RNA syntheses in S-180 tumor [40].

Homopterocapin separated from the roots of *Glycyrrhiza pallidiflora* was proved to be an effective ingredient for inhibiting (low concentration) or killing (high concentration) human's throat cancer cell (HEP-2) in cultural condition [41].

Pedicularis striate is used in folk medicine as cardio-tonics for treatment of collapse, exhaustion and senility, and is usually called "pseudo-ginseng" by local inhabitant of northwestern China. Several phenylpropanoid glycosides, isoverbascoside, verbascoside, pedicularioside A, echinacoside, cistanoside D(5), were obtained from this plant. They showed scavenging effect on superoxide and antioxidation effect [42]. Those glycoside exhibited antitumor effects on three different tumor. cell lines SMMC-7721, L342 and MGC-803 [43].

Platelet Activating Factor (PAF) is a lipid medium that is relative to rheumatism, asthma and inflammation. The possible way of finding new medicine for the above disease will be tried through searching antagonist of PAF receptor.

In a continuing search for PAF antagonists, nine bicyclo (3, 2, 1) octanoid and benzofuran neolignan have been isolated from the aerial pert of *Piper kadsura (ohoisy) ohwi*, a Chinese traditional drug used for the treatment of inflammation and rheumatic disease. Among them, kadsurenin K, kadsurenin L, (-)-denudatin B, kadsurenin M are new neolignans, and demonstrated significant PAF antagonistic activity in ^{3}H-PAF receptor binding assay [44,45].

Other species of the same genus, Piper polysyphorum C.DC. is indigenous to the southern part of China. In the course of screening for inhibitors of PAF, the norpolar fraction was found to exhibit PAF inhibitory activity. Four new neolignans polysyphorin, (+)-virolongin, (+)-grandisin, (+)-lancifolin D were isolated for preliminary pharmacological tests. They exhibited PAF antagonistic activity in ^{3}H-PAF receptor binding assay and inhibition of PAF induced platelet aggregation [46].

A new steroidal saponin, anemarsaponin B obtained from *Anmarrhena asphodeloides Bunge* could inhibit PAF-induced rabbit platelet aggregation in vitro [47].

Phytolecca acinosa polysaccharide I (PAP-I) and Esculentoside A were obtained from the roots of *phytolecca acinosa van Houtte* which, as relieve ocdema's drug in Chinese traditional medicine.PAP-I, is an acidic polysaccharide consisted of galacturonic acid, galactose, arabinose and rhamnose, molecular weight 10 kDa, enhanced the production of inteleukin-2 and interleukin-3 in vitro and increased the cytotoxicity of macrophages and its production of tumor necrosis factor and interleukin-1. It augmented the immunological functions in vivo [48,49].

Esculentosin A (ESA) markedly lowered the increase of vascular permeability induced by acetic acid and the swelling of murine ears induced by zylene in mice. It suppressed the swelling of rat hind pows induced by carrageenan, there were no marked effects on adrenal weight. It also showed suppressing effect on adrenalectomised rat. This suggested that the antiinflammatory properties of ESA was not depended on the pituitary-adrenal system [50]. ESA was shown to inhibit significantly phagocytic activity intracellular and extracellular production of interleukin I by thioglycollate primed murine peritoneal macrophages in vitro and markedly decreased serum hemolysin concentration in sensitized mice challenged with sheep red blood cells in vivo. Inhibition of antibody production by B lymphocytes, phagocytosis and the production of inflammatory mediators by macrophages may partially explain the wide and strong antiinflammatory effect of ESA [51].

Edulinine was isolated from the Chinese herb *Zanthoxylum simulens Hance* used for treatment of stomachic and rheumarthritis. Because edulinine is as a minor constituent, ($\pm$)-edulinine was synthesized. Analgesic effects of the ($\pm$)-edulinine was demonstrated in rats by using the tail flick method and by using the hot plate and writhing analgesic assays ($\pm$)-edulinine was also shown to have anticonvulsive activity [52]. The anticonvulsive action of edulinine is related to 5-HT and DA system in the brain [53]. It exhibited inhibition of bicuculline-induced epileptiform activity of pyramidal cells in rat hippocampal slices [54]. Preliminary clinical trials have shown that the symptoms of epileptic patients were controlled or improved on oral administration of the ($\pm$)-edulinine. In order to enhance its pharmacological activity, various analogues were synthesized. Some compounds of them, such as 6-chloro or 6-bromo-($\pm$)edulinine had more potent activity in central depressant action and had more potent anticonvulsant than the parent alkaloid. [55]

Osthole, a coumarin, obtained from fruits of *Cnidium monnieri (L) Cusson*. It showed a hypotensive and an antiarrhymic activity. It possessed calcium antagonistic effect, which would contribute to the antiarrhythmic action [56, 57].

Guan-fu base A, G, I were isolated from the tuber root of *Aconitum corenum(Lew) Raips* which was used for treatment of migraine and facial paralysis in Chinese folk medicine. Animal tests suggested these alkaloids had antiarrhythmic action. Guan-fu base A, G, I, iv 133, 175, 266.2 mg/kg respectively counteracted the VF (93%, 90%, 90%) induced by chloroform in mice. The ED_{50} were 81.87$\pm$26.2 (base A), 9.53$\pm$0.14 (base G) and 189.9$\pm$26.2 (base I) mg/kg. Guan-fu base A 10 mg/kg, base G 25.5 mg/kg, base I 10 mg/kg increased the dose of aconitine or beiwutine for inducing cardiac arrhythmias in anesthetized rats respectively. Base G 29.5 μmol/L, base A 32.2 μmol/L could antagonized electrical stimulation-induced ventricular fibrillation in isolated guinea pig heart. Base G and A could reduce heart rate and prolong P-R, QRS interval. In the isolated guinea pig left atria, the functional refractory period could be prolonged by base G and base A. In mice, the acute LD50 of base A and G were 582.2 mg/kg, 185.5 mg/kg by ip, and 134 mg/kg, 35 mg/kg by iv respectively. [58] Clinic test of antiarrhythmic effect of Guan-fu base A is under progress. Guan-fu base A has been chosen as a lead compound for chemical modification. Screening test of 23 cogenerous compounds indicated four phenylproanediolamine derivatives markedly antagonized chloroform-induced arrhythmias in rats, and appeared to be more potent than the parent alkaloid [59].

Chinese traditional medicines have been mostly used in the form of prescriptions with thousands of years of clinical practice and its unique theoretical system. Its extraordinary special characteristic is that the patients are treated with the whole system under the doctors' consideration. What is more, the Chinese traditional medicine holds that all the systems, organs of a human are closely linked together and therefore certainly affected by one another when the person is sick. And on the contrary, the affected body will have great effect on some parts or organs of the body itself.

Therefore, Chinese traditional medical workers pay great attention to the practical use and balance of the different herbs and so make them become more effective in treatments. In China, many processed Chinese drugs have been manufactured based on numerous medically effective traditional prescriptions. These drugs make great contributions to the health of the Chinese and the abroad people.

However, we have to accept the fact that which constituents of the drugs are medically effective and their chemical functions cannot be fully determined at present. And some compounds are found effective, but the actual therapeutic effects do not represent the medical effects originally thought to have in the prescriptions. For example, indurubin isolated from Danggui luwei pill have been used to treat chronic granulocytic leukemia, but its therapeutic effect is not as good as Danggui luwei pill. So, it is necessary to consider the synergism, antagonism, potentiation and addition of all the constituents in the prescription when they may affect each other in use. We have already begun researching on those very effective Chinese traditional medicine complexe in chemical, pharmacological and biochemical ways. The main effective principles isolated from those Chinese traditional medicine complex were optimized with computer. So it will be certain that safer, more effective and well-balanced in clinic. Some stable new generations Chinese traditional medicines will come into being in the future.

References

1. China cooperative Research Group on Qinghaosu and Its Derivatives as Antimalarials. J. Tradl. Chi. Med., 1982, 2: 17-24
2. D.Y. Zhu, B.S. Huang, Z.L. Chen, M.L. Yin. Acta Pharm. Sin., 1980, 15: 509-512
3. D.Y. Zhu, B.S. Huang, Z.L. Chen, M.L. Yin. Acta Pharmacol. Sin., 1983, 4: 194-197
4. Ibid. J. Tradl. Chi. Med., 1982, 2: 45-50

5. J.S. Liu, C.M. Yu, Y.Z. Zhou. Acta Chim. Sin., 1986, 44: 1035
6. Y.E. Wang, D.X. Yue, X.C. Tang. Acta Pharmacol. Sin., 1986, 7: 110-113
7. X.C. Tang, P. Desarno, K. Sugaya, et al. J. Nearosici. Res., 1989, 24: 276-285
8. X.C. Tang, Y.F. Han, X.P. Chen, et al. Acta Pharmacol. Sin., 1986, 7: 507-511
9. B.C. Qian, M. Wang, Z.F. Zhou, et al. Acta Pharmacol. Sin., 1995, in press.
10. Y.S. Chen, C.Z. Lu, Z.L.Ying. New Drugs and Clinical Remedies 1986, 5: 197-199
11. R.W. Zhang, X.C. Tang, Y.Y. Han, et al. Acta Pharmacol. Sin., 1991, 12: 250-252
12. S.S. Xu, Z.X. Gao, Z. Wang. Acta Pharmacol. Sin., 1995, in press.
13. D.Y. Zhu, X.Z. Tang, J.Q. Seng, et al. One patent have been filed in China office. 1995
14. Beijing Institute Pharmacentical Industry. Chin. J. Med., 1977, 57: 467-469
15. J.X. Yin. Chinese J. of Integrated Traditional and Western Medicine 1993, 13: 32
16. Z.H. Zhang, S.Z. Yu, Z.T. Wang. Acta Pharmacol. Sin., 1994, 15: 229-231
17. J.R. Liu, S.B. Ye. Chin. J. Pharmacol. and Toxicol., 1993, 7: 149-152
18. Z.Y. Wang, Y.C. Chen, Y. He, C.G. Ruan. Acta Pharmacol. Sin., 1993, 14: 148-151
19. Q.S. Zhou, Y.M. Zhao, X. Bai, P.X. Li, C.G. Ruan. Acta Pharmacol. Sin., 1992, 13: 239-242
20. X.P. Song, P.P. Chen, X.S. Chai. Acta Pharmacol. Sin., 1988, 9: 55-58
21. X.S. Chai, Z.X. Wang, P.P. Chen , L.Y. Wang . Acta Pharmacol. Sin., 1985, 6: 166-168
22. J.L. Shi, Q.G. Huang, X.M. Shang, Z.R. Li, X.M. Jia. Chin. J. Pharmacol. and Toxicol 1992, 6: 223
23. J.H. Lin, W. Cai, X.C. Tang. Acta Pharmacol. Sin., 1987, 8: 301-305
24. X.C. Tang, X.J. Liu, W.H. Lu. Acta Pharm. Sin., 1986, 21: 886-891
25. X.C. Tang, X.J. Liu, J. Feng. Acta Pharmacol. Sin., 1986, 7: 413-418
26. J. Liu, Y.P. Liu, C.D. Klaassen. Acta Pharmacol. Sin., 1995, 16: 97-102
27. Y. Dai , B.Q. Hang, Q.Y. Meng, S.P. Ma, L.W. Tan. Acta Pharmacol. Sin., 1988, 9: 562-565
28. Y. Dai, B.Q. Hang, P.Z. Li, L.W. Tan . Acta Pharmacol. Sin., 1989, 10: 381-384
29. F.H. Kong, D.Y. Zhu, R.S. Xu, Z.C. Fu, L.Y. Zhou, T. Twashita. Tetrahedron Lett 1986, 27: 576-579
30. R. Kasai, M. Miyakoshi, K. Matsumoto, R.L. Nis, J. Zhou, T. Morita and Tanaka. Chem. Pharm. Bull., 1986, 34: 3974
31. R.D. Ma, L.J. Yu, Y.Q. Wang, H. Nishino, J. Takayasu. Chin. Sci. Bull., 1992, 602-606
32. L.J. Yu, R.D. Ma, Y.Q. Wang, H.N. Shino. Plant Medica, in press.
33. L.J. Yu, R.D. Ma, Y.Q. Wang, H. Nishino, J. Takayusu, W.Z. He, et al. Int. Cancer, 1992, 50: 635- 638
34. L.J. Yu, R.D. Ma, S.B. Jiang. Acta Pharmacol. Sin., 1994, 15: 103-106
35. B.X. Xu, G.Y. Han, X.P.W, M.Z. Liu, X.Y. Xu, L.N. Meng, Z.L. Chen, D.Y. Zhu. Kexue Tong bao 1980, 25: 444-445
36. J.Z. Yao, H.Q. Liang, S.X. Liao. Acta Pharm. Sin., 1994, 29: 845-850
37. D.Z. Wang, G.E. Ma, R.S. Xu. Acta Pharm. Sin., 1992, 27: 173-177
38. Y.M. Ye, C.X. Xu, R.H. Sui, J.Y. Guo and G.J.Cui. Acta Pharm. Sin., 1995, 30: 12-16
39. C.X. Xu, L. Lin, R.H. Sun, X. Liu and R. Han. Acta Pharm. Sin., 1990, 25: 330-335
40. Z.J. Zhu, Z.C. Zhong, X.Y. Luo and Z.Y.Xiao. Acta Pharm. Sin., 1990, 25: 898-903
41. Y.M. Kan, Q. Zhu, L.Chen, R.D. Wang, X. Li, M.F. Hong. Chin. J. Pharm., 1994, 29: 608-609
42. J. Li, R.L. Zheng, Z.M. Liu, Z.J. Jia. Acta Pharmacol. Sin., 1992, 13: 427-430
43. L. Ji, Y.Zheng, R.L. Zheng, Z.M. Liu, Z.J. Jia. Chin. J. Pharm., 1995, 30: 269-271
44. Y. Ma, G.Q. Han, Z.J. Liu. Acta Pharm. Sin., 1993, 28: 207-211
45. Y. Ma, G.Q. Han, Y.Y. Wang. Acta Pharm. Sin., 1993, 28: 370-373
46. Y. Ma, Q. Han, C.L. Li, J.R. Cheng, B.H. Arison, S.B. Hang. Acta Pharm. Sin., 1991, 26: 345-350
47. J.X. Dong, G.Y. Han. Acta Pharm. Sin., 1992, 27: 26-32
48. H.B. Wang, Q.Y.Zheng, D.H. Qian, J. Fang, D.W. Ju . Acta Pharmacol. Sin., 1993, 14: 243-248
49. J.P. Zhang , D.H. Qian , Q.Y. Zheng. Acta Pharmacol. Sin., 1990, 11: 375-377
50. Q.Y. Zheng, K. Mai, X.F. Pan, Y.H. Yi. Chin. J. Pharmacol. and Toxicol., 1992, 6: 221-223
51. D.W. Ju, Q.Y. Zheng, H.B. Wang , X.J. Guan, J. Fang, Y.H. Yi. Acta Pharm. Sin., 1994, 29: 252-255
52. Z.Q. Cheng, G.X. Hong, Z.Teng, L.G.Qian, G.S. Li, K.J. Gu, R.Y. Ji. Chin. J. Pharmacol. and Toxicol., 1988, 2: 109-112
53. T. Zhong, G.X. Hong, Z.Q. Chang. Chin. J. Pharmacol. and Toxicol., 1989, 3: 247-250
54. Y.K. Li, S.R. Yang, X.L. Dai, Z.Q. Chang. Chin. J Pharmacol. and Toxicol., 1992, 6: 201-203
55. L.G. Qian, K.J. Gu, R.Y. Ji. Acta Pharm. Sin., 1991, 26: 572-577
56. L. Li, F.E. Zhang, C.L. Zhang, G.S. Zhao D.K. Zhao. Chin. J. Pharmacol. and Toxicol., 1995, 9: 108-112
57. L. Li, F.E. Hang, L. Yang, C.L. Zhang, G.S. Zhao, D.K. Zhao. Acta Pharmacol. Sin., 1995, 16: 251- 254
58. D.H. Chen, L.Y. Mu, W.R. Xu. J. China Pharm. Uni., 1987, 18: 268-272
59. R.B. Wang, S.X. Peng, W.Y. Hua. Acta Pharm. Sin., 1993, 28: 583-593

An integrated approach to exploiting molecular diversity

W.H. Moos, S.C. Banville, J.M. Blaney, E.K. Bradley, R.A. Braeckman, A.M. Bray, E.G. Brown, M.C. Desai, G.D. Dollinger, M.V. Doyle, J.A. Gibbons, D.A. Goff, R.J. Goodson, V.D. Huebner, D.E. Johnson, S.E. Kaufman, L.A. McGuire, N.J. Maeji, E.J. Martin, H.Y. Min, S. Ng, J.M. Nuss, L.S. Richter, S. Rosenberg, K.R. Shoemaker, K.L. Spear, D.C. Spellmeyer, G.B. Stauber, J.R. Stratton-Thomas, L. Wang, J. Winter, G.H.I. Wolfgang, A.K. Wong, R. Yamamoto, R.J. Zimmerman and R.N. Zuckermann

Chiron Technologies, Chiron Corporation, 4560 Horton Street, Emeryville, CA 94608, USA

Abstract: A new field of research, "molecular diversity", represents a paradigm shift for drug discovery in the '90s. In the identification of drug candidates, the automated, permutational, and combinatorial use of chemical building blocks now allows the generation and screening of unprecedented numbers of compounds. Drug discovery— better, faster, cheaper? Note that diversity-based discovery programs are currently producing many, many potential leads every year. Indeed, more compounds have been made and screened in the '90s than in the previous century of pharmaceutical research. Of course, diversity is much more than a game of numbers. Key elements of variety, complexity, spatial features, and multiple physicochemical parameters contribute to diversity. And, combinatorial synthesis begs for new assay schemes to be established, including affinity selection techniques, tagging methodologies, deconvolution tactics, and other brute-force strategies. However, the resulting demands placed on informatics can be staggering. Massive data sets are now amassed in a period of weeks to months, and innovative tools for data handling and analysis are critical. Lead discovery is often followed by lead optimization, although diversity libraries based on well-known medicinal pharmacophores have the potential to produce development candidates directly. Where optimization is needed, diversity can also play a role in speeding and enhancing the research process, yielding better drug candidates, more quickly. Discovery and development should be contrasted in relation to diversity libraries. One might predict that any highly diverse set of building blocks might yield potent and selective leads. In contrast, the need for orally available therapies can restrict the viable chemistries to those producing low molecular weight compounds, with stability in appropriate biological milieus, suitable safety characteristics, etc. Furthermore, there are noteworthy ramifications of the diversity game beyond targeted pharmaceutical research, for example, in novel ways of representing multi-dimensional physicochemical parameter space in our limited three- to five-dimensional world (3-dimensions, x, y, and z, plus color and time), in biophysics and spectroscopy applied to assays, for example, mass spectrometric identification of leads and fluorescence-activated cell sorting techniques, and in other disciplines too. Thus, an integrated approach to exploiting molecular diversity, taking into account both discovery and development considerations, plus multiple scientific disciplines, is ultimately required for optimal success.

INTRODUCTION

Molecular diversity approaches, including combinatorial chemistry, represent a recent addition to the toolbox of medicinal chemists (for background, see ref. 1-3, and citations therein). Together with paradigm shifts of the past several decades (TABLE 1), combinatorial discovery and development technologies promise to revolutionize the way drug hunters practice their trade during the 1990s. This is not to say that the "old fashioned" way of discovering drugs will be completely overturned. Indeed, molecular diversity is but one of many tools, taking its rightful place alongside structure-based design, biotechnology, and other modern machinery—but a powerful tool it is!

137

A fundamental underpinning of combinatorial technologies is the ability to create large, sometimes random, pools of compound mixtures, followed by selection methods to identify the active component(s) of a pool. This "shotgun" approach has been well known in molecular biology for some time now, but only became commonplace in pharmaceutical chemistry circles in this decade.

The need for better, faster, and cheaper ways of discovering new drugs is accentuated by the extreme cost, high failure rate, and long time frame associated with the R&D and regulatory approval process (Fig. 1; see also ref. 4). Molecular diversity provides an opportunity to generate more information at each stage of preclinical R&D, in both the discovery and optimization of leads, hopefully allowing better decisions to be made along the way, thus possibly also resulting in higher success rates. Moreover, with advantages of speed and numbers, combinatorial chemistry may allow the advancement of worthwhile drug candidates in record time, with much higher productivity per employee than usual.

TABLE 1. Paradigm shifts in the pharmaceutical industry in recent decades.

1960s
solid-phase peptide synthesis (SPPS)
routine nuclear magnetic resonance (NMR) spectroscopy

1970s
affinity chromatography
radioligand receptor binding

1980s
biotechnologies
structure-based design

1990s
molecular diversity
solid-phase organic synthesis (SPOS)

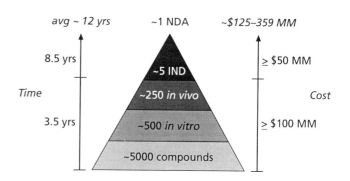

Fig. 1. Risk in biomedical R&D—long time frames, high costs, high failure rates.

Thus, through smart application of molecular diversity technologies, forward-looking academic laboratories and start-up biotechnology ventures have been able in recent years to compete head-to-head in research and discovery with much larger pharmaceutical rivals. Whether the brute force strength of major pharma will now begin to overpower such smaller programs remains to be seen.

COMBINATORIAL CHEMISTRY

At the heart of combinatorial chemistry is the concept of combining readily available chemical building blocks like "beads on a chain". This concept was first popularized by Merrifield, who received a Nobel Prize for his work on solid-phase peptide synthesis (SPPS) in the 1960s (Fig. 2). The SPPS concept was broadened to multiple synthesis by Geysen in the 1980s (see ref. 5-7, and citations therein), and further to peptoids (ref. 8-11), robotics (ref. 12), and other chemistries in the 1990s. (While combinatorial approaches may eventually extend to worthwhile solution phase synthetic approaches, the seminal workers in this field have invariably used solid phase strategies.) It is noteworthy that it took two to three decades from the first reports of SPPS before the general organic and medicinal chemistry communities began to embrace the automated solid-supported synthetic methodologies that today represent such a large portion of leading drug discovery programs.

Many classical and modern day organic transformations have been shown to work well on the solid phase, much to the surprise of a number of chemists (Fig. 3). General solid phase organic synthesis (SPOS) requires appropriate access to and knowledge of the characteristics of base polymers, supports, linkers, spacers, and the like. With the basics in hand, chemists have the ability to prepare both mixture libraries, using techniques such as "mixed resin", and multiple sets of individual compounds, using "multipin" methodologies, for example. Efficient synthetic approaches that are simultaneously convergent and divergent have been developed for mixture library synthesis (Fig. 4), and libraries can be

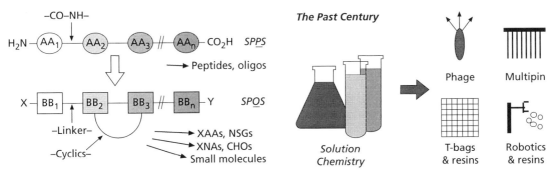

Fig. 2. The building block ("BB") concept applied to SPPS and SPOS.

Fig. 3. From "pot boiling" solution chemistry to solid-phase organic synthesis.

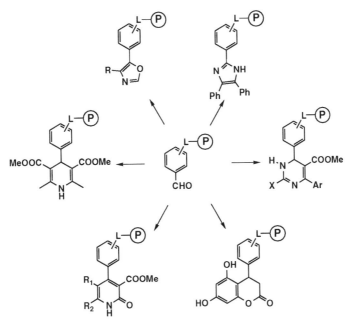

Fig. 4. Efficient combinatorial synthesis approaches from readily available starting materials.

prepared from existing libraries using "post-modification" approaches (ref. 13). Furthermore, several automated, robotics-based, and semi-automated systems have been developed.

Biological approaches also exist, such as the "biological pin", namely, bacteriophage libraries, wherein a genetic insert encodes random or structured peptide or protein sequences that are then expressed on the surface of the phage (ref. 14). The strength of the biological method is its ability to prepare and screen literally more than 100 million peptide sequences in a single experiment. A weakness is that only the 20 standard amino acids are readily available at this time as building blocks.

ANALYTICAL PLUS PROCESS CHEMISTRY AND MOLECULAR DIVERSITY

Quality assurance and quality control ("QX") have an important place in the preparation of libraries of compounds, since it is often difficult or impossible to determine whether each compound has been prepared in a given library. Of course, if at the end of the day a lead has been identified, the concerns about completeness of a library are less of an issue. In any event, it behooves the combinatorial chemist to establish that trial reactions across a range of reactants run successfully under the conditions used to create the library, factoring in electron donating and withdrawing substituents, steric or charge interactions, stability of protecting groups, etc. Once the reaction characteristics are defined, the user has greater expectation that most of the compounds planned will indeed be present in the library. Otherwise, combinatorial chemistry can take on too much of the undesirable side of natural products research.

Reaction progress can be monitored in a variety of ways, some of which do not require formal cleavage from the solid phase. The various methods include: colorimetric techniques; gel phase nuclear magnetic resonance (GP-NMR) and infrared (IR) spectroscopies; liquid chromatography (LC-MS) and gas chromatography (GC-MS) mass spectrometries.

A strong process chemistry effort assists combinatorial discovery through developing new synthetic methods, testing reaction conditions, preparing quantities of building blocks, and often "turning the crank" in the production of libraries or deconvolutions. With success, rapid scale-up to multi-gram quantities facilitates further study of biological activity.

BIOTECHNOLOGY AND MOLECULAR DIVERSITY

Chemistry by itself is not enough in our industry, and the ability to clone, sequence, express, produce, and purify human biological targets is critical to the optimal use of combinatorial diversity. Having the human sequence of a given protein is important because non-human protein sequences can lead to several orders of magnitude difference in affinity. This may at least partly explain certain clinical trial failures, where preclinical work in rodents, dogs, or other experimental animals identified agents that in fact had little affinity for the human protein homologues.

With the human protein reagents in hand, molecular diversity researchers must then develop assays and strategies to uncover active "hit pools" and lead series (see, for example, ref. 15-17). If libraries are screened as pools of mixtures, high throughput screening (HTS) is often not necessary, but nonetheless, a number of HTS approaches have been developed to screen from one to a few dozen compounds per well. A variety of approaches are available to identify hits in pools, including "deconvolution" (Fig. 5) of hit pools into smaller and smaller pools until individual compounds are identified, selection of stained beads under a microscope, affinity chromatography, encoding approaches (Fig. 6; see, for example, ref. 18), and so on. The most powerful method developed to date uses affinity selection plus size exclusion chromatography, followed by mass spectroscopy (AS-MS), to identify leads that bind to the target of interest. This method is particularly useful, as it eliminates the need to separately encode compounds on beads, using instead a code intrinsic to all molecules, namely, molecular weight (MW). (One AS-MS consideration worth mentioning is the possible redundancy of some MWs within a library of compounds.)

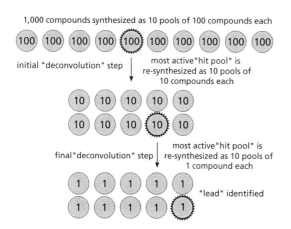

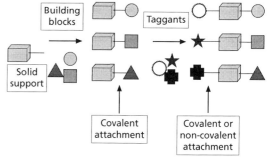

▲Fig. 6. Encoded library concept.

◄Fig. 5. Example of a "deconvolution" process.

COMPUTATIONAL CHEMISTRY AND MOLECULAR DIVERSITY

Some have called molecular diversity simply a "game of numbers" (Fig. 7). Numbers or "variety" are certainly part of the story, but we believe that the complexity or "diversity" of libraries are perhaps more important (Fig. 8). As a result, we have developed several ways of calculating and visualizing diversity, using statistical, "fragment", and graphic means. Calculation of structural properties, followed by statistical selection methods, generates non-redundant libraries that optimally sample property space. Our "flower plots" (Fig. 9; see ref. 19) have captured the interest of many groups for their ability to represent the many dimensional physicochemical parameter space in our limited dimensional world. We are in the process of adding true 3-dimensional descriptors to our analyses, as well as coupling genetic

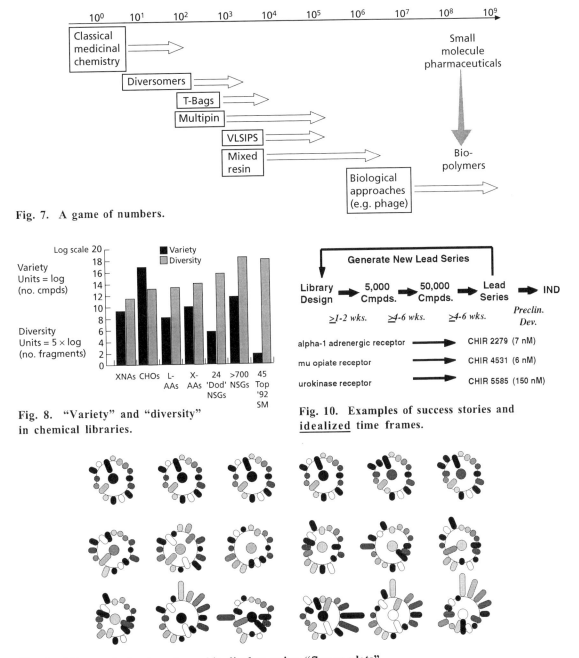

Fig. 7. A game of numbers.

Fig. 8. "Variety" and "diversity" in chemical libraries.

Fig. 10. Examples of success stories and <u>idealized</u> time frames.

Fig. 9. Library design-based graphic display, using "flower plots".

algorithms with docking programs to enable the automated design of libraries that fit the spatial requirements of crystal structures or pharmacophoric hypotheses. New informatics approaches are also necessary in order to handle the overwhelming amount of data and compounds resulting from molecular diversity approaches (see, for example, ref. 20).

COMBINATORIAL DISCOVERY VS. COMBINATORIAL DEVELOPMENT

There are now a number of success stories in discovery or early development phases (Fig. 10; see, for example, ref. 21-22). We have found it, at times, relatively straightforward to uncover novel, nanomolar hits to a wide variety of biological targets. (In some cases, for reasons not yet fully understood, it is clearly not so easy!) We have also been able to use the chemical methods described above to enhance the activity or potency of initial leads. Thus, combinatorial chemistry is finding a significant role in both lead finding and lead optimization.

With increasing success in lead discovery, the "bottleneck" shifts back to preclinical development. While the idea of studying multiple compounds simultaneously via *in vivo* models may seem daunting, if not impossible, even as equimolar mixtures, we have found that at least some pharmacokinetic parameters may potentially be studied with the assistance of ultra-sensitive detection methods like "XMS" (accelerator mass spectrometry, better known as "AMS"). There are also useful *in vitro* methods that can be employed to study absorption parameters, including gastrointestinal cell monolayers (e.g., Caco-2). In vitro methods can further be used to study molecular (e.g., stress gene reporter systems like those available from Xenometrix, Inc., Boulder, CO, USA)., cellular, and organ-specific toxicity.

CONCLUSIONS

Combinatorial chemistry has joined the arsenal alongside structure-based design and other tools of the recent decades, and it is here to stay. The new methods have the potential to underwrite a cheaper, faster, and better preclinical phase of study, but to make the most of it, one needs the full menu of preclinical technologies (Fig. 11). Only then will the super-additive complementarity of technological progress earn its keep in this most competitive arena.

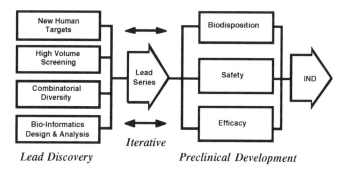

Fig. 11. Full menu of preclinical R&D technologies.

ACKNOWLEDGMENTS

We thank our many colleagues past and present for their numerous contributions to this work.

REFERENCES

1. M. Desai, R. N. Zuckermann and W. H. Moos. *Drug Dev. Res.* **33**, 174-188 (1994).
2. W. H. Moos, G. Green and M. R. Pavia. *Annu. Rep. Med. Chem.* **28**, 315-323 (1993).
3. M. R. Pavia, T. K. Sawyer and W. H. Moos. *Bioorg. Med. Chem. Lett.* **3**, 387-396 (1993).
4. J. A. Weisbach and W. H. Moos. *Drug Dev. Res.* **34**, 243-259 (1995).
5. N. J. Maeji, R. M. Valerio, A. M. Bray and R. A. Campbell. *Reactive Polymers* **22**, 203-212 (1994).
6. R. M. Valerio, A. M. Bray and N. J. Maeji. *Int. J. Pep. Prot. Res.* **44**, 158-165 (1994).
7. H. M. Geysen and T. J. Mason. *Bioorg. Med. Chem. Lett.* **3**, 397-404 (1993).
8. S. M. Miller, R. J. Simon, S. Ng, R. N. Zuckermann, J. M. Kerr and W. H. Moos. *Drug Dev. Res.* **35**, 20-32 (1995).
9. R. J. Simon, E. J. Martin, S. M. Miller, R. N. Zuckermann, J. M. Blaney and W. H. Moos. *Tech. Prot. Chem.* **5**, 533-539 (1994).
10. R. J. Simon, R. S. Kania, R. N. Zuckermann, V. D. Huebner, D. A. Jewell, S. Banville, S. Ng, L. Wang, S. Rosenberg, C. K. Marlowe, D. C. Spellmeyer, R. Tan, A. D. Frankel, D. V. Santi, F. E. Cohen and P. A. Bartlett. *Proc. Natl. Acad. Sci. USA* **89**, 9367-9371 (1992).
11. R. N. Zuckermann, J. M. Kerr, S. B. H. Kent and W. H. Moos. *J. Am. Chem. Soc.* **114**, 10646-10647 (1992).
12. R. N. Zuckermann, J. M. Kerr, M. A. Siani and S. C. Banville. *Int. J. Peptide Prot. Res.* **40**, 498-507 (1992).
13. Y. Pei and W. H. Moos. *Tetrahedron Lett.* **35**, 5825-5828 (1994).
14. J. Winter. *Drug Dev. Res.* **33**, 71-89 (1994).
15. S. E. Kaufman, S. Brown and G. B. Stauber. *Anal. Biochem.* **211**, 261-266 (1993).
16. J. M. Kerr, S. C. Banville and R. N. Zuckermann. *Bioorg. Med. Chem. Lett.* **3**, 463-468 (1992).
17. R. N. Zuckermann, J. M. Kerr, M. A. Siani, S. C. Banville and D. V. Santi. *Proc. Natl. Acad. Sci. USA* **89**, 4505-4509 (1992).
18. J. M. Kerr, S. C. Banville and R. N. Zuckermann. *J. Am. Chem. Soc.* **115**, 2529-2531 (1993).
19. E. J. Martin, J. M. Blaney, M. A. Siani, D. C. Spellmeyer, A. K. Wong and W. H. Moos. *J. Med Chem.* **38**, 1431-1436 (1995).
20. M. A. Siani, D. Weininger and J. M. Blaney. *J. Chem. Info. Comp. Sci.* **34**, 588-593 (1994).
21. R. N. Zuckermann, E. J. Martin, D. C. Spellmeyer, G. B. Stauber, K. R. Shoemaker, J. M. Kerr, G. M. Figliozzi, D. A. Goff, M. A. Siani, R. J. Simon, S. C. Banville, E. G. Brown, L. Wang, L. S. Richter and W. H. Moos. *J. Med. Chem.* **37**, 2678-2685 (1994).
22. R. J. Goodson, M. V. Doyle, S. E. Kaufman and S. Rosenberg. *Proc. Natl. Acad. Sci. USA* **91**, 7129-7133 (1994).

Structures and functions of prostanoid receptors

Atsushi Ichikawa and Manabu Negishi

The Department of Physiological Chemistry, Faculty of Pharmaceutical Sciences, Kyoto University, Yoshida, Sakyo-ku, Kyoto 606, Japan

Abstract

Prostaglandin (PG) E_2 exhibits a wide spectrum of physiological and pathological actions in diverse tissues and cells through four subtypes of PGE receptors. We recently isolated cDNAs for four mouse PGE receptor subtypes, and demonstrated that they differed in signal transduction pathways; EP1 was coupled to Ca^{2+}-permeable channels, EP2 and EP4 were coupled to stimulation of adenylate cyclase, and EP3 was coupled to inhibition of adenylate cyclase. Among these PGE receptor subtypes, EP3 had several isoforms produced through alternative mRNA splicing. They differ only in the carboxy-terminal tail. They had almost identical ligand binding properties, but they differ in G protein coupling or desensitization. The carboxy-terminal tails of EP3 determine not only the G protein selectivity but also regulatory systems of receptor activation. The diversity of the cellular responses to PGE_2 is based on the existence of functionally different PGE receptor subtypes and isoforms.

Introduction

Prostanoids comprise PGs and thromboxanes, are synthesized via the cyclooxygenase pathway from arachidonic acid by a variety of cells in response to various physiological or pathological stimuli (1). They exert a wide spectrum of actions in the body, which are mediated by specific receptors on plasma membranes. Among the prostanoids, PGE_2, in particular, exhibits a variety of actions in diverse tissues and cells. The characteristic features of PGE_2 are versatile and opposite actions and wide distribution of its receptors in the body. PGE_2 can cause the contraction or suppression of the neurotransmitter release, and inhibition or stimulation of sodium and water reabsorption in the kidney. These opposite actions of PGE_2 are due to PGE receptor subtypes being coupled to a variety of signal transduction pathways. PGE receptors are pharmacologically divided into four subtypes, EP1, EP2, EP3 and EP4, on the basis of their responses to various agonists and antagonists (2). We recently revealed the molecular structures of four subtypes of mouse PGE receptor and characterized their biochemical properties and distribution in the body.

Correspondence to: Atsushi Ichikawa,
The Department of Physiological Chemistry, Faculty of Pharmaceutical Sciences, Kyoto University, Sakyo-ku, Kyoto 606, Japan
Tel. +81-75-753-4527
Fax: +81-75-753-4557
E-mail: aichikaw@pharmsun.pharm.kyoto-u.ac.jp

Structure of Prostanoid Receptors

Fig. 1 shows the amino acid sequence alignment of cloned mouse prostanoid receptors, including TP, IP, FP, DP, EP1, EP2, EP3 and EP4 receptors. They have hydrophobic putative transmembrane domains, and thus belong to a rhodopsin-type receptor superfamily. Although the overall homology of amino acid sequences among the prostanoid receptors ranges from approximately 20-30%, the prostanoid receptors show significantly higher homology to each other than other rhodopsin-type receptors, and thus they constitute a new subfamily within the rhodopsin-type receptor superfamily. There are several regions conserved specifically among the prostanoid receptors. The highly conserved regions were found in the third and seventh transmembrane domains and in the second extracellular loop. Prostanoids have common structural features such as carboxylic acid, a hydroxyl group at position 15, and two aliphatic side chains. These structures are thought to play an important role in the binding to the receptors and activation. The conserved sequences of the receptors noted above may involve counter binding sites for the groups of prostanoid structure.

Fig. 1. Comparison of the amino acid sequences of mouse prostanoid receptors. The deduced amino acid sequences of the mouse prostanoid receptors are aligned to optimize homology. Identical amino acid residues in four or more sequences are indicated by bold characters. The approximate positions of the putative transmembrane regions are indicated above the amino acid sequences.

The prostanoid receptors can be grouped into three categories according to the coupled signal transductions. The IP, EP2, EP4 and DP receptors are coupled to stimulation of adenylate cyclase, the TP, EP1 and FP receptors are coupled to Ca^{2+} mobilization, and the EP3 receptors are coupled to inhibition of adenylate cyclase. Sequence homology among these functionally related receptors is higher than that between the two groups. Among the prostanoid receptors, subtypes of EP receptors are not significantly closer to each other than to other prostanoid receptors, but are even lower, except for between EP2 and EP4 receptors, and appear not to form a subfamily among prostanoid receptors. The subtypes of EP receptors appear to evolve from an ancestral prostanoid receptor responding to PGE_2, and then these functionally different subtypes might evolve to other functionally similar receptors responding to other prostanoids.

EP1 Receptors

EP1 receptors mediate the contraction of smooth muscle in various tissues including the gastrointestinal tract, respiratory tract, vas deferens, myometrium and iris sphincter muscle. In addition, EP1 receptors modulate neurotransmitter release. We isolated a cDNA for mouse EP1 receptor from the mouse kidney library (3). Among various mouse tissues, the receptor was abundantly expressed in the kidney and in a lessor amount in the lung and stomach. *In situ* hybridization study revealed that EP1 receptor was specifically localized to the collecting ducts from the cortex to the papilla in the kidney, where PGE_2 attenuates the vasopressin-induced osmotic water permeability through Ca^{2+} mobilization. PGE_2 induced rapid extracellular Ca^{2+} influx in CHO cells expressing the mouse EP1 receptor, followed by subsequent stimulation of phospholipase C (PLC) as a result of the secondary effect of Ca^{2+} mobilization, suggesting that the EP1 receptor activates a novel type of Ca^{2+}-permeable channel. On the other hand, the receptor was not coupled to either stimulation or inhibition of adenylate cyclase.

EP2 and EP4 Receptors

EP2 receptors mediate the relaxation of trachea and ileum circular muscles, and the vasodilatation of various blood vessels. EP4 receptors induce relaxation of pig and dog saphenous vein and rabbit ductus arteriosus. Among the four PGE receptor subtypes, EP2 and EP4 receptors are coupled to the same signal transduction pathway, stimulation of adenylate cyclase, but they differ in the activity induced by certain ligands. While EP2 receptor is sensitive to butaprost, an EP2 agonist, EP4 receptor is insensitive but sensitive to AH23848B, an EP4 antagonist.

We recently isolated cDNAs for mouse EP2 and EP4 receptors, and demonstrated that they were exclusively coupled to stimulation of adenylate cyclase (4, 5). EP2 and EP4 receptors showed a wide distribution in mouse tissues, abundant expression being detected in uterus, spleen, lung, thymus, ileum and stomach. The expression level of the EP4 receptor was much higher than that of the EP2 receptor in most of the mouse tissues, except for the liver, where only EP2 receptor was expressed. As based in *in situ* hybridization study, the EP4 receptor was localized to the mesangial cells of glomeruli in the kidney. In cultured mesangial cells, PGE_2 is known to elicit cAMP formation and attenuate the contractility induced by various vasoconstrictors. This relaxant action of

PGE_2 may be mediated by the EP4 receptor, and the receptor may be involved in regulation of glomerular filtration.

EP3 Receptors

EP3 receptors mediate the contraction of the uterus, modulation of neurotransmitter release, inhibition of gastric acid secretion, lipolysis in adipose tissue, sodium and water reabsorption in kidney tubules, glycogenolysis and fatty acid oxidation in hepatocytes, and stimulation of catecholamine release from adrenal chromaffin cells. We isolated a cDNA for mouse EP3 receptor, and demonstrated that it was functionally coupled to Gi (6). In addition, the receptor induced PLC activation and subsequent Ca^{2+} mobilization mainly from extracellular medium via Gi (7). This Ca^{2+} mobilization would be mediated by IP_3-sensitive Ca^{2+}-permeable channel as the result of PLC activation by $\beta\gamma$ subunits liberated from the EP3 receptor-activated Gi proteins. PGE_2-induced uterus contraction and inhibition of a housekeeping Cl^- channel in gastric parietal cells appear to be mediated by this Ca^{2+} signaling.

The EP3 receptor was expressed in the kidney at the highest level among mouse tissues. The *in situ* hybridization study has revealed that the EP3 receptor was present in the tubules in the outer medulla and in the distal tubules in the cortex, where PGE_2 attenuates vasopressin-induced sodium and water reabsorption.

The molecular and biological characteristics of the four PGE receptor subtypes, as currently defined, are summarized in Table 1.

Table 1. Comparison of the molecular and biochemical characteristics of the cloned mouse four PGE receptor subtypes.

Subtypes	EP1	EP2	EP3	EP4
Potency of PGs	$E_2>I>E_1>F>D$	$E_2=E_1>>I>F>D$	$E_2=E_1>I>F>D$	$E_2=E_1>I>F>D$
Agonists	17-phenyl PGE_2	butaprost	M&B28767	none
Antagonists	AH6809	none	none	AH23848B
Effectors	Ca^{2+} channel	AC	AC , Ca^{2+}	AC
G proteins	unknown	Gs	Gi, Go	Gs
Tissues	kidney, stomach	uterus, thymus	kidney, uterus	uterus, ileum
M. W.	43 k	56 k	40 k	40 k
Amino acids	405	362	365	513
K_d	21	27	2.9	11

AC: adenylate cyclase; M. W.: molecular weight

EP3 Receptor Isoforms

We have found three isoforms of mouse EP3 receptor (EP3α, β and γ), which are produced through alternative splicing and differ only in the carboxy-terminal tail (8, 9). They have almost identical ligand binding properties, but they differ in G protein coupling. EP3α and β were exclusively coupled to inhibition of adenylate cyclase via Gi, but EP3α caused adenylate cyclase inhibition more efficiently than EP3β. On the other hand, EP3γ was coupled to both Gi and Gs, causing the inhibition of adenylate cyclase at low concentrations of PGE$_2$ and stimulation at high concentrations. Furthermore, we have found four isoforms of bovine EP3 receptor (EP3A, B, C and D), which are also produced through alternative splicing and differ only in the carboxy-terminal tail (10). They were coupled to different G proteins; EP3A was coupled to adenylate cyclase inhibition, EP3B and C were coupled to adenylate cyclase stimulation, and EP3D was coupled to both stimulation and inhibition of adenylate cyclase and stimulation of phosphoinositide metabolism in a pertussis toxin-insensitive manner. EP3B and C was further coupled to Go, but they inhibited the GTPase activity of Go (11). The carboxy-terminal tails of the EP3 receptors determine not only the G protein selectivity but also regulatory systems of G protein activity. G protein coupling selectivity of the EP3 receptor isoforms are summarized in Fig. 2.

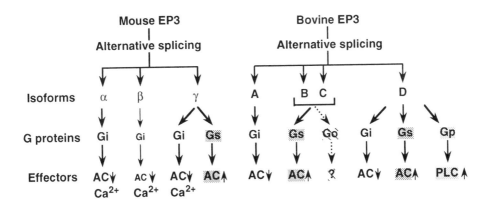

Fig. 2. Signal routing in mouse and bovine EP3 receptor isoforms - G protein networks.

We also demonstrated that the EP3 receptor isoforms differed in agonist-induced regulation of receptor function. EP3α underwent both short- and long-term agonist-induced desensitization, whereas EP3β did not undergo either short- or long-term desensitization (12). Furthermore, EP3α and β isoforms showed different efficiencies of enhancement of adenylate cyclase stimulation after prolonged agonist exposure and subsequent removal (i. e., sensitization of adenylate cyclase stimulation, representing an attempt by cells to maintain cAMP homeostasis on prolonged inhibition of adenylate cyclase by Gi-coupled receptor agonists); EP3α caused enhancement of adenylate cyclase stimulation more efficiently than EP3β (13). The carboxy-terminal tails add the EP3 receptors distinct properties with respect to agonist-induced regulation of receptor function.

Agonist Structure-dependent Regulation of G Protein Coupling

The positively charged arginine residue within the seventh transmembrane domain, which is conserved in all of the prostanoid receptors (Fig. 1), was proposed to be the binding site of the negatively charged carboxyl group of prostanoid molecules by analogy to the retinal binding site of rhodopsin, in which the hydrocarbon chain of retinal was shown to be attached to Lys-296 in the seventh transmembrane domain. A variety of PGE analogues with potent agonist activity have been developed, but some EP3 agonists have the structural features of substitution at C-1, where carboxylic acid is replaced by a variety of esters or methanesulfonamido groups, its negative charge being blocked. Chemical structures of EP3 agonists are shown in Fig. 3.

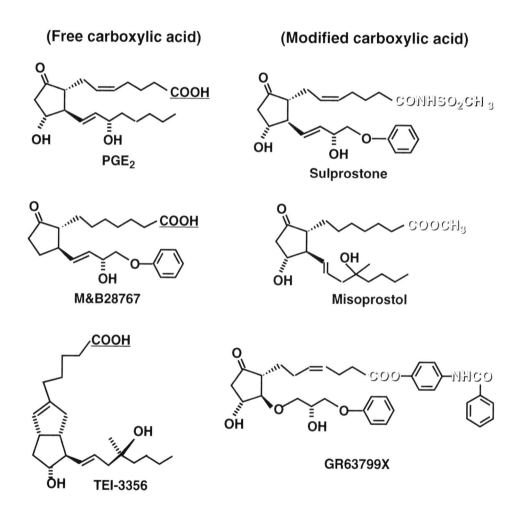

Fig. 3. Chemical strucures of various EP3 agonists.

We mutated the Arg residue in the seventh transmembrane domain to the noncharged Gln in EP3D, which is coupled to Gi, Gs and Gp, and examined the roles of the interaction of carboxylic acid of PGE_2 and the Arg residue in receptor-G protein coupling (14). Among various EP3 agonists, carboxylic acid-unmodified agonists showed Gi and Gs activities and stimulation of phosphatidylinositol hydrolysis, but the modified agonists showed only Gi activity in wild type EP3D receptor. In the mutated receptor, both modified and unmodified agonists showed only Gi activity. Thus, the interaction between the carboxylic acid of PGE_2 and the Arg residue of the receptor plays an important role for the Gs and Gq activation but not for the Gi activation. This finding provides the idea that receptors form multiple conformations, which are differentially coupled to multiple G proteins. In EP3 receptors, carboxy-terminal tail determines the selectivity of G protein coupling. The seventh transmembrane domain is directly connected with the carboxy-terminal tail. Thus, a conformational change of the carboxy-terminal tail may regulate the selectivity of G protein coupling, and the EP3D receptor may be able to form two types of conformation, which can selectively be associated with either Gi or Gs/Gq, depending on the interaction of the carboxylic acid and the Arg residue. We show the hypothetical model of coupling of EP3D receptor to G proteins depending on the interaction of the carboxylic acid and the Arg residue in Fig. 4.

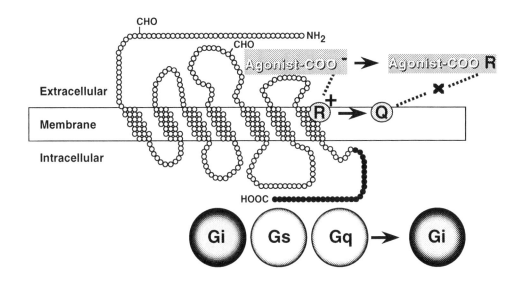

Fig. 4. Selective coupling of EP3D receptor to Gi, Gs and Gq through interaction of the carboxylic acid of agonists and the Arg residue.

Conclusion

We here showed molecular and biochemical properties of four PGE receptor subtypes and EP3 receptor isoforms. These findings strongly suggest that the diverse and exquisite actions of PGE$_2$ are based on the existence of functionally different PGE receptor subtypes and isoforms. These studies will contribute not only to understanding the diverse PGE$_2$ actions, but they will also help to elucidate the molecular mechanism of receptor-mediated activation of G proteins. In addition, we demonstrated ligand binding site of the receptor and the responding ligand structure in the receptor-induced selection and activation of G proteins. Functional elucidation of the ligand binding structures of the receptors will facilitate the development of subtype-specific and signal pathway-specific agonists and antagonists for potential therapeutic uses.

References

1.   Coleman, R. A., Kennedy, I., Humphrey, P. P. A., Bunce, K., and Kumley, P. (1990) in Comprehensive Medicinal Chemistry (Hansch, C., Sammes, P. G., Taylor, J. B., and Emmett, J. C. eds.), Vol. 3, 643-714. Pergammon Press, Oxford.

2.   Negishi, M., Sugimoto, Y., and Ichikawa, A. (1993) Prog. Lipid Res. 32, 417-434.

3.   Watabe, A., Sugimoto, Y., Honda, A., Irie, A., Namba, T., Negishi, M., Ito, S., Narumiya, S., and Ichikawa, A. (1993) J. Biol. Chem. 268, 20175-20178.

4.   Honda, A., Sugimoto, Y., Namba, T., Watabe, A., Irie, A., Negishi, M., Narumiya, S., and Ichikawa, A. (1993) J. Biol. Chem. 268, 7759-7762.

5.   Nishigaki, N., Negishi, M., Honda, A., Sugimoto, Y., Namba, T., Narumiya, S., and Ichikawa, A. (1995) FEBS Lett. 364, 339-341.

6.   Sugimoto, Y., Namba, T., Honda, A., Hayashi, Y., Negishi, M., Ichikawa, A., and Narumiya, S. (1992) J. Biol. Chem. 267, 6463-6466.

7.   Irie, A., Segi, E., Sugimoto, Y., Ichikawa, A., and Negishi, M. (1994) Biochem. Biophys. Res. Commun. 204, 303-309.

8.   Sugimoto, Y., Negishi, M., Hayashi, Y., Namba, T., Honda, A., Watabe, A., Hirata, M., Narumiya, S., and Ichikawa, A. (1993) J. Biol. Chem. 268, 2712-2718.

9.   Irie, A., Sugimoto, Y., Namba, T., Harazono, A., Honda, A., Watabe, A., Negishi, M., and Ichikawa, A. (1993) Eur. J. Biochem. 217, 313-318.

10.   Namba, T., Sugimoto, Y., Negishi, M., Irie, A., Ushikubi, F., Kakizuka, A., Ito, S., Ichikawa, A., and Narumiya, S. (1993) Nature 365, 166-170.

11.   Negishi, M., Namba, T., Sugimoto, Y., Irie, A., Katada, T., Narumiya, S., and Ichikawa, A. (1993) J. Biol. Chem. 268, 26067-26070.

12.   Negishi, M., Sugimoto, Y., Itie, A., Narumiya, S., and Ichikawa, A. (1993) J. Biol. Chem. 268, 9517-9521.

13.   Harazono, A., Sugimoto, Y., Ichikawa, A., and Negishi, M. (1994) Biochem. Biophys. Res. Commun. 201, 340-345.

14.   Negishi, M., Irie, A., Sugimoto, Y., Namba, T., and Ichikawa, A. (1995) J. Biol. Chem. 270, 16122-16127.

Allylic nitro-oximes:
an original moiety pattern
to store and release nitric oxide

J.-C. MULLER, P. BELLEVERGUE, G. LASSALLE, D. LOYAUX,
P.H. WILLIAMS, M. FONTECAVE*, J.L. DECOUT* and B. ROY*.

SYNTHELABO Recherche, 31 av. P. Vaillant-Couturier, BP 110
F - 92225 BAGNEUX (France)

* Laboratoire d'Etudes Dynamiques et Structurales de la Sélectivité, BP 53X
F - 38041 GRENOBLE (France)

Abstract: Amongst the various categories of NO donors, allylic nitro-oximes behave in a rather peculiar way. Like organic nitrates, nitrothiols, sydnonimines, furoxans and NONOates, they relax rabbit aorta through an activation of guanylate cyclase. The study reported was designed to prove that FK 409, an allylic nitro-oxime derivative isolated from *Streptomyces griseosporeus*, directly releases NO.

Unlike most NO donors, FK 409 spontaneously and rapidly decomposes without activation, in aqueous solution at physiological pH to release nitric oxide. Decomposition products were identified through chemical and physicochemical experiments. The presence of NO was clearly and fully established. A plausible mechanism for this NO liberation will be discussed.

Since the pionnering discovery by FURCHGOTT and ZAWADSKI (1), in 1980, proving that the endothelial cell was able to release a potent vasorelaxant factor which they named EDRF, numerous papers dealing with this topic have appeared. In 1987, evidence was presented indicating that EDRF is nitric oxide, NO$^\bullet$ (2), a conclusion that has been largely documented in recent years. Nitric oxide is a unique biological messenger which mediates various physiological processes (3). Conventional neurotransmitters are enzymatically synthesized, stored in vesicles and released upon excitation or depolarisation. Nitric oxide, on the other hand, is synthesized on demand by the enzyme NO synthase and exerts its biological activity through its chemical reactivity rather than its molecular structure.

The study of compounds susceptible to release nitric oxide is therefore of current importance and allows a novel insight in the various mechanisms underlying the actions of NO$^\bullet$.

We decided to investigate the decomposition of ($\pm$)-(3E)-4-Ethyl-2[(E)-hydroxyimino]-5-nitro-3-hexeneamide, FK 409, **1**, a new type of vasodilator isolated from the fermentation broth of *Streptomyces griseosporeus*. FK 409 shows strong vasorelaxant effects and inhibits platelet aggregation (4,5). In order to understand the molecular mechanism by which allylic nitro-oximes, such as FK 409, behave as a vasodilator, we questioned whether the drug is transformed into NO$^\bullet$.

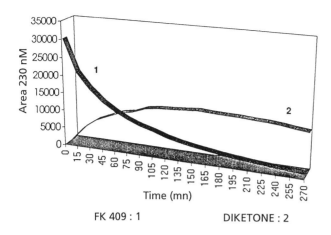

FK 409 : 1   DIKETONE : 2

Figure 1 : Respective Peak Areas of FK 409 and diketon 2 observed by LC / UV (230 nm absorbance), pH = 7.5

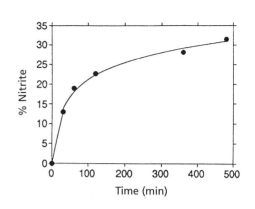

Figure 2 : Nitrite formation during FK 409 decomposition

<u>Decomposition of FK 409</u> :

FK 409 decomposes spontaneously in aqueous buffer solution. The decomposition was pH dependent with the highest rates observed at alkaline pH. At pH 7.5, 85% of parent compound was lost after 180 minutes (figure 1). By mass spectrometry, the decomposition product was identified as diketone **2**.

As shown in figure **2**, during decomposition, FK 409 produced large amounts of nitrite NO_2^-.

<u>NO release from FK 409</u> :

To demonstrate that $NO^•$ was an intermediate product during the decomposition of **1** into nitrite, two scavengers of $NO^•$ and EPR spectroscopy were used. The first $NO^•$ trap was the Fe^{2+}-diethyldithiocarbamate (DETC) complex, dissolved in yeast membranes at pH 7.5 (6). After addition of **1**, the hyperfine triplet signal characteristic of the paramagnetic mononitrosyl-Fe^{2+} (DETC), stable complex shown in figure **3**, was recorded. The second assay was based on the ability of $NO^•$ to reversibly scavenge the tyrosyl radical of the small subunit of ribonucleotide reductase named R_2 (7), (scheme 1).

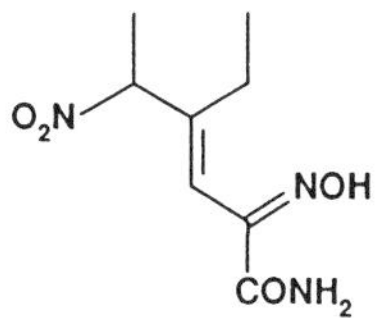

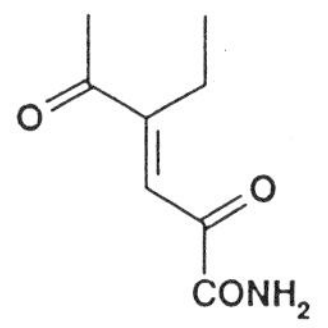

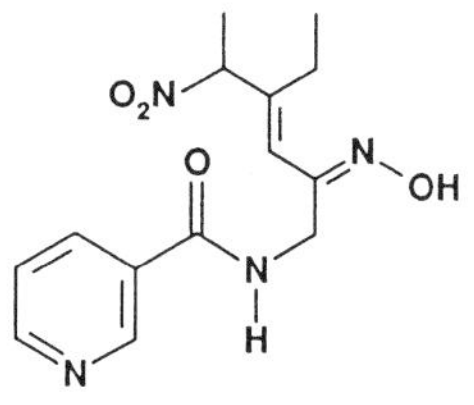

FK 409 : <u>1</u>   <u>2</u>   FR 144 420 : <u>3</u>

Scheme 1

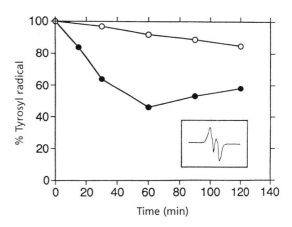

Figure 3 : EPR Spectrum of the paramagnetic NO-Fe^{2+} (DETC) complex

Figure 4 : Loss of tyrosyl radical signal during incubation of FK 409 with protein R2

This reactivity is an intrinsic property of phenoxyl radicals and is very specific for NO$^{•}$. When a pure preparation of protein R$_2$, displaying a characteristic stable tyrosyl radical, was incubated with FK 409, this signal disappeared and later reappeared (figure 4). This is best explained by the formation of the EPR - silent nitrosotyrosine (scheme 1). This reversible reaction is driven backwards when the free NO$^{•}$ reacts with molecular oxygen and its concentration decreases.

This study confirms, through physicochemical studies, what had been evidenced by biochemical studies (8), that **1** produces NO$^{•}$ when spontaneously decomposing.

1 contains a nitro and an oxime moiety which might be transformed into NO$^{•}$. Chemical difficulties to prepare and isolate a large series of targeted analogues of this original allylic nitrooxime have hampered SAR studies. FR 14420, **3**, a derivative of **1**, was recently described by KITA et al. (9). These authors disclosed that **3** released NO$^{•}$ at at half the rate of **1** and was less potent as a vasorelaxant. However, *in vivo*, the duration of hypotensive effects induced by **3** was longer than those triggered by **1**. Therefore, it is believed that **3** is more stable and releases NO$^{•}$ more slowly than does **1**.

We have tried to clarify the molecular mechanism for the spontaneous decomposition of **1** into NO$^{•}$ and diketone **2** :

- Nuclear magnetic resonance spectroscopy studies indicated that the allylic proton, in the α position with regard to the nitro group, was very labile at low pH.
- Mass spectrometry experiments allowed to identify diketone **2** as the only stable by product of the decomposition.
- Generation of nitrite anion is not very reliable indicator for NO$^{•}$ production

The proposed mechanism would have to include the three following key features :

- both the nitro and the oxime groups are needed for the generation of NO•

- both groups are hydrolysed into the corresponding ketones.

- most probably NO• and nitrite anions are formed simultaneously in the decomposition of **1**.

An attractive possibility is that **1** is the site of an unusually mild Nef reaction. The intermediately formed NO – would be a direct or indirect precursor of NO• (10). Further studies to understand the fine tuning of this mechanism will request specific ^{15}N labelling of the nitro and/or oxime moieties of a more stable derivative of **1**.

References

1.   R.F. Furchgott and J.V. Zawadski   *Nature* **288**, 373-376 (1980)

2.   R.M.J. Palmer, A.G. Ferrige and S. Moncada   *Nature* **327**, 524-526 (1987)

3.   T.M. Dawson and V.L. Dawson   *The Neuroscientist* **1**, 9-20 (1995)

4.   S. Shibata, N. Satake, N. Sato, M. Matsuo, Y. Koibuchi and R.K. Hester.
       J. Cardiovasc. Pharmacol. **17**, 508-518 (1991)

5.   T. Isono, Y. Koibuchi, N. Sato, A. Furuichi, M. Niishi, T. Yamamoto, J. Mori,
       M. Kohsaka and M. Ohtsuka   *Europ. J. Pharmacol.* **246**, 205-212 (1993)

6.   P. Mordvintcev, A. Mulsch, R. Busse and A. Vanin   *Anal. Biochem.* **199**,
       142-146 (1991)

7.   B. Roy, M. Lepoivre, Y. Henri and M. Fontecave   *Biochemistry*   in press

8.   Y. Kita, Y. Hirasawa, K. Maeda, M. Nishio and K. Yoshida.
       Europ. J. Pharmacol. **257**, 123-130 (1994)

9.   Y. Kita, K. Ohkubo, Y. Hirasawa, Y. Katayama, M. Ohno, S. Nishino, M. Kato and
       K. Yoshida   *Europ. J. Pharmacol.* **275**, 125-130 (1995)

10.   J.L. Decout, B. Roy, M. Fontecave, J.C. Muller, P.H. Williams and D. Loyaux.
       Bioorganic and Medicinal Chemistry Letters **5**, 973-978 (1995)

Design and synthesis of novel potent, non-peptide and orally active renin inhibitors

J. Maibaum,* V. Rasetti, H. Rüeger, N.C. Cohen, R. Göschke, R. Mah, J. Rahuel, M. Grütter, F. Cumin, and J. M. Wood

CIBA-GEIGY Ltd. Pharmaceuticals Division, Research Department, Klybeckstrasse 220, CH-4002 Basel, Switzerland

Abstract: A new concept used to design a novel, completely non-peptide class of potent, orally active human renin inhibitors of low molecular weight is described. The rationale was based on modeling studies and on the subsequent discovery of the non-substrate binding pocket S_3^{sp} by X-ray crystallography of enzyme-inhibitor complexes. The truncated dipeptide transition state mimetics with an extended hydrophobic $(P_3$-$P_1)$-moiety lack the P_4-P_2 spanning peptide-like backbone of all previously reported renin inhibitors. The additional P_3^{sp} residue attached to the P_3 heterocycle provided enhanced binding affinities up to the sub-nanomolar range. In vitro SAR data for two series of compounds with structurally different P_3-P_1 side chains as well as the pronounced blood pressure lowering effects of inhibitors $\underline{16}$ and $\underline{17}$ in marmosets is described.

INTRODUCTION

The renin-angiotensin system (RAS) plays a central role in the regulation of blood pressure, fluid volume and electrolyte balance (1), and is well established as a target for cardiovascular disease therapy. Inhibitors of the angiotensin-converting enzyme (ACE), blocking the formation of the potent vasopressor hormone angiotensin II, are widely used for the treatment of hypertension and congestive heart failure (2). The metabolic degradation of bradykinin and other members of the kinin family by ACE may be responsible for some of the adverse effects of ACE inhibitors (2). Considerable efforts have therefore been invested in the development of agents that would more specifically interfere with the RAS cascade, such as antagonists of the angiotensin II receptor (3) and inhibitors of renin which possesses high substrate specificity in catalysing the first and rate-limiting step in the formation of angiotensin II (4).

Poor oral absorption and rapid biliary uptake, due to unfavourable lipophilicity and molecular size, contribute to the limited bioavailabilty of most of the potent peptide-like renin inhibitors that have been reported to date, including CGP 38560 (5) (Fig. 1). Only recently have peptide-based renin inhibitors with good oral bioavailability and efficacy in several species been discovered (6). We have developed a completely different concept for the design of smaller non-peptide inhibitors with altered physico-chemical properties in order to improve oral potency (7, 8). A preliminary account of the structure activity relationship (SAR) of two series of dipeptide transition state isosteres, which contain a substituted P_3 moiety directly linked to the P_1 side chain and which lack the P_4-P_2 portion, is reported here. Furthermore, an unexpected binding mode of these compounds to renin as discovered by X-ray crystallography is described.

RATIONALE FOR THE DESIGN OF NON-PEPTIDIC PROTOTYPE RENIN INHIBITORS

In the quest for structurally different small molecule renin inhibitors, we investigated the binding mode of inhibitors such as CGP 38560 ($\underline{1}$) within the active site (9) by using the computational model of human renin constructed by Sibanda, Blundell et al. (10). These studies indicated that in the predicted bioactive extended conformation of $\underline{1}$ both the P_1 cyclohexyl side chain and the P_3 phenyl are in close spatial proximity, occupying the contiguous hydrophobic binding site of the S_3 and S_1 subpockets (cf. Fig. 2). Therefore, it appeared to be feasible to append the P_1 residue of a transition state mimic directly with a lipophilic P3 moiety via a spacer group. At the outset, we also considered the possibility to enforce the relative binding contribution of the tethered side chain by optimisation of its van der Waals interactions to the huge surface of the S_3/S_1 'superpocket', which would in turn allow truncation of the N-terminal inhibitor peptide backbone and thus an overall reduction of molecular size (Fig. 1). A similar topographical approach has been employed recently for the synthesis of transition state isosteres with modified (P_3-P_1)-moieties, that contain the P_2-P_4 portion with no P_3 phenyl side chain (11).

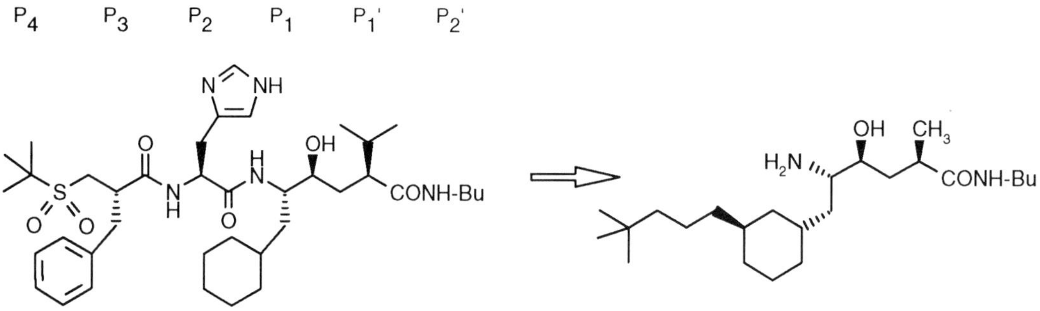

Fig. 1   Schematic presentation of the P_3-P_1 approach leading to the non-peptide inhibitor 2 (7).

The non-peptidic prototype inhibitors 2 and 3a were found to inhibit purified human renin in the sub-micromolar range, being >100-fold more potent than the corresponding non-tethered analogues (7). We extended these findings in the present study to a related class of inhibitors by appending a P_3 tetrahydroquinoline moiety via an amide linkage to the P_1 side chain of the transition state mimetic. For compound 4, a more than 10-fold improvement in the *in vitro* activity was observed over the parent inhibitor 3a. The increase in binding affinity of 4 vs. 3a could be mainly rationalized by the reduced conformational flexibility of the hydrophobic P_3-P_1 moiety due to the amide carbonyl of the spacer. Moreover, the amide linkage allowed a broad SAR study at the P_3 position within this series, starting from a suitably protected key acid intermediate. Subsequent variation of the N-containing ring size in 4 or the spacer group (cf. reduction of the carbonyl) did not further increase *in vitro* potency.

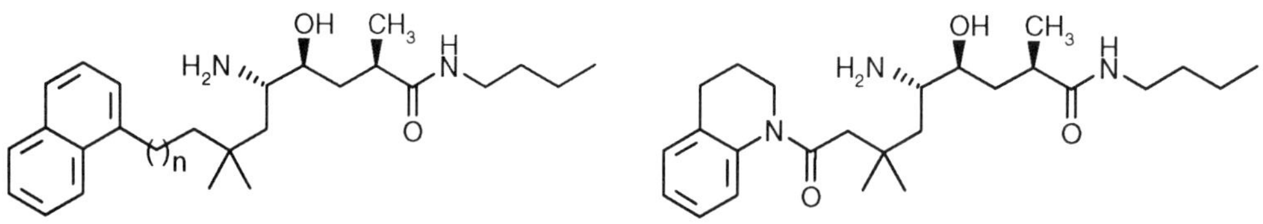

3a: n = 1 (IC$_{50}$ = 0.7 μM)

3b: n = 2 (IC$_{50}$ = 2.1 μM)

4 (IC$_{50}$ = 0.05 μM)

The energy minimized conformation of 4 docked into the binding cleft of human renin indicated a very close fit of the hydrophobic P_3-P_1 moiety to the complementary S_3-S_1 'tandem pocket' of the enzyme. It appeared therefore difficult to enhance the binding affinity of inhibitors just by further optimisation of the van der Waals interactions to the hydrophobic binding site without increasing the synthetic complexity within this series. Instead, emphasis was directed towards substitution of the P_3 heterocycle in $\underline{4}$ with the aim to induce additional hydrogen bonding interactions to appropriate amino acids of the enzyme aligning the S_3 site (7). We reasoned that a hydrogen bond with optimal distance and geometry might lead to tighter binding either by its free energy contribution or by inducing a more precise alignment of the ligand within the active site (12). The computational model of human renin-bound inhibitor $\underline{4}$ showed the 'upper' portion of the P_3 heterocycle oriented towards the amide bond of Ser219. In particular, this model suggested that an additional carboxylic ester group introduced to the 3'-position of the ring would be in a similar position as the P_3/P_2 amide carbonyl of CGP 38560 ($\underline{1}$) in its predicted binding mode, and thus within reasonable H-bond distance to the amide NH of Ser219, as depicted in Fig. 2. A conserved hydrogen bond between the carbonyl of the inhibitor P_3/P_2 backbone amide bond and the main chain NH of Ser/Thr219 (human renin numbering) has been found in the crystal structures of several peptide-like inhibitors complexed to various aspartic acid proteases (12, 13) and recently for CGP 38560 bound to human renin (14).

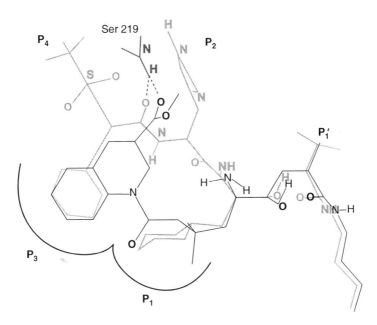

Fig. 2   Illustration of the concept used in the design of inhibitors $\underline{5}$-$\underline{15}$ of Table 1. The bound conformation of $\underline{1}$ (gray lines) within the model of the human renin active site is overlapped with $\underline{5a}$ (black lines); H-bonds to the amide NH of Ser219 are indicated by dotted lines.

The binding affinities of the diastereomeric carboxylic esters $\underline{5a}$ and $\underline{5b}$ (Table 1) as the first members of a series of P_3-substituted compounds were determined against purified human renin as described previously (6). Isomer $\underline{5a}$ revealed sub-nanomolar enzyme inhibition (IC_{50}=0.8 nM) and was more than 50-fold more potent as compared to the non-substituted analogue $\underline{4}$. On the other hand, epimer $\underline{5b}$ was markedly less active than $\underline{5a}$ in this assay, being basically equipotent to $\underline{4}$. The increase in binding affinity of $\underline{5a}$ vs. $\underline{5b}$ and $\underline{4}$ corresponds to about 2 kcal/mol , which would be in agreement with a single hydrogen bond interaction. The 3,4-dihydro-2H-1,4-benzothiadiazine derivative $\underline{7}$, prepared as a mixture of C-2' isomers, was 2-3 times more active than its congener $\underline{5a}$. These results clearly demonstrated the importance of the correct spatial orientation of the alkoxycarbonyl residue for optimal binding, with the binding affinity being affected even by subtle changes in the geometry of the P_3 moiety. As predicted by the model, the C-2' ester analogue of $\underline{5a}$ was a much weaker inhibitor of human renin (IC_{50} = 2.5 μM).

OPTIMIZATION AND SAR OF P3 APPENDED DIPEPTIDE ISOSTERES

<u>Substituted 1,2,3,4-Tetrahydroquinolines and Related Heterocycles at P3</u>

Although compounds **5a** and **7** were very effective at inhibiting human renin *in vitro*, they lacked any significant oral activity in marmosets. This was rationalized by a rapid metabolic hydrolysis of the carboxylic esters to the free acid derivatives which were more than three orders of magnitude less active *in vitro*. We therefore envisaged a variety of potential H-bonding acceptor and/or donor groups as replacements for the labile ester function, the SAR of which is highlighted in Table 1.

Within a series of carboxamide analogues only secondary amides such as **8** showed low nanomolar potency against the purified human enzyme. However, the N-monoalkyl carboxamides as well as 'reversed amides' such as **9** were generally at least one order of magnitude less active than esters **5** and **7**. As is also evident from the data in Table 1, a reduction in potency by a factor of about 10 was observed for both the ester and amide derivatives **5-8** in the physiologically more relevant human plasma renin assay (5,15). Such discrepancies of IC$_{50}$ values measured in the presence vs. absence of plasma, as well as a larger variability of these values, has been reported previously in the case of peptide-like renin inhibitors, the magnitude of which also being affected by only minor structural modifications (15). Despite its relative moderate *in vitro*

TABLE 1. *In vitro* activities of (P$_3$-P$_1$)-dipeptide transition state mimetics modified at P$_3$

Compd[a]	R	X	IC$_{50}$ (nM), Human Renin Purified	Plasma
<u>5a</u> [b]	CO$_2$Me	CH$_2$	0.8	6
<u>b</u> [c]			34	ND[f]
<u>6</u>	CO$_2$Et	O	2	40
<u>7</u>	CO$_2$Me	S	0.3	3
<u>8</u>	CONHMe	S	4	30
<u>9</u>	NHCOMe	CH$_2$	19	54
<u>10a</u> [c]	NHCO$_2$Me	CH$_2$	0.5	1.0
<u>b</u> [b]			6	25
<u>11</u>	NHCO$_2$Et	CH$_2$	1	26
<u>12</u>	CH$_2$OMe	CH$_2$	2	23
<u>13</u>	CH$_2$OCH$_2$OMe	CH$_2$	0.8	2
<u>14a</u> [d]	CH$_2$OCH$_2$OMe	O	0.1	2
<u>b</u> [e]			2	74
<u>15</u>	(phenyl)	CH$_2$	2	39

a) Ca. 1:1-mixtures of C-2' or C-3'-diastereomers; *b)* Pure C-3'(S) diastereomer; *c)* Pure C-3'(R) diastereomer; *d)* Pure C-2'(S) isomer; *e)* Pure C-2'(R) isomer; *f)* Not determined.

potency in the presence of plasma, compound $\underline{8}$ was found to be the first inhibitor in this series that induced a significant blood pressure lowering effect in marmosets when given orally at 10 mg/kg, hence providing encouragement in continuing with this approach.

Substitution of the P3 heterocycle with a 'reversed' methyl carbamate residue maintained the excellent *in vitro* potency seen for the ester analogues. Thus, the more active 3'(R)-configured $\underline{10a}$ showed a dramatic improvement in binding affinity over the reversed amide $\underline{9}$. More importantly, $\underline{10a}$ appeared to be basically equipotent in the purified enzyme and in the plasma renin assay. On the other hand, the ethyl analogue $\underline{11}$ lost more than 20-fold in potency in the plasma vs. the purified renin assay (compare also methyl ester $\underline{5a}$ vs. ethyl ester $\underline{6}$). Thus, even subtle structural changes at the P3 residue may have a marked impact on the *in vitro* activity when measured in the presence of plasma. An increase in the partial lipophilicity of the P3 side chain tended to increase the difference of IC_{50} values determined under the two different conditions.

Introduction of an alkyl ether residue into P3 also gave very potent inhibitors *in vitro*, such as compounds $\underline{12}$-$\underline{14}$, being consistent with the strong H-bonding acceptor potential of the ether oxygen (16). Numerous analogues of $\underline{12}$ were synthesised by varying the length of the side chain and incorporating one or two heteroatoms at different positions in order to optimize potency in the plasma assay. In this series, derivatives $\underline{13}$ and $\underline{14a}$ bearing a MOM ether side chain showed favourable *in vitro* profiles (Table 1).

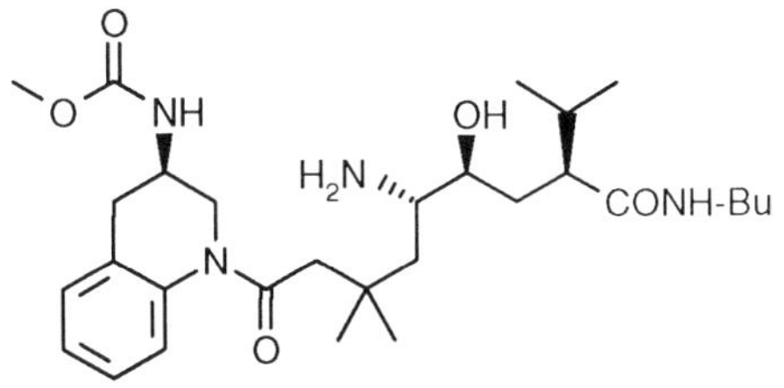

CGP 55128A ($\underline{16}$) 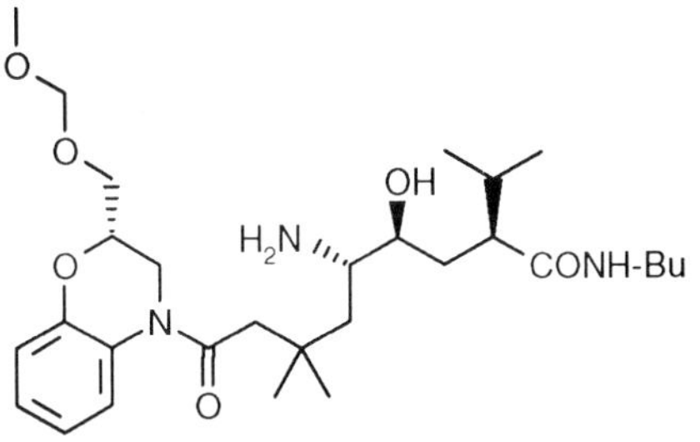   CGP 56346A ($\underline{17}$)

Compd	Human Plasma Renin IC_{50} (nM)	Marmoset Plasma Renin IC_{50} (nM)
$\underline{16}$	0.6	2.0
$\underline{17}$	0.5	5.0

Fig. 3 *In vitro* activity and species specificity of inhibitors $\underline{16}$ and $\underline{17}$.

The influence of the P_1' position in the transition state mimic moiety on enzyme inhibition for a number of analogues was also investigated. Eventually, the P_1' isopropyl substituted tetrahydroquinoline $\underline{16}$ and the 3,4-dihydro-2*H*-1,4-benzoxazine derivative $\underline{17}$ emerged with sub-nanomolar IC_{50} values towards human renin and high *in vitro* potency in the marmoset plasma renin assay (Fig. 3). Furthermore, both $\underline{16}$ and $\underline{17}$ were found to be highly selective for renin with no inhibition activity against the related aspartic acid proteases pepsin and cathepsin D at concentrations <100 μM. Both compounds were also selective for primate renin and less effective against dog or rat plasma renin.

N-Benzyl Indole-3-carboxamides at P_3

A second series of inhibitors of human renin was derived from the weakly active lead $\underline{3b}$ by: i) replacement of the P_1 dimethyl group with mono-methyl at position C-7; and ii) incorporation of a synthetically flexible carboxylic ester group into the three carbon spacer of $\underline{3b}$. The 2-benzyl benzoate $\underline{18}$ (mixture of four diastereomers, Fig. 4) was found to be the most active inhibitor of a series of *ortho* or *meta* substituted analogues *in vitro*. Docking analysis of the energy minimized conformation of $\underline{18}$ in the enzyme active site suggested a hydrogen bond between the ester carbonyl of the spacer and the amide NH of Ser219, whereas the *ortho*-benzyl would fit into the S_3 binding pocket. Based on this model, further structural modifications of the P_3 moiety led to the N-benzyl indole-3-carboxylic ester $\underline{19}$ as the pure (2R,4S,5S,7S)-configured diastereomer (Fig. 4) showing a more than 100-fold increase in binding affinity compared to $\underline{18}$. Isosteric replacement of the potentially labile ester linkage afforded the *in vitro* equipotent indole-3-carboxamide $\underline{20a}$ which appeared to be >20 times more active *in vitro* than the corresponding C-7(R)-epimer $\underline{20b}$. Finally, increasing the steric bulk of the alkyl residue at P_1 by replacing methyl with isopropyl induced a more than 5-fold enhancement in binding affinity in the plasma assay, compound $\underline{21}$ being the most potent inhibitor prepared in this series.

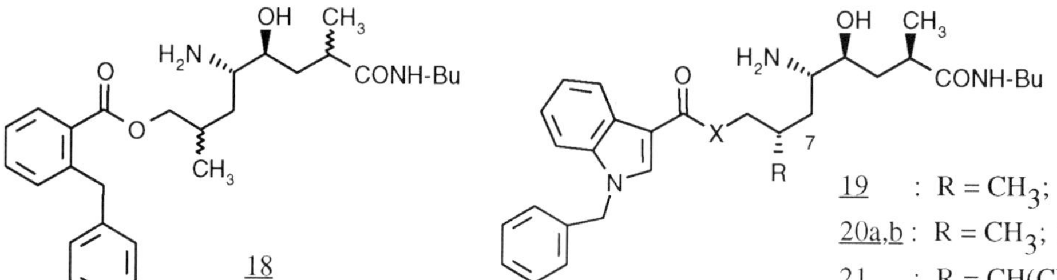

Compd		Purified Human Renin IC$_{50}$ (nM)	Human Plasma Renin IC$_{50}$ (nM)
$\underline{18}$		800	ND[b]
$\underline{19}$		3	34
$\underline{20a}$	(7S)[a]	3	28
$\underline{20b}$	(7R)[a]	72	ND[b]
$\underline{21}$		1	4

[a] Absolute configuration at C-7;   [b] Not determined.

Fig. 4   *In vitro* potency of $\underline{18}$ and N-benzyl indole-3-carboxylic acid derivatives.

X-RAY CRYSTAL STRUCTURE OF HUMAN RENIN - INHIBITOR COMPLEXES

Inhibitor $\underline{5a}$ was soaked into preformed cubic crystals of rh-renin (14) and data to 2.8 Å resolution were collected. The hydroxyethylene mimic including the P_1'-P_2' moiety occupied roughly the same position in the enzyme binding cleft as CGP 38560 ($\underline{1}$). Although a slight relative shift of the transition state mimicking OH-group was seen in the complex with $\underline{5a}$, the hydroxyl is still involved in a hydrogen bonding network with the two catalytic aspartic acid residues Asp32 and Asp215 of the active site, similar to that observed for $\underline{1}$ (10). Accordingly, the P_1'-P_2' amide bond interacted with Ser76 and Gly34 of the enzyme backbone. The primary amino group was in hydrogen bonding distance to the carboxyl group of Asp215 as well as the amide carbonyl of Gly219. In contrast, however, to what was predicted from molecular modeling, the P_3 heterocycle was rotated about 60° from the presumed position within the S_3 pocket in such a way that the ester carbonyl became exclusively hydrogen bonded to the side chain hydroxyl of Ser219 rather than to the Ser219 backbone NH. Most unexpectedly, the methoxy residue of the carboxylic ester reached into a hitherto unobserved subpocket of human renin, which normally appeared to be partially filled by a water molecule (14). This newly discovered binding site, which we named the S_3^{sp} subpocket, is a distinct binding site formed by amino acids Tyr14, Tyr155 (bottom of the pocket), Ala303, Thr216, Gly217 and Ala218. The absolute configuration of 5a was determined to be C-3'(S) from the above crystal data.

X-ray data of the enzyme complex with the indole-3-carboxylic ester 19 (soaking in method, 3.1 Å resolution) showed similar binding interactions between the transition state mimic of 19 and the binding cleft as observed for 5a. Surprisingly, the indolyl residue was positioned in the same plane as the heterocycle of 5a within the S_3 binding site wheras the N-benzyl group filled the S_3^{sp} subpocket. Thus again, the data of the crystal structure were in contrast to the bound conformation of 19 predicted by modeling (*vide infra*). An additional hydrogen bond was observed between the spacer ester carbonyl and the Oγ of Thr77 in the flap region. These unexpected findings, as well as the extensive SAR data, demonstrated that the new S_3^{sp} subsite common to two structurally distinct series of inhibitors can accommodate a variety of functional hydrogen bonding groups as well as hydrophobic residues, and that these additional interactions of the inhibitor with this subpocket may lead to a dramatic increase of the overall binding affinity. In line with these results, the tetrahydroquinoline derivative 15 substituted with phenyl at C-3' was prepared and found to be a very potent inhibitor against purified renin, albeit potency dropped 20-fold in the presence of plasma due to the lipophilicity of the aromatic P_3^{sp} residue.

IN VIVO ACTIVITIES

The *in vitro* potent renin inhibitors were tested for oral activity in conscious, sodium-depleted marmosets freely moving in their home cages. Mean arterial blood pressure (MAP) was measured by telemetry and the effects of compounds monitored over a 24 hour period after administration (17). Both inhibitors 16 and 17 induced a pronounced and long-lasting reduction in MAP in a dose dependent manner when given as a single dose at 1 and 3 mg/kg p.o. The maximum fall in MAP developed within 1-2 hours after administration with a maximum decrease in MAP of -30 mmHg, lasting for at least 6 hours after a 3 mg/kg p.o. dose. Blood pressure was still significantly reduced after 20 hours, as compared to the control group, with ΔMAP= -10 mmHg for 16 and -15 mmHg for 17. In parallel to the effect on blood pressure, plasma renin activity (PRA) was completely blocked at 1.5, 3 and 6 hours after dosing, and was still inhibited up to 95% after 24 hours. The N-benzyl indole-3-carboxamide 21 also induced a blood pressure lowering effect in marmosets when given orally in a dose of 10 mg/kg. However, the maximum response was less pronounced (ΔMAP=-20 mmHg) compared to compounds 16, 17, and duration of action was shorter. The novel renin inhibitors CGP 55128A (16) and CGP 56346A (17) appeared to be the most active compounds of this series *in vivo* in our experimental primate model. Both analogues have a much higher oral potency and a longer duration of action on blood pressure in comparison to the known peptide-based renin inhibitors RO-42-5892 (18) and A-72517 (6).

CONCLUSIONS

We have discovered two series of novel, completely non-peptidic and small molecule renin inhibitors by rational design based on the concept of directly linking the P_3-P_1 pharmacophores. This approach has been extended towards a previously unknown non-substrate binding site S_3^{sp} which was unveiled by X-ray analyses of inhibitor-human renin complexes. These structurally unique inhibitors are highly potent *in vitro* and very specific for primate renin. *In vivo*, several compounds show pronounced blood pressure lowering effects through complete inhibition of plasma renin in marmosets after oral administration with a long duration of action. These findings may lead to the development of a new generation of renin inhibitors which could be useful for the treatment of hypertension.

ACKNOWLEDGEMENTS

We thank T. Blundell et al. (Birkbeck College, University of London, UK) for providing to us the coordinates of their model of human renin. We would like in particular to thank Dr. W. Fuhrer and Prof. K. Hofbauer for their generous support and encouragement during the course of this program. Mr. Ch. Schnell and Mr. H.P.Baum are acknowledged for their technical assistance.

REFERENCES

1. Sealey, J.E.; Laragh, J.M. *Hypertension: Pathophysiology, Diagnosis and Management*; Laragh, J.M.; Brenner, B.M., Eds.; Raven Press, New York, 1995; pp 1763-1796.
2. Waeber, B.; Nussberger, J.; Brunner, H.R. *Hypertension: Pathophysiology, Diagnosis and Management*; Laragh, J.M.; Brenner, B.M., Eds.; Raven Press, New York, 1995; pp 2861-2876.
3. Nelson, E.B.; Harm, S.C.; Goldberg, M.; Shahinfar, S.; Goldberg, A.; Sweet, C.S. *Hypertension: Pathophysiology, Diagnosis and Management*; Laragh, J.M.; Brenner, B.M., Eds.; Raven Press, New York, 1995; pp 2895-2916.
4. For reviews: (a) Kleinert, H.D. *Exp. Opin. Invest. Drugs* **3**, 1087-1104 (1994). (b) Greenlee, W. *J. Med. Res. Rev.* **10**, 173-236 (1990).
5. Bühlmayer, P.; Caselli, A.; Fuhrer, W.; Göschke, R.; Rasetti, V.; Rüeger, H.; Stanton, J. L.; Criscione, L.; Wood, J. M. *J. Med. Chem.* **31**, 1839-1846 (1988).
6. Kleinert, H.D.; Rosenberg, S.H.; Baker, W.R.; Stein, H.H.; Klinghofer, V.; Barlow, J.; Spina, K.; Polakowski, J.; Kovar, P.; Cohen, J.; Denissen, J. *Science* **257**, 1940-1943 (1992). (b) Rosenberg, S.H.; Spina, K.P.; Condon, S.L.; Polakowski, J.S.; Yao, Z.; Kovar, P.; Stein, H.H.; Cohen, J.; Barlow, J.L.; Klinghofer, V.; Egan, D.A.; Tricarico, K.A.; Perun, T.J.; Baker, W.R.; Kleinert, H.D.; *J.Med.Chem.* **36**, 460-467 (1993).
7. Rasetti, V.; Cohen, N.C.; Rüeger, H.; Göschke, R.; Maibaum, J.; Cumin, F.; Fuhrer, W.; Wood, J.; paper manuscript in preparation.
8. For a related preliminary account of our design efforts, see: Hanessian, S.; Raghavan, S. Bioorg. Med. Chem. Lett. **4**, 1697-1702 (1994).
9. Cohen, N. C. *Trends in Med. Chem.* '88; van der Goot, H.; Dománi, G.; Pallos, L.; Timmerman, H., Eds.; Elsevier Science Publishers: Amsterdam, 1989; pp. 13-28.
10. Sibanda, B.L.; Blundell, T.; Hobart, P.M.; Fogliano, M.; Bindra, J.S.; Dominy, B.W.; Chirgwin, J.M. *FEBS Letters* **174**, 102-111 (1984).
11. a) Plummer, M.; Hamby, J. M.; Hingorani, G.; Batley, B. L.; Rapundalo, S. T. Bioorg. Med. Chem. Lett. **3**, 2119-2124 (1994); b) Plummer, M.S; Shahripour, A.; Kaltenbronn, J.S.; Lunney, E.A.; Steinbaugh, B.A.; Hamby, J.M.; Hamilton, H.W.; Sawyer. T.K.; Humblet, C.; Doherty, A.M.; Taylor, M.D.; Hingorani, G.; Batley, B.L.; Rapundalo, S.T. *J.Med.Chem.* **38**, 2893-2905 (1995).
12. Sali, A.; Veerapandian, B.; Cooper, J. B.; Foundling, S. I.; Hoover, D. J.; Blundell, T. L. *The EMBO J.* **8**, 2179-2188 (1989).
13. See for example: (a) Metcalf, P.; Fusek, M. *The EMBO Journal* **12**, 1293-1302 (1993). (b) Suguna, K.; Padlan, E.A.; Bott, R.; Boger, J.; Parris, K.D.; Davies, D.R. *PROTEINS: Structure, Function, and Genetics* **13**, 195-205 (1992).
14. Rahuel, J.; Priestle, J. P.; Grütter, M. G. *J. Struct. Biol.* **107**, 227-236 (1991).
15. Rosenberg, S.H.; Kleinert, H.D.; Stein, H.H.; Martin, D.L.; Chekal, M.A.; Cohen, J.; Egan, D.A.; Tricarico, K.A.; Baker, W.R. *J.Med.Chem.* **34**, 469-471 (1991); and references therein.
16. Abraham, M.H.; Duce, P.P.; Prior, D.V.; Barratt, D.G.; Morris, J.J.; Taylor, P.J. *J.Chem.Soc Perkin Trans II* 1355-1375 (1989).
17. Schnell, C.R.; Wood, J.M. *Am.J.Physiol.* **264**, H1509-H1516 (1993).
18. Fischli, W., Clozel, J.P., Elamrani, K., Wostl, W., Neidhart, W., Stadler,H. and Branca, Q. *Hypertension* **18**, 22-31 (1991).

Inhibition of HIV proteinase

Joseph A Martin

Roche Research Centre, 40 Broadwater Road, Welwyn Garden City,
Herts AL7 3AY, England

Abstract: Potent and selective inhibitors of HIV proteinase have been developed
in a rational design based approach. Optimisation of the lead structure which
contained the hydroxyethylamine transition-state mimetic afforded saquinavir.
This potent and highly selective inhibitor of HIV proteinase has excellent
antiviral activity in cell culture, is not cytotoxic and is synergistic with a range of
other anti-HIV agents. In rat and in human volunteers oral doses of saquinavir
gave good plasma levels of the drug which is currently in Phase III clinical trials.

Introduction

The elucidation of processes involved in the replication of human immunodeficiency virus (HIV)
identified a number of molecular targets suitable for therapeutic intervention in the treatment of HIV
infection and AIDS. One of the most extensively studied approaches is inhibition of the HIV
proteinase, an enzyme that is essential for maturation of new virus particles. From an analysis of its
amino acid sequence (1) and its inhibition by pepstatin (2), HIV proteinase was classified as a
member of the aspartic family of enzymes. Inhibition of aspartic proteinases, for example renin, has
been achieved with peptide analogues where the P1-P1' residues of an appropriate substrate are
replaced by a non-hydrolysable isostere which mimics the transition-state of hydrolysis (3).

Inhibitor design and lead generation

HIV proteinase cleaves the Gag and Gag-Pol polyproteins at eight different sites which have been
classified into three different types (4). Of these, we were particularly interested in the Tyr-Pro and
Phe-Pro cleavage sites because amide bonds N-terminal to proline are not cleaved by mammalian
endopeptidases and we reasoned that inhibitors based on these sequences should lead to high
selectivity for the viral enzyme.

Transition-state mimetics

Reduced amide

PheΨ[CH$_2$NH]Pro

Hydroxyethylamine

PheΨ[CH(OH)CH$_2$N]Pro

Of the various transition-state mimetics that we considered, the reduced amide and
hydroxyethylamine isosteres most readily accommodate the imino acid moiety of the Tyr-Pro and
Phe-Pro cleavage sites. Inhibitors based on the former transition-state mimetic (compounds 1 - 3)

gave only weak inhibition, whereas corresponding analogues containing the hydroxyethylamine moiety gave markedly more active compounds (Table 1).

Table 1. Generation of lead structures

Compound	Structure	IC50 (nM)
1	Cbz.PheΨ[CH2N]Pro.O^{t}Bu	> 100,000
2	Cbz. Asn.PheΨ[CH2N]Pro.O^{t}Bu	> 10,000
3	Cbz.Asn.PheΨ[CH2N]Pro.Ile.NH^{i}Bu	50,000
4, 5	Cbz.PheΨ[CH(OH)CH2N]Pro.O^{t}Bu	6,500 & 30,000
6, 7	Cbz.Asn.PheΨ[CH(OH)CH2N]Pro.O^{t}Bu	140 & 300
8, 9	Cbz.Leu.Asn.PheΨ[CH(OH)CH2N]Pro.O^{t}Bu	600 & 1,100
10, 11	Cbz.Asn.PheΨ[CH(OH)CH2N]Pro.Ile.NH^{i}Bu	130 & 2,400

Compounds 4 - 11 were prepared to determine the minimum size required to give a level of enzyme inhibition that hopefully could be enhanced by structural modification. In each case both diastereoisomers were prepared because it was not possible to predict with certainty the stereochemical requirement of the hydroxyl group in the transition-state mimetic for optimal activity. Thus, compounds 4 and 5 showed weak enzyme inhibitory activity but extension at the N-terminus gave markedly more active compounds (6 and 7). Further extension at either the N-terminus (8 and 9) or at the C-terminus (10 and 11) did not increase potency. The most active tripeptide derivative was compound (6) in which the hydroxyl group has the *R* configuration, therefore we focused our optimisation studies on this compound.

Optimisation of activity

Initially, we studied variations in the protecting groups. Replacement of the N-terminal benzyloxy-carbonyl group in 6 by smaller groups such as acetyl (compound 12) reduced activity. Interestingly, both the dihydrocinnamoyl analogue 13 and the cinnamoyl derivative 14 retained good potency. The introduction of conformational restriction, via the beta-naphthoyl derivative 15, gave an increase in potency which was enhanced further with the quinoline-carbonyl analogue 16, one of the most potent inhibitors in the series thus far (Table 2).

Next, we investigated modifications of the C-terminal ester protecting group (Table 3). Replacement of the *tert*-butyl ester with either *tert*-butylamide 17 or benzylamide 19 afforded no loss in potency, whereas other modifications such as methylamide (18) resulted in reduced activity. For further development of the structure-activity relationships the choice between the *tert*-butyl ester and *tert*-butylamide was made according to ease of synthesis of the individual compound.

Replacement of asparagine at P2 in 6 resulted in loss of activity. Similarly, at the P1 position no improvement was found by modification of the benzyl side-chain derived from the phenylalanine of the natural substrate. However, the most marked improvements in potency were found by modification of the proline residue in the P1' position (Table 4).

Table 2. Optimisation of P3 residue

R.Asn.PheΨ[CH(OH)CH2N]Pro.OtBu

Compound	R	IC50 (nM)
6	PhCH2OCO (Cbz)	140
12	CH3CO	8,600
13	PhCH2CH2CO	240
14	PhCH:CHCO	240
15	beta-Naphthoyl	46
16	Quinoline-2-CO	23

Table 3. Optimisation of C-terminal residue

Cbz.Asn.PheΨ[CH(OH)CH2N]Pro.**R**

Compound	R	IC50 (nM)
6	OtBu	140
17	NHtBu	210
18	NHMe	670
19	NHCH2Ph	160

Table 4. Optimisation of P1' residue

Cbz.Asn.PheΨ[CH(OH)CH2N].**R**CO.NHtBu

Compound	R	IC50 (nM)
17		210
20		18
21		8.4
22		5.6
23		2.7

A significant improvement in potency was found with the pipecolic acid and the thiazole derivatives 20 and 21 respectively. Early molecular modelling studies, prior to X-ray crystallographic work, indicated that the S1' subsite could accommodate a large residue. Indeed, this was found to be so with the bicyclic systems in compounds 22 and 23. The most potent inhibitor thus far was compound 23, and interestingly the *S, S, S* stereochemistry for the bicyclic systems in both 22 and 23 was essential for high activity.

Ro 31-8959 (Saquinavir)

(24)

Having identified optimal residues at each subsite in the lead structure, we prepared analogues which contained combinations of the preferred side-chains. It was satisfying to find that the effect of individual residues was additive, the most potent compound was 24, Ro 31-8959 (Saquinavir). Furthermore, an excellent correlation exists between enzyme inhibitory activity and antiviral activity, with none of the compounds showing cytotoxicity at concentrations markedly in excess of that required for antiviral activity (5). Consequently, saquinavir was elevated to development status and a more detailed evaluation of its biological properties was undertaken.

Selectivity

A key feature of our design process focused on the Phe-Pro cleavage sequence in the viral polyproteins and the potential of derived inhibitors to exhibit high selectivity for the viral enzyme. It was not unexpected to find that saquinavir did not inhibit a range of mammalian aspartic proteinases as well as representatives of serine, cysteine and metallo enzymes (4).

In vitro antiviral activity

An extensive evaluation of the antiviral properties of saquinavir has been carried out. In brief, it is highly active in a variety of cell lines using different endpoints against a range of virus strains, including clinical isolates, AZT resistant strains as well as HIV-2 (6,7). Typically, the antiviral IC_{50} is in the 2-20 nanomolar range. Furthermore, saquinavir has been shown by electron microscopy to prevent maturation of virions produced in chronically infected cells (8). Saquinavir produces a synergistic effect with other anti-HIV agents such as AZT and ddC which are reverse transcriptase inhibitors (9).

Pharmacokinetics

A single dose of 10mg/kg of saquinavir in rat gave plasma levels that remained above the antiviral IC_{50} for six hours. Pharmacokinetic and tolerence studies in healthy volunteers dosed at 600mg tid for seven days gave plasma drug concentrations in excess of the *in vitro* antiviral IC_{90} throughout the dosing period (10).

Clinical studies

Clinical studies have shown saquinavir is well tolerated in patients at all dose levels upto 600mg tid. In a dose-ranging Phase II monotherapy clinical trial conducted in London a beneficial effect was observed in patients on the highest dose as measured by immunological and antiviral laboratory markers of disease (11). Similarly, in a Phase II study in Italy the effect of saquinavir alone, AZT alone and combinations of both drugs was compared. A reduction in virus load was observed in each group as measured by viral RNA using PCR, with the best effect seen in the high dose combination group. In another Phase II clinical study in the USA (ACTG 229) the triple combination of saquinavir, AZT and ddC was significantly better than either AZT and ddC or AZT plus saquinavir. Long term Phase III clinical studies to demonstrate the efficacy of saquinavir as measured by clinical end points have commenced.

References
1. L. Ratner *et al. Nature* **313**, 277 (1985)
2. A. D. Richards et al. *FEBS Lett.* **247**, 113 (1989)
3. N. W. Hamilton *et al. Annual Rep. Med. Chem.* **24**, 51 (1989)
4. J. A. Martin. *Antiviral Res.* **17**, 265 (1992)
5. N. A. Roberts *et al. Science* **248**, 358 (1990)
6. S. Galpin *et al. Antiviral Chem. Chemother.* **5**, 43 (1994)
7. H.C. Holmes *et al. Antiviral Chem. Chemother.* **2**, 287 (1991)
8. J. C. Craig *et al. Antiviral Res.* **16**, 295 (1991)
9. J. C. Craig *et al. Antiviral Chem.* Chemother. **5**, 380 (1994)
10. G. J. Muirhead *et al. Brit. J. Clin. Pharm.* **34**, 170 (1992)
11. V. S. Kitchen *et al. The Lancet* **345**, 952 (1995)

Drug metabolism and toxicokinetics

R. Kato

Department of Pharmacology, School of Medicine, Keio University,
Shinjuku-ku, Tokyo 160, Japan

Abstract

Metabolisms of drugs produce a variety of metabolites having different toxicological potency. Most of the metabolites show lower toxicological activities, but some metabolites show higher toxicological activities than parent compounds. Especially, reactive intermediates which covalently bind to protein and nucleic acids cause drug-induced allergic cell toxicity, mutagenesis and carcinogenesis. Toxicokinetics is an important science for interpretation of toxicological findings and analysis of the mechanism of drug toxicities. For extrapolation of data obtained from rats to humans, the use of female rats is highly recommendable, instead of use of males, since the drug-metabolizing enzyme activities in male rats are androgen-dependent and regulated peculiarly and species specifically by growth hormone. Drug metabolism studies with human liver enzymes are a potential approach for better interpretation and extrapolation to humans of data obtained from toxicokinetics and drug metabolism studies with experimental animals.

1. Drug metabolism and drug toxicity

Metabolisms of a variety of drugs and chemicals usually cause a loss or decrease in their pharmacological activities, but sometimes they increase the toxicological activities. In particular, some drugs cause severe drug toxicities through the formation of reactive intermediates which covalently bind to the high molecular weight of cellular components, such as proteins and nucleic acids.

Chemicals bound covalently to proteins produce specific antibodies and cause immunoreaction-related drug toxicities, such as drug-induced hepatitis, skin eruption and agranulocytosis (1). On the other hand, chemicals bound covalently to DNA cause mutation and carcinogenesis. The formation of reactive intermediates is species-dependent among experimental animals and humans (1-3). Drug metabolism study, therefore, is an essential method for clarification of the mechanism of drug toxicity and of species differences in drug toxicity.

2. Toxicokinetics

Originally, toxicokinetics was a component of pharmacokinetics. Toxicokinetics is now defined as pharmacokinetic studies which are carried out during drug toxicity studies to interpret toxicological results and to assist analysis of the mechanism of drug toxicity. I should like to discuss here two kinds of toxicokinetics: Regulatory toxicokinetics and academic toxicokinetics. Regulatory toxicokinetics has been established by ICH 2 (International Conference on Harmonization 2) and is an obligatory part of new drug application for the evaluation of drug toxicity in relation to plasma (tissue) levels of drugs in order to assess systemic exposure and to evaluate risk and safety in humans in relation to clinical data.

On the other hand, academic toxicokinetics is mainly involved in analysis and clarification of the mechanism of drug toxicity. In academic toxicokinetics, therefore, we have to determine not only plasma (tissue) levels of administered drugs, but also several metabolites in plasma and tissues including tiny amounts of metabolites when they are considered to be related to the drug toxicity.

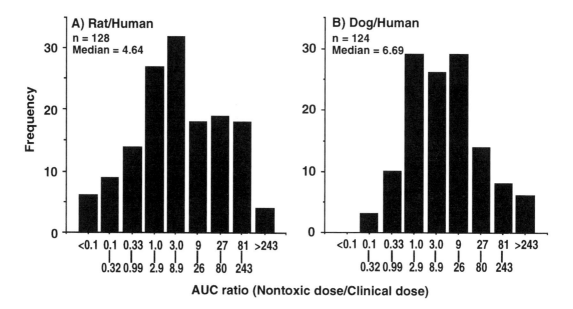

Fig. 1. Distrribution frequency of AUC ratio between nontoxic dose and clinical dose. 150 new and investigated orally used drugs were surveyed. Rapidly metabolized drugs (the ratio of parent drug/all metabolite+parent drug is less than 0.15 at peak plasma concentration in rats) were excluded.

To assure the drug safety, plasma levels of administered drugs in humans, in principle, should be lower than those in experimental animals treated repeatedly with the highest nontoxic dose. However, we observed that the drug plasma levels in humans are often higher than those in experimental animals as shown in Fig. 1 and 2 (4). In such cases, in general, we have to consider a possible presence of toxic metabolite(s), which dominates overall drug toxicity in the bodies. Unfortunately, in most cases, such toxic compounds are not easily determined during drug toxicity studies.

We have to consider, therefore, some important differences between pharmacokinetic data and toxicokinetic data. Most critical points come from the difference in the dose used and duration of treatment between both studies. Firstly, long terms of treatments cause high degrees of induction in drug-metabolizing enzymes. Secondly, high-dose treatments enhance drug interaction, such as induction and inhibition of drug metabolism. Third, high amounts of metabolites cause inhibitions in metabolism of parent drugs. Fourth, the metabolic pathway and the enzyme involved in the same metabolic pathway may be different, depending on the substrate concentration (administered dose) as shown later. Fifth, saturation kinetics are often observed after high-dose administrations in absorption, protein binding, metabolism and excretion of drugs. The contribution of metabolites to drug toxicity has not usually been evaluated, but in future we need to evaluate the contribution of each metabolite and parent compound causing whole drug toxicity.

3. Female rats are a more useful model for extrapolation to humans

In toxicological studies both sexes of animals are used, but data from male rats are usually employed for extrapolation to humans. However, male rats are very unique animals which show clear sex related-differences in some drug metabolisms depending on cytochrome P450 and sulfotransferase (Table 1 and Fig. 2) (5).

Table 1   Sex-related differences in hepatic level of cytochrome P450 (CYP) in rats.

P450 forms	Difference (fold)		P450 forms	Difference (fold)	
Male specific			Female specific		
2C11	m>f	20 × ↑	2C12	m<f	20 × ↑
3A2	m>f	10 × ↑			
2A2	m>f	10 ×			
2C13	m>f	10 × ↑			
2C22	m>f	10 × ↑			
Male dominant			Female dominant		
2B1[a]	m>f	3~10 ×	2A1	m<f	2 ×
2B2[a]	m>f	2 ×	2C7	m<f	2 ×
3A1	m>f	5 ×	2E1	m<f	1.5 ×
			1A2	m<f	2 ×
No sex difference					
2C6	m=f				
1A1	m=f				

[a]Levels in non-induced male rats are low (0.6-4 pmol/mg protein).

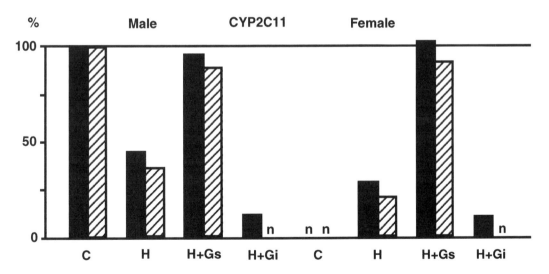

Fig. 2. Regulation of CYP2C11 and CYP2C11 mRNA levels by growth hormone in male and female rats. C: Control; H: hypophysectomy; Gs: growth hormone injected subcutaneously twice a day for 7 days; Gi: growth hormone infused subcutaneously using a mini-pump for 7 days; n: not detectable; filled columns: proteins; dashed columns: mRNA.

These sex-related differences are species-specific and clearly related to the secretion pattern of growth hormone. The metabolic activities in male rats are usually markedly higher than those in humans (female and male), but in female rats the activities are only slightly higher than those in humans. The high activities in male rats, therefore, easily decreased under a variety of pathophysiological conditions, but most of the changes are not observed in other species of experimental animals and humans (Table 2). Thus the use of data from female rat for extrapolation to humans is highly recommendable (6).

Table 2 Species differences in changes of hepatic microsomal drug-metabolizing enzyme in pathological states.

	Male rats	Male mice	Male rabbits	Human
1) Fasting	↓↓	↓ ~ ⇄	→	⇄
2) Diabetes	↓↓	⇄ ~ ↑	⇄ ~ ↑	→
3) Morphine intoxication	↓↓	⇄	→	→
4) Hyperthyroidism	↓↓	→	↑	⇄ ~ ↑
5) Adrenectomy	↓↓	↓ ~ ⇄	→	⇄ ~ ↓

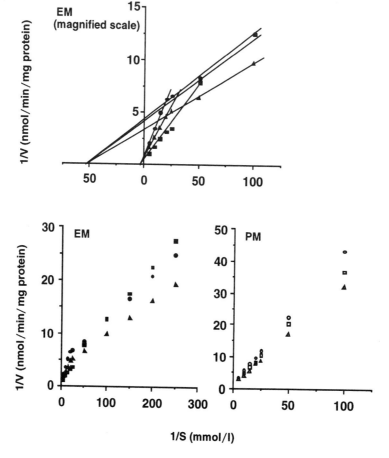

Fig. 3. Lineweaver-Burk plots for diazepam N-demethylation in liver microsomes from EMs and PMs. The closed and open symbols indicate EM and PM, respectively.

4. Drug metabolism studies with human livers

It is well known that there are marked species differences in drug metabolism between experimental animals and humans. Therefore, uses of human liver in an early stage of drug development are highly recommendable for drug metabolism studies and interpretation of toxicokinetics and toxicological findings. However, the metabolic pathway and the molecular form of enzyme involved in the same pathway may be different depending on the substrate concentrations. For example, the major metabolic pathways of diazepam by human liver microsomes are dose-related. The 3-hydroxylation to form temazepam is a major pathway at higher substrate concentrations, whereas, at lower substrate concentrations, the ratio of the N-demethylation (formation of medazepam) to 3-hydroxylation was clearly increased. However, liver microsomes derived from poor metabolizers (PM) of *S*-mephenytoin 4'-hydroxylation also showed low diazepam N-demethylation; thus the ratio of N-demethylation to 3-hydroxylation was not markedly increased with microsomes from PM.

Lineweaver-Burk plots for diazepam N-demethylation were biphasic in the liver microsomes from the extensive metabolizer (EM) of *S*-mephenytoin 4'-hydroxylation, suggesting that at least two forms of cytochrome P450 (CYP) are involved in the N-demethylation (Fig. 3). However, the liver microsomes from PM showed a monophasic pattern. Thus, the formation of nordazepam in liver microsomes from EM appeared to be mediated by low and high K_m enzymes. On the other hand, in liver microsomes from PM the formation of nordazepam appeared to be catalyzed only by high K_m enzyme as shown in Table 3. Thus, in the low substrate concentrations, nordazepam formation is catalyzed by low K_m enzyme in liver microsomes from EM.

Table 3 Michaelis-Menten parameters for diazepam N-demethylation in human liver microsomes from EM and PM for *S*-mephenytoin 4'-hydroxylation.

Liver microsomes	Diazepam N-demethylation			
	K_m1	K_m2	$V_{max}1$	$V_{max}2$
	(μM)		(nmol/min/mg protein)	
EM	19.4 ± 0.4^a	346 ± 34	0.27 ± 0.04	1.82 ± 0.63
PM	——	319 ± 30	——	1.49 ± 0.62

[a]Mean ± SD.

Diazepam 3-hydroxylations in liver microsomes from EM and PM were not inhibited by anti-human CYP2C antibody at either high (0.2mM) or low (0.02mM) substrate concentration, but the activities were inhibited by anti-human CYP3A antibody. On the other hand, diazepam N-demethylation was inhibited by anti-human CYP2C antibody at 0.02mM, but not 0.2mM substrate concentration in liver microsomes from EM (Fig. 4). In contrast, diazepam N-demethylation was not inhibited by the same antibody at either 0.02 and 0.2mM substrate concentration in liver microsomes from PM.

These results indicate that the metabolic pathway of diazepam is concentration-dependent and at clinical dose-relevant hepatic concentration the N-demethylation may be a major pathway. The N-demethylation of diazepam is catalyzed by CYP2C (probably CYP2C19) and the 3-hydroxylation is catalyzed by CYP3A. These results also strongly support the notion that the use of substrate concentration relevant to hepatic concentration after administration of clinical dose is necessary for determination of the major metabolic pathway and the molecular form of cytochrome P450 responsible for the metabolic pathway (8).

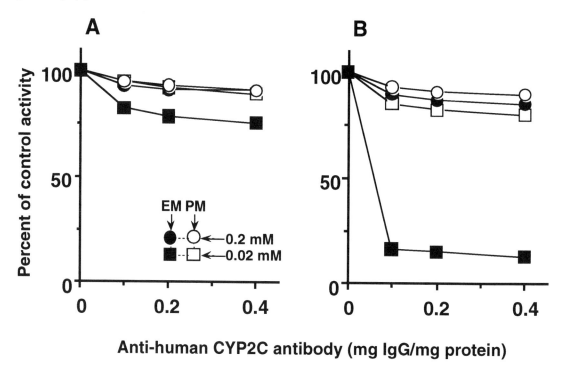

Anti-human CYP2C antibody (mg IgG/mg protein)

Fig. 4. Inhibition by anti-human CYP2C antibody of diazepam 3-hydroxylation and N-demethylation in liver microsomes of EMs and PMs. A: 3-hydroxylation; B: N-demethylation. The squares and circles indicate the mean values from two microsomal samples at substrate concentrations of 0.02 and 0.2 mM, respectively.

5. Drug metabolism studies with heterologous expressed human liver enzymes

To determine major enzymes responsible for each metabolic pathway, the use of heterologous expressed human enzymes is highly recommendable. Commonly used hosts for the expression are yeast (*Saccharomyces cerevisiae*), *E. Coli*, and cultured mammalian cell lines, depending on the nature of the metabolic enzymes and purpose of utilization (9, 10). For expressions in the cultured mammalian cell lines, the Baculovirus expression system, Vaccinia virus expression system and lymphoblastoid cells expression system are usually used.

We recently have used *Schizosaccharomyces pombe* which produce a 10-fold larger amount of cytochrome P450 in comparison with the conventional yeast (Table 4) (11). Approximate contents of CYP2C9 and CYP2C19 in human livers are 0.15 and 0.01 pmole/mg protein, respectively. The activity for *S*-mephenytoin 4'-hydroxylation was very high in CYP2C19-expressed *Schizosaccharomyces pombe*. These expressed systems are also utilized as alternative materials for human livers by using them as a single component and as an enzyme cocktail. These metabolic studies are potentially useful to predict the efficacy and toxicity of new drugs in humans.

Table 4 Tolbutamide hydroxylase and *R*- and *S*-mephenytoin 4'-hydroxylase activities in yeast microsomes.

	P450 content	Tolbutamide hydroxylation	*R*-Mephenytoin 4'-hydroxylation	*S*-Mephenytoin 4'-hydroxylation
	nmol/mg protein	nmol/min/mg protein		
S. pombe leu-32h-[pTLCYP2C9]	1.26	2.61 ± 0.06[a]	0.59 ± 0.08	6.75 ± 0.27
S. cerevisiae MT8-1[pTF450]	0.12	0.24 ± 0.01	0.06 ± 0.01	0.65 ± 0.03
S. pombe leu-32h-[pTLCYP2C19]	0.78	1.77 ± 0.36	3.07 ± 0.53	187.98 ± 12.36

[a]Values are the results of experiments with 1 unit of NADPH-cytochrome P450 reductase in the incubation mixture. Substrate concentrations of tolbutamide and mephenytoins were 0.1 and 0.2 mM, respectively.

References

1)Boelsterli, U. A.: Specific targets of covalent drug-protein interactions in hepatocytes and their toxicological significance in drug-induced liver injury. Drug Metab. Rev., 25: 395-451, 1993.

2)Kato, R.: Metabolic activation of mutagenic heterocyclic aromatic amines from protein pyrolysates. CRC Crit. Rev. Toxicol., 16: 307-348, 1986.

3)Kato, R. and Yamazoe, Y.: Metabolic activation of N-hydroxylated metabolites of carcinogenic and mutagenic arylamines and arylamides by esterification. Drug Metab. Rev., 26: 413-430, 1994.

4)Kato, R.: Clinical dose and nontoxic dose in repeated administration of a variety of drugs in relation to their pharmacokinetics. Jpn. J. Clin. Pharmacol. Ther., 24: 595-602, 1993.

5)Kato, R. and Yamazoe, Y.: Hormonal regulation of cytochrome P450 in rat liver. In Handbook of Experimental Pharmacology, Vol. 105, "Cytochrome P450", pp. 447-459, Eds. by J. B. Schenkman and H. Greim, Springer-Verlag, Berlin, 1993.

6)Kato, R. and Yamazoe, Y.: Sex-specific cytochrome P450 as a cause of sex- and species-related differences in drug toxicity. Tox. Lett., 64/65: 661-667, 1992.

7)Yasumori, T., Li, Q.-H., Yamazoe, Y., Ueda, M., Tsuzuki, T. and Kato, R.: Lack of low K_m diazepam N-demethylase in livers of poor metabolizers for S-mephenytoin 4'-hydroxylation. Pharmacogenetics, 4: 323-331, 1994.

8)Kato, R. and Yamazoe, Y.: The importance of substrate concentration in determining cytochrome P450 therapeutically relevant in vivo. Pharmacogenetics, 4: 362, 1994.

9)Yasumori, T., Murayama, N., Yamazoe, Y., Abe, A., Nogi, Y., Fukasawa, T. and Kato, R.: Expression of a human P-450IIC gene in yeast cells using galactose-inducible expression system. Mol. Pharmacol., 35: 443-449, 1989.

10)Kato, R., Yasumori, T. and Yamazoe, Y.: Characterization of human P450IIC isozymes by using yeast expression system. In "Methods in Enzymology", Vol. 206, pp. 183-190, Eds. by M. R. Waterman and E. F. Johnson, Academic Press, Orlando F. L., 1991.

11)Yasumori, T., Tohda, H., Giga-Hama, Y., Kumagai, H., Li, Q.-H., Yamazoe, Y. and Kato, R.: Expression of human cytochrome P450 in *Schizosaccharomyces pombe*: Comparison with the expression in *Saccharomyces cerevisiae*. submitted.

Calcium-mediated mechanisms in chemically induced cell death

S. Orrenius

Institute of Environmental Medicine, Karolinska Institutet, Box 210,
S–171 77 Stockholm, Sweden

Abstract: There is now convincing evidence that the calcium ion can play a critical role in cell killing in various tissues. Recent research has established some of the biochemical mechanisms by which intracellular Ca^{2+} overload can trigger either necrotic or apoptotic cell death, and a number of studies have shown that prevention of Ca^{2+} overload by pretreatment with either Ca^{2+} chelators, receptor antagonists, or channel blockers can rescue cells that would otherwise die. This overview summarizes current evidence for an association between a perturbation of intracellular Ca^{2+} homeostasis and cytotoxicity and discusses various Ca^{2+}– mediated mechanisms that may lead to the death of cells.

The role of the calcium ion as intracellular regulator of many physiological processes is now well established. Thus, the effects of a variety of hormones and growth factors have been found to be mediated by transient increases in the level of cytosolic Ca^{2+}, which frequently assume oscillatory patterns (1,2). Most often, the Ca^{2+} increase is initiated by the release of Ca^{2+} from intracellular stores followed by the stimulation of influx of extracellular Ca^{2+}. Most regulatory effects of Ca^{2+} are mediated by Ca^{2+}–binding proteins (e.g. calmodulin) and achieved by alterations of the phosphorylation state of target proteins.

Along with this knowledge has come the understanding that Ca^{2+} can also play a determinant role in a variety of pathological and toxicological processes. It has long been recognized that Ca^{2+} accumulates in necrotic tissue, and more recent work has revealed that a disruption of intracellular Ca^{2+} homeostasis is frequently associated with the early development of cell injury (3–5). This led to the formulation of the calcium hypothesis of cell injury, proposing that perturbation of intracellular Ca^{2+} homeostasis may be a common step in the development of cytotoxicity. Support for this hypothesis has come from a large number of studies demonstrating that the calcium ion plays a critical role in cytotoxicity and cell killing in many tissues, notably the central nervous system and the immune system (ref. 6 for review).

Intracellular Ca^{2+} Homeostasis and Signalling

Studies using selective indicators have shown that the Ca^{2+} concentration in the cytosol of unstimulated cells is maintained between 0.05 and 0.2 μM (1). Extracellular Ca^{2+} levels are approximately four orders of magnitude higher (~1.3 mM). This produces a large, inwardly directed electrochemical driving force that is primarily balanced by active Ca^{2+} extrusion through the plasma membrane and by the coordinated activity of Ca^{2+}-sequestering systems located in the mitochondrial, endoplasmic reticular, and nuclear membranes (Fig. 1). In excitable tissues, different types of voltage–operated Ca^{2+} channels have been identified and characterized as well as receptor–operated channels that are involved in Ca^{2+} entry during hormone stimulation. It is still unclear whether voltage–operated Ca^{2+} channels exist in nonexcitable cells, whereas receptor–operated channels mediate Ca^{2+} influx in both excitable and nonexcitable tissues (7).

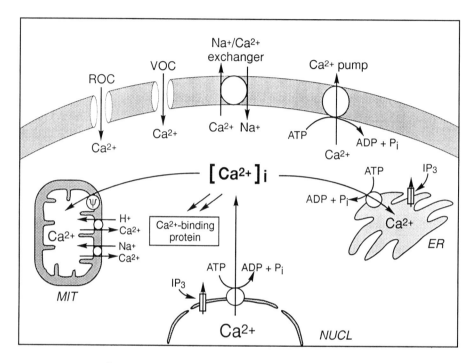

Fig. 1. Intracellular Ca^{2+} transport processes. ROC, receptor–operated channel; VOC, voltage.operated channel; IP3, inosotol 1,4,5–trisphosphate; ER, endoplasmic reticulum; MIT mitochondrion; NUCL, nucleus

Although isolated mitochondria can accumulate large amounts of Ca^{2+}, the affinity of the uniport carrier for Ca^{2+} uptake is low, and the mitochondria appear to play a minor role in buffering cytosolic Ca^{2+} under normal conditions. Electron probe X–ray microanalysis of rapidly frozen liver sections has shown that mitochondria contain little Ca^{2+} in situ (about 1 nmol Ca^{2+} per mg protein) (8), whereas the endoplasmic reticulum represents the major intracellular Ca^{2+} store. However, it should be noted that several physiological constituents (e.g. polyamines) can potentially increase the affinity of the uniport carrier for Ca^{2+}, at least to the level that is reached in the cytosol during agonist stimulation (about 400 nM) (9). Thus, the role of mitochondria in the modulation of the agonist–stimulated Ca^{2+} transients deserves further consideration.

Studies in our laboratory have shown that liver nuclei possess an ATP–stimulated Ca^{2+} uptake system involved in intranuclear Ca^{2+} accumulation (10). The more recent finding that the nucleus has a high Ca^{2+}–buffering capacity and that Ca^{2+} can be released from a nuclear compartment in response to intracellular messengers (11), suggests the possibility that the nucleus may have self–regulating mechanisms to control its Ca^{2+} level and to modulate intranuclear Ca^{2+} responses to hormones and growth factors.

The mechanisms whereby Ca^{2+}–mobilizing hormones elicit Ca^{2+} transients have been extensively studied in recent years (12). The signal transduction pathway leading to an elevation of cytosolic Ca^{2+} can be summarized as follows: upon binding of the agonist to its plasma membrane receptor, a specific phospholipase C is activated via stimulation of a G protein, resulting in the hydrolysis of phosphatidylinositol 4,5–bisphosphate and the generation of two second messengers, inositol 1,4,5–trisphosphate ($Ins(1,4,5)P_3$) and diacylglycerol. Diacylglycerol is a potent activator of protein kinase C, whereas $Ins(1,4,5)P_3$ binds to specific receptors and stimulates Ca^{2+} release from nonmitochondrial intracellular stores.

In addition to mobilizing Ca^{2+} from intracellular stores, hormones can stimulate Ca^{2+} influx from the extracellular compartment through specific receptor–operated Ca^{2+} channels (13). With the development of techniques to study Ca^{2+} changes in single cells, it has become possible to study the spatial and temporal distribution of transients within the cell. This led to the observation that at low, close–to–threshold concentrations of Ca^{2+}–mobilizing hormones, many cells respond to these agents with relatively rapid, oscillating spikes (14). It has been suggested that such oscillatory patterns may carry a frequency–encoded message (2). However, the possible implications of these phenomena to cell regulation have not been identified.

Interference with Cell Signalling.

Xenobiotics can interfere with signal transduction at different levels with a resulting loss of normal Ca^{2+} responses to hormones and growth factors. Chemical, bacterial and viral toxins can interact with receptors, G proteins and other enzymes involved in cell signalling, or can directly affect intracellular Ca^{2+} homeostasis by interfering with Ca^{2+} pumps or Ca^{2+} channels. Toxic chemicals can also inhibit the generation of inositol polyphosphates (15), and may cause either stimulation or inhibition of protein kinase C (16). Although the implications of these effects for cell survival have yet to be established, it is becoming increasingly apparent that the inhibition of hormonal responses may result not only in the loss of a trophic stimulus, but also in the activation of a program for cell self–deletion in some instances (17).

Other environmental toxins which may interfere with cell signalling include metal ions (see ref. 6 and references therein). Many metals have been shown to interfere with intracellular Ca^{2+} transport systems and with Ca^{2+} channels and to compete with Ca^{2+} for Ca^{2+}–binding proteins, including calmodulin. Divalent metals, such as Cd^{2+} and Hg^{2+}, can also interact with protein thiol groups and inhibit Ca^{2+} transport systems within the cell and thus prevent Ca^{2+} efflux and release of Ca^{2+} from intracellular stores. Moreover, experiments using synaptosomal preparations have indicated that the effects of certain metals within a given class on intracellular Ca^{2+} levels can be directly correlated with their neurotoxic effects *in vivo*. For example, the neurotoxic organotin derivative, trimethyltin, is much more potent than the closely related non–neurotoxic mono– and dimethyltin in causing increases in intracellular Ca^{2+}(18).

Effects of Ca²⁺Overload

Thus, intracellular Ca^{2+} homeostasis is controlled by the concerted operation of plasma membrane Ca^{2+} translocases and intracellular compartmentalization systems. Disturbances of these processes during cell injury can result in release of Ca^{2+} from intracellular stores, enhanced Ca^{2+} influx, and inhibition of Ca^{2+} extrusion at the plasma membrane, which can lead to an uncontrolled rise in intracellular Ca^{2+} concentration. In ischemia–reperfusion injury the Ca^{2+} overload is normally preceded by intracellular Na^+ accumulation caused by inhibition of the Na^+/K^+–pump and activation of Na^+/H^+ exchange and Na^+–HCO_3 co–transport due to ATP depletion and acidosis, respectively (Fig. 2). Reversal of the Na^+/Ca^{2+} exchanger would then contribute to intracellular Ca^{2+} accumulation (19). Such sustained increases in intracellular Ca^{2+} will obliterate the transient Ca^{2+} responses normally evoked by hormone or growth factor stimulation, compromise mitochondrial function and cytoskeletal organization and, ultimately, activate degradative processes. Since Ca^{2+} is an activator of several enzymes involved in the catabolism of proteins, phospholipids and nucleic acids, a sustained increase in cytosolic free Ca^{2+} concentration above the physiological level can result in uncontrolled breakdown of macromolecules of vital importance for the maintenance of cell structure and function.

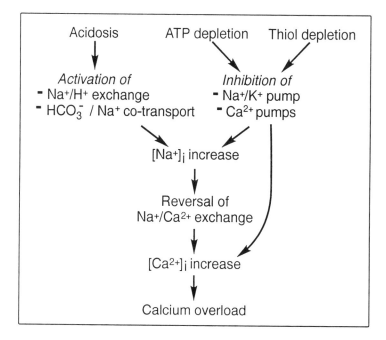

Fig. 2. Mechanisms proposed to contribute to intracellular Ca^{2+} overload in ischemic/oxidative injury.

Mitochondrial damage. Work from several laboratories has indicated that mitochondrial damage may represent a common event in cell injury caused by toxic agents and other insults, e.g. ischemia. Mitochondrial damage is initially manifested by a decrease in the mitochondrial membrane potential followed by ATP depletion. Calcium ions can be actively transported into mitochondria via an electrophoretic uniporter. The driving force for the continuous Ca^{2+} pumping is provided by the transmembrane potential. However, studies performed in isolated mitochondria have demonstrated that during Ca^{2+} uptake the membrane potential decreases and the extent of the decrease is proportional to the amount of Ca^{2+} taken up by the mitochondria (20). Thus, it appears that, under conditions which cause massive amounts of Ca^{2+} to accumulate in the mitochondria, their membrane potential would collapse.

The existence of different Ca^{2+} uptake and release pathways in mitochondria provides a basis for Ca^{2+} cycling. This process continuously utilizes energy which is supplied by the membrane potential. Oxidation of intramitochondrial NAD(P)H can activate the release route and accelerate Ca^{2+} cycling across the mitochondrial membrane (21). This condition is associated with a decrease in the mitochondrial membrane potential that parallels the rate of Ca^{2+} cycling. Evidence that this mechanism is also operational in intact cells has been obtained using video imaging analysis in cultured hepatocytes loaded with Rhodamine 123 (22).

Cytoskeletal alterations. One of the early signs of cell injury is the appearance of multiple surface protrusions (blebs) (4). The events leading to bleb formation have not yet been fully elucidated and several mechanisms may independently contribute to their formation. However, it is generally accepted that a perturbation of cytoskeletal organization and of the interaction between the cytoskeleton and the plasma membrane plays an important role. Evidence for this assumption is provided by the observation that agents which disrupt the organization of the cytoskeleton, such as cytochalasins and phalloidin, stimulate bleb formation and by the recent demonstration that the bundles of actin filaments present at the base of the bleb appear to be totally dissociated from the bleb–forming portion of the plasma membrane (23). The finding that treatment of cells with a Ca^{2+} ionophore was able to induce similar blebbing, and that this was prevented by the omission of Ca^{2+} from the incubation medium, led to our proposal that Ca^{2+} is involved in the cytoskeletal alterations associated with the formation of surface blebs during cell injury (4). A number of studies have clearly demonstrated the importance of various Ca^{2+}–dependent events in the control of cytoskeletal organization and function. Ca^{2+}–dependent cytoskeletal alterations include modification of the association between actin and actin–binding proteins, disruption of microtubular organization, and Ca^{2+}–dependent proteolysis of actin–binding proteins (23,24). Thus, perturbation of intracellular Ca^{2+} homeostasis is one important factor contributing to plasma membrane blebbing during cell injury.

Role of Ca^{2+} in Apoptotic Cell Killing

Most often, Ca^{2+}–mediated cell killing is of the necrotic type and characterized by cell and organelle swelling and cytolysis. However, there is accumulating evidence that the calcium ion may also play an important role in the triggering of apoptotic cell death. Apoptosis is a form of programmed cell death which occurs during fetal development as well as in adult life and which can also be induced by physiological and chemical agents and in certain disease conditions. Typical early morphological changes occur in apoptotic cells, including cell shrinkage, widespread plasma and nuclear membrane blebbing, compacting of organelles, and chromatin condensation (25). Activation of proteases and endonucleases results in the cleavage of cell chromatin into oligonucleosome–length fragments.

The results of several recent studies have shown that Ca^{2+} overload can trigger DNA fragmentation. The Ca^{2+} ionophore A23187 stimulates apoptosis in thymocytes, and characteristic endonuclease activity in isolated nuclei is dependent on Ca^{2+} (26). More recently, we found that exposure of human adenocarcinoma cells to tumor necrosis factor causes intracellular Ca^{2+} accumulation and endonuclease activation (27). Interestingly, in many cells the initial Ca^{2+} increase was found in the nucleus, suggesting that a selective elevation of the nuclear Ca^{2+} concentration may be sufficient to stimulate DNA fragmentation. However, the implications of this finding remain to be elucidated. Thus, recent observations suggest that the endonuclease–catalyzed internucleosomal cleavage is secondary to the formation of high molecular weight DNA fragments of 300 kbp and 50 kbp corresponding to hexameric rosette structures (300 kbp) and single loops (50 kbp), respectively (28). It now appears that both protease and endonuclease activities are required for the multistep chromatin cleavage process, and that Ca^{2+} is required for the formation of both the 50 kbp and oligonucleosome–length fragments (29). The mechanism(s) responsible for the initial chromatin changes in apoptosis are currently under study in our laboratory.

We and others have shown that intracellular Ca^{2+} buffers and extracellular EGTA can inhibit both DNA fragmentation and death in apoptotic cells, suggesting that sustained Ca^{2+} elevations are required for both responses (6). Independent evidence for the involvement of Ca^{2+} influx has come from studies with specific Ca^{2+} channel blockers, which abrogate apoptosis in the regressing prostate following testosterone withdrawal (30) and in pancreatic β cells treated with serum from patients with type I diabetes (31). Further evidence for a critical role of Ca^{2+} in triggering apoptosis comes from a study by Dowd et al. (32), demonstrating that the stable overexpresssion of calbindin–D28K in a thymoma cell line protected the cells from apoptosis normally produced by exposure to dexamethasone, forskolin and calcium ionophore A23187. Thus, in several models of apoptosis, the calcium ion plays a critical role in triggering this process. However, this is not always true. There are several experimental systems which fail to show any Ca^{2+} dependency (33). In fact, there are examples where elevations of cytosolic Ca^{2+} have been found to block apoptosis. Therefore, further studies are required to define in which cases the triggering of apoptosis is mediated by Ca^{2+} and how this is achieved.

Ca^{2+} Overload in Neuronal Cell Death

The evidence for a role of Ca^{2+} in cell killing is particularly strong in the central nervous system. Much recent research has focused on glutamate–induced excitotoxicity and its contribution to brain damage in various disease conditions (34). The ability of glutamate or related compounds to induce neuronal death by receptor overstimulation has long been recognized. The calcium ion plays a critical role in this process, and intracellular Ca^{2+} overload appears to mediate the lethal effects of NMDA receptor overactivation. This mechanism is responsible not only for the brain damage induced by certain neurotoxins, but excitotoxicity is also strongly implicated in neuronal death following insults such as ischemia and trauma.

Recent studies in our laboratory have shown that glutamate can trigger the onset of both necrosis and apoptosis in cerebellar granule cells. Necrosis occurred during the early phase of exposure; and the type of cell death was related to glutamate concentration; necrosiswas predominant at higher doses (35). Radical scavengers and agents which inhibit the generation of nitric oxide by nitric oxide synthase were ineffective in preventing apoptosis in this system, whereas NMDA receptor/channel blockers prevented both necrosis and apoptosis in glutamate-treated cerebellar granule cells. These data suggest that Ca^{2+} overload mediated through NMDA receptor-operated channels is sufficient to induce either necrosis or apoptosis in cerebellar granule cells in vitro.

There are several recent studies suggesting that cell killing in neurodegenerative disorders may occur by apoptosis. Thus, MPP^+ (1-methyl-4-phenylpyridinium), the neurotoxic metabolite of the Parkinsonian agent MPTP, can trigger internucleosomal DNA cleavage and apoptosis in cerebellar granule neurons (36). Further, Forloni et al. (37) have reported that the neurotoxicity induced by chronic application of a synthetic peptide homologous to residues 25–35 of the amyloid β protein to rat hippocampal neurons occurs by apoptosis, suggesting that amyloid fibrils may induce neuronal death by apoptosis in Alzheimer's disease. Evidence supporting a role for apoptosis in focal cerebral ischemia in rats has recently been published (38), whereas the proto-oncogene bcl-2 has been found to prevent neuronal cell death induced by glutathione depletion by decreasing the net cellular generation of reactive oxygen species (39). Of considerable interest in this context is the recent observation that the β amyloid peptide kills various cell lines as well as cortical cultures by an H_2O_2-mediated mechanism and that the type of cell death is necrotic rather than apoptotic (40).

Our understanding of the mechanisms responsible for cell death in various neurodegenerative disorders has improved during recent years but is still far from complete. It is clear that intracellular Ca^{2+} overload is an important factor in various in vitro models, and results from clinical studies appear to support this hypothesis. The Ca^{2+} overload seems to result from increased Ca^{2+} influx through both receptor-operated and L-type Ca^{2+} channels as indicated by the neuroprotective effects of glutamate receptor antagonists and L-type channel blockers. Various Ca^{2+}-dependent degradative processes have been found to contribute to Ca^{2+}-mediated cell killing in in vitro studies and, although their relative contribution is unclear; it appears that perturbation of cytoskeletal organization and impairment of mitochondrial function may be of particular importance. Finally, although cell death in neurodegenerative disorders has been assumed to be of the necrotic type, recent studies have indicated that it may also occur by apoptosis. In either case, it appears that the calcium ion may play a determinant role in the killing process.

References

1. E. Carafoli. *Annu. Rev. Biochem.* **56,** 395 (1989).
2. M.J. Berridge. *J. Biol. Chem.* **265,** 9583 (1991).
3. F.A.X. Schanne *et al. Science* **206,**702 (1979).
4. S.A. Jewell *et al. Science* 217,1257 (1982).
5. A. Fleckenstein *et al.*, *Mechanisms of Hepatocyte Injury and Death.* p 321, MTP Press Limited, Lancaster, (1983).
6. P. Nicotera. *et al. Annu. Rev. Pharmacol. Toxicol.* **32,** 449 (1992).
7. C. Fasolato *et al. J. Biol. Chem.* **265,** 20351 (1990).
8. A.P. Somylo *et al. Nature* **314,** 622 (1985).
9. H. Rottenberg and M. Marbach. *Biochim. Biophys. Acta* **1016,** 77 (1990).
10. P. Nicotera *et al. Proc. Natl. Acad. Sci. USA* **86,** 453 (1989).
11. P. Nicotera *et al. Proc. Natl. Acad. Sci. USA* **87,**6858 (1990b).
12. M.J. Berridge. *Annu. Rev. Biochem.* **56,**159 (1987).
13. G.J. Barritt *et al. J. Physiol.* **312,** 9 (1981).
14. N.M. Woods *et al. Nature* **319,** 600 (1986).
15. G. Bellomo *et al. J. Biol. Chem.* **262,** 1530 (1987).
16. G.E.N. Kass *et al. Biochem. J.* **260,** 499(1989).
17. D.S. Barnes. *J. Biol. Chem.* **262,** 1530 (1988).
18. H. Komulainen and S.C. Bondy *Toxicol. Appl. Pharmacol.* **88,**77(1987).
19. R. Carini *et al. Biochem. Biophys. Res. Commun.* **202,** 360 (1994).
20. T.E. Gunther and D.R. Pfeiffer. *Am. J. Physiol.* **258,** c755 (1990).
21. A.L. Lehninger *et al. Proc. Natl. Acad. Sci. USA* **75,** 1690 (1978).
22. P. Nicotera *et al. Chem. Res. Toxicol.* **3,** 484 (1990a).
23. P.C. Phelps *et al. Lab. Invest.* **60,** 630 (1989).
24. F. Mirabelli *et al. Arch. Biochem. Biophys.* **270,** 478 (1989).
25. M.J. Arends *et al. Am. J. Pathol.* **136,** 593 (1990).
26. D.P. Jones *et al. J. Biol. Chem.* **264,** 6398 (1989).
27. G. Bellomo *et al. Cancer Res.* **52,** 1342 (1992).
28. G.M. Cohen *et al. Biochem. J.* **286,** 331 (1992).
29. B. Zhivotovsky *et al. FEBS Lett.* **351,** 150 (1994).
30. P. Martikainen and J. Isaacs. *Prostate* **17,** 175 (1990)
31. L. Juntti–Berggren *et al. Science* **261,** 86 (1993).
32. D.R. Dowd *et al. Mol. Endocrinol.* **6,** 1843 (1992).
33. R. Iseki *et al. J. Immunol.* **151,** 5198 (1993).
34. D.W. Choi. *Science* **258,** 241 (1992).
35. M. Akarcrona *et al. Neuron* In press.
36. B. Dipasquale *et al. Biochem. Biophys. Res. Commun.* **181,** 1442 (1991).
37. G. Forloni *et al. NeuroReport* **4,** 523 (1993).
38. M.D. Linnik *et al. Stroke* **24,** 2002 (1993).
39. D.J. Kane *et al. Science* **262,** 1274 (1993).
40. C. Behl *et al. Cell* **77,** 817 (1994).

Targeted drug delivery to the brain using chemical delivery systems

Nicholas Bodor

Center for Drug Discovery, University of Florida, Health Science Center, Box 100497, Gainesville, Florida 32610, USA

Abstract

A general approach for brain targeting of a wide variety of biologically active compounds and drugs was developed based on a dihydropyridine-pyridinium salt redox targetor system. Subsequent advancement and modifications of the system allowed brain targeting of such complicated molecules as various neuropeptides.

INTRODUCTION

Targeting drugs to specific receptors or specific organs is very attractive goal of medicinal and pharmaceutical research, since this provides one of the most significant potential ways to improve the therapeutic index (TI) of the drugs (1). This is because when a drug is delivered exclusively or preferentially to the site of the action, it will spare the rest of the body by virtue of this desired differential distribution, and thus it will significantly reduce the overall toxicity while maintaining the therapeutic benefits. In this context, drug targeting is defined in the broadest sense. In other words the objective is to optimize a drug therapeutic index by strictly localizing it pharmacological activity to the site or organ of action. This is a very important point, since previously the objective of drug discovery was to improve affinity, fit and binding to the specific receptor, which ultimately would trigger the pharmacological activity. However, in many instances, the general distribution of the receptors, which does not generally follow the various disease states, places serious limitations on the success of targeting only to the receptor. Actually, most of the time, drug toxicity is receptor related and receptor mediated. Improving intrinsic drug affinity and activity, based on increasing receptor binding does not improve the therapeutic index. However, localizing an active drug at the target site can have dramatic improvement in the treatment of diseases, particularly if the localized delivery is also controlled in time.

In principal there are three basically different approaches for site specific drug delivery. The most obvious is the physical or mechanical approach, which is based on effectively formulating a drug in a delivery device, which by physical localization will allow differential release of the drug, that is higher drug concentrations at the site the device is localized, while the drug concentration in the rest of the body is very much diminished, due to simple dilution factor. Of course, metabolism and pharmacokinetic parameters also contribute to the success of this still rather limited approach. The prime obvious limitation is due to the fact that many desired target sites are simply not available for the physical approach.

The second basic approach is the biological one, where a drug is targeted by a biological carrier that would have specific affinity for certain receptor sites, organs or other biological targets. This kind of approach, for example using monoclonal antibody-drug conjugates, erythrocyte carriers, or molecular carriers, have again a number of limitations presented by stoichiometry, control of the processes related to releasing the drugs from the biological carriers, in addition to frequent biological incompatibility of the carriers.

The third approach is the chemical-enzymatic approach, that is the use of site-specific chemical targeted delivery systems (CSD'S), (1,2), which provide a wide variety of possibilities for site enhanced, site-specific, targeted drug delivery.

CHEMICAL DRUG DELIVERY SYSTEM

The concept of chemical delivery systems evolved from the prodrug concept (3,4). A prodrug is a pharmacologically inactive compound which results from chemical modification of biological active species. The chemical change is designed to improve some deficient chemical-physical, kinetic or other property of the drug, however, after administration, the prodrug provides more desired overall distribution and by one chemical or enzymatic step will be converted to the active drug. Most of the time, the objective is to improve membrane permeability by transiently increasing lipophilicity of the drug. Generally, this approach does not lead to improved targeting since changes in distribution are reflected essentially everywhere in the body. In this way, many times non target site toxicities are also increased.

A chemical delivery system (CDS) is defined as a biologically inert molecule which requires several steps in its conversion to the active drug and which enhances drug delivery to a particular organ or site (5). The concept of chemical delivery systems constitutes one of the major approaches in the retrometabolic drug design (5). Strictly speaking, CDS-s are inactive compounds produced by one or more chemical modifications of the drug (D), where the attached moieties are comparable in size or smaller than the drug. In general, multi-step enzymatic transformations *in vivo* produces the targeted drug. There are two types of covalent attachments to the drug. The most important one is the "targetor" moiety (T), which in some cases could have multiple, bioreversible forms. The other class of moieties are protector functions (F), which have the objective to protect certain other functions from premature metabolism and/or optimize overall physical-chemical and transport properties. Chemical delivery systems were successfully designed to target drugs to various organs such as the eye (6) to the lung and other organs. One of the most important general targetor systems involves the brain. Here a combination of chemical-enzymatic design with respect to the important biological barrier, the blood-brain-barrier (BBB) led to successful brain targeting.

DRUG TARGETING TO THE BRAIN BY REDOX CHEMICAL DELIVERY SYSTEMS

The blood-brain-barrier provides a very efficient protection of the brain against a wide variety of blood borne compounds (7,8). The basic morphological feature of the BBB is the presence of epithelial-like, high resistance-type junctions that fuse brain capillaries and epithelial together into a continuous cellular layer, separating blood and brain interstitial space. Due to the lack of fenestrae and vesicular traffic in the brain capillary endothelial cells, the free flow between brain interstitial space and the blood is restricted. Accordingly, this lipid-like barrier prevents penetration into the brain of hydrophilic compounds, unless they are transported into the brain by an active transport system. In addition the BBB contains highly active enzyme systems, which further enhance the already protective function. Besides the molecular size and lipophilicity, the affinity of the substances to various blood proteins, specific enzymes in the blood or the BBB, will also very much influence the amount of drug reaching the brain. For example neurotransmitters, when injected into the blood system will not pass the BBB. On the other hand, when the very same compounds are synthesized *in situ* in the brain, they are basically confined to the brain due to participation of the barrier function of the BBB. The recognition that the BBB should act as a barrier against the afflux of hydrophilic molecules formed in situ in the brain (9) led to the idea of developing a brain enhanced or brain specific drug delivery system (10). The basic concept is a simple physical-chemical one. Accordingly, the basic inactive CDS is a lipophilic molecule containing the drug, as being derivatized by a targetor (T) and appropriate protective functions (F_1-F_n). This lipophilic molecule can penetrate the BBB. Subsequently, the carefully chosen targetor moiety will undergo enzymatic conversion to a charged specie. Clearly, the most dramatic change that can occur in the physical-chemical properties of a molecule is to acquire a permanent charge. This will lead generally to about four log unit change in the partition coefficient, providing a truly dramatic difference in membrane permeability of the compound. According to the CDS concept, this change in physical-chemical properties should be localized on the targetor moiety, thus the appropriate enzymatic modifications still would be able to release the intact active drug. The new, hydrophilic form of the CDS, still inactive, is now effectively "locked-in" in the brain. Subsequent enzymatic reactions then cleave the other protective functions, providing a direct precursor, which is essentially a covalent combination of the active drug and the charged targetor moiety. In the final step the targetor is removed enzymatically, allowing the active drug to be effective. Targetor moiety has to be chosen in such a way to be non-toxic after the release, and also to be quickly eliminated from the brain. That is, why the drug-targetor charged precursor should be locked-in the brain, the charged hydrophilic targetor upon cleavage, should be easily eliminated from the brain. The process is illustrated in a general form in Scheme 1.

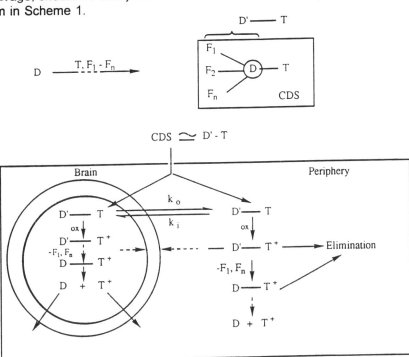

A Brain-targeting chemical delivery system.

Scheme 1.

An example shows the conversion of the lipophilic T form of the targetor into a positively charged T+ form. Alternate methods can use negatively charged targetor systems as well.

In order to satisfy the above described requirements, we have concentrated on a dihydropyridine-quaternary pyridinium salt - type redox system, which is analogous to the ubiquitous NAD^+-NADH coenzyme system, as illustrated on Fig. 1.
The idea was that if we can introduce a covalently bound, but bioreversible redox moiety of this type, the ubiquitous conversion of the initial lipophilic CDS, to the very hydrophilic charged species is assured. The N_1-substituted nicotinic acid amides and esters were the first choice, since the finally released positively charged targetor, trigonelline is a natural compound of virtually no toxicity and which is also eliminated very fast from the brain, apparently by an active transport mechanism (9,11). It is important to recognize that the conversion to the hydrophilic quaternary salt occurs virtually with the same rate everywhere in the body. This further enhances the targeting value of this system. While the quaternary form formed in the brain is locked-in, the oxidation in the peripheral tissues leading to the peripheral quaternary

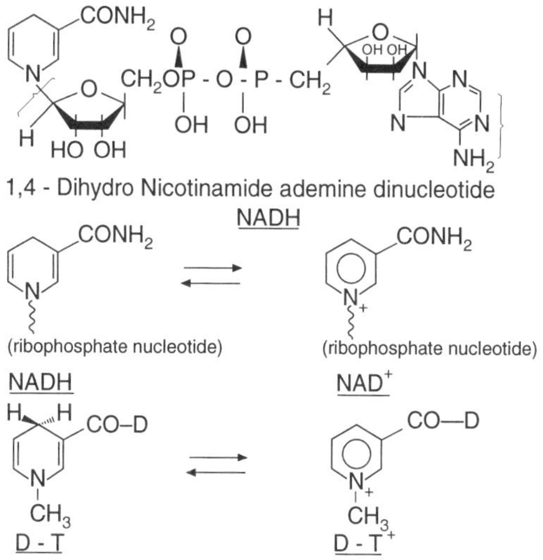

Fig. 1. The redox targetor ($T \rightleftharpoons T^+$) system and its analogy to the ubiquitous NADH $\rightleftharpoons$ NAD$^+$ coenzyme system.

species accelerate elimination and thus after a relatively short time there is a dramatic difference in the concentration of the quaternary precursor concentrations in the brain and the rest of the body. Actually, if the rate of administration is compatible with the rate of oxidation, one can enhance constantly the concentrations in the brain, against the concentration gradient, while maintaining relatively low, practically steady state concentrations in the periphery. The conjugate which is trapped behind the BBB will then yield the active drug in a sustained released manner. The amount of drug which is formed in the periphery is very low and thus systemic toxicities are very much reduced. In addition, since the drug in the CNS is present mostly in the form of inactive conjugates, central toxicities are also reduced.

APPLICATION OF THE REDOX CDS TO BRAIN TARGETING OF VARIOUS DRUGS

The above redox targetor system can have wide applicability. Actually, there are two types of drugs, biologically active compounds which would benefit from this system. The most obvious application is to biologically active compounds, such as neurotransmitters, peptides, anti-viral agents, and others which have very low penetration to the brain. In this case, not only can such a drug be delivered to the brain, but maintained there for a prolonged time. On the other hand compounds which do penetrate the blood-brain barrier can also significantly benefit from this system, by dramatically increasing brain/blood concentration ratios. Examples for application of the redox brain targeting systems are listed in Table 1.

Table 1. Application of the Redox Brain Targeting System to Neurotransmitters, Hormones and Drugs

Compound	Reference
Dopamine	Bodor and Farag, 1993 (12); Bodor and Simpkins, 1983 (13); Simpkins et al., 1985 (14)
Tryptophan	Bodor et al. , 1986 (15)
γ-Aminobutyric acid	Anderson et al., 1987 (16)
Trifluorothymidine	Rand et al., 1986 (17)
Azidothymidine	Little et al., 1989 (18); Brewster et al., 1993 (19); Pop et al., 1992 (20)
CCNU	Raghavan et al., 1987 (21)
Berberine	Bodor and Brewster, 1983 (22)
Phenytoin	Shek et al., 1989 (23)
Propofol	Pop et al., 1991 (24)
Benzyl Penicillin	Pop et al., 1989 (25)
Naproxen	Phelan and Bodor, 1989 (26)
Chlorambucil	Bodor et al., 1989 (27)
Dexamethasone	Anderson et al., 1989 (28)
Testosterone	Bodor and Farag, 1984 (29)
Estradiol	Bodor and Abdelalim, 1986 (30); Brewster et al., 1986 (31); Bodor et al., 1987 (32); Brewster et al., 1990 (33)

Among the many examples, two important and two distinctly different applications are presented. The first one is the brain targeting of an endogenous hormone, estradiol (E$_2$), which obviously, easily passes the BBB, thus there is no problem in actual brain delivery. The issue here is <u>brain targeting</u>, that is, differential distribution to the brain. This is a good example for brain targeting of molecules, which otherwise do go to the brain. The main issue here is that estrogen being circulated in the periphery has poor retention in the brain. Accordingly, it would be required frequent doses to maintain therapeutically significant CNS levels. Unfortunately, constant peripheral exposure to estrogen has been related to a number of pathological conditions, including cancer, hypertension and altered metabolism (34, 35). The brain targeting of E$_2$ is depicted on Fig. 2.

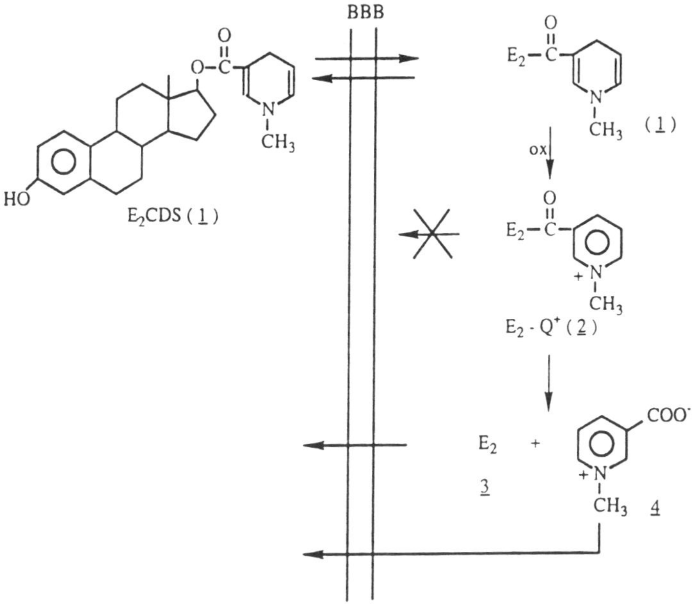

Fig. 2. Brain targeting of estradiol (E$_2$) by E$_2$CDS.

Estradiol is a lipophilic molecule (octanol/water logP = 3.76). The corresponding dyhydrotrigonellinate (E$_2$-CDS, <u>1</u>) is somewhat more lipophilic, being a 17-ester, its logP is 4.50. As such, <u>1</u> is easily distributed from the periphery to the brain. The earlier described oxidation of the dihydrotrigonellinate to the positively charged trigonelline ester, E$_2$-Q$^+$(<u>2</u>), introduces a dramatic difference in the physical properties of the molecule. The logP of the quaternary salt <u>2</u> is -0.14. Accordingly, the quaternary salt is 8,000 times more hydrophilic then the parent estradiol and more than 44,000 times more hydrophilic than the E$_2$-CDS. These facts clearly describe why the system works so well. The E$_2$-Q$^+$ is "locked-in" in the brain, and provides a slow, sustained release of the estradiol, in addition to the trigonelline. Both these species, compounds <u>3</u> and <u>4</u> are now easily released from the brain. Pharmacologically, an estrogen chemical delivery system could be used to reduce the secretion of luteinizing hormone releasing hormone (LHRH), and hence the luteinizing hormone (LH) and gonadal steroids (36). As such, the E$_2$-CDS could be employed to achieve contraception and to reduce growths of peripheral steroid dependent tumors and to treat endometriosis (37). Additionally, brain enhanced delivery of estradiol could be useful in stimulating male and female sexual behavior (38,39), in treatment of menopausal syndrome (40), and possibly most importantly, the treatment of depression and of various types of dementias, including Alzheimer's disease (41,42). *In vivo* studies clearly support the functioning of the E$_2$-CDS. After i.v. administration of one single dose of E$_2$-CDS to rats, a significant selectivity was observed for the CNS, as brain tissue half life was between 5-9 times greater than that observed in the kidney, heart, lung, testes, eye, or peritoneal fat (43). This highly differential distribution of the "locked-in" form, E$_2$-Q$^+$ was reflected in concentration of estradiol released over time from depot E$_2$-Q$^+$ (44). It was found that the levels of E$_2$ in brain tissue after E$_2$-CDS administration are elevated about five times longer than after simple

E_2 treatment. In addition, the ratio of concentration of E_2 in brain and plasma after E_2-CDS treatment was approximately twelve. Pharmacological studies confirmed the differential estradiol distribution. Accordingly, E_2-CDS suppressed LH secretion at 12,18, and 24 days at 88, 86, and 66%, respectively, while equivalent doses of estradiol or the isolipophilic estradiol 17-valerate, did not show any difference from the control at these time points. Other pharmacological studies such as the effect on cyclicity and ovulation (44) or on sexual behavior in castrate male rats (45), were again consistent with the functioning of the brain targeted estradiol chemical delivery system. It is important to underline that various toxicological evaluation of E_2-CDS resulted no neurotoxicological side effects (46). Subsequent human studies demonstrated similar efficacy and lack of toxicity (47). All these clearly demonstrate the potential importance of brain targeted estradiol chemical delivery system in a variety of clinical application, among which its use in Alzheimer's disease could be of major importance.

PEPTIDE DELIVERY TO THE BRAIN BY SEQUENTIAL METABOLISM

Neuropeptides in general are in a class of compounds in which brain delivery could be highly desired. Although transport of some peptide molecules across the BBB cannot be ruled out, it is unlikely that endogenous peptides pass the BBB in physiologically significant amounts. The various attempts to deliver peptides to the brain have recently been reviewed (48), at this point we are going to summarize the most successful approach, an extension of the redox brain targeting system named "molecular packaging". Accordingly, it was recognized (49), that simple CDS is not sufficient to allow brain delivery and targeting of peptides. Due to the high metabolic, peptidase activity in the BBB, even apparently lipophilic peptide conjugates containing the dihydrotrigonelline targetor, fail to deliver the peptide to the brain and produce the expected activity. The molecular packaging approach places the neuropeptide to be delivered in an environment which is not recognized by the peptidase, as the peptide represents a perturbation of a large molecule. Accordingly, the first example involved the brain delivery of a Leu-enkephalin (Tyr-Gly-Gly-Phe-Leu, YGGFL). This target peptide is prone to cleavage and deactivation by peptidase at the Tyr and at the Gly-Phe position. The use of D-Ala^2 and D-Leu^5 analog renders the enkephalin more stable. In order to effect the ultimate release, a spacer, strategically designed to be a substrate to dipeptidal peptidase is inserted between the targetor redox component and the peptide. In addition, the carboxy terminus was derivatized by a large lipophilic function, such as cholesterol, accomplishing the packaging of the peptide. Thus, the peptide in the delivery system appears as a perturbation on the bulky molecule dominated by L and T. A simple schematic representation of the brain targeting of the neuropeptide is described on Scheme 2.

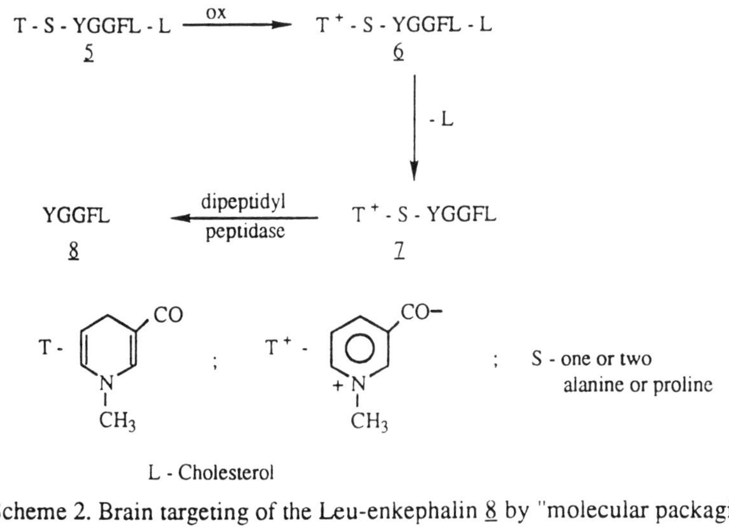

Scheme 2. Brain targeting of the Leu-enkephalin 8 by "molecular packaging".

Upon administration, the modified peptide (5) simply enters the brain because of its lipoidal nature.   The oxidation of the dihydrotrigonellinate function to the hydrophilic, membrane impermeable trigonellinate ion produces a (6), which remains trapped behind the BBB. Oxidation of 5 in the periphery, on the other hand, results in its rapid secretion from the body. The removal of the cholesterol (L) by esterase or lipase occurs subsequent two or simultaneously with the brain targeting enzymatic oxidation.   The targetor-peptide conjugate (7) is then formed and could be detected at significant amounts in the brain for many hours (49).   *In vivo* pharmacological studies, demonstrated sustained and statistically significant increase in the tail flick response, a measure of the spinal cord mediated analgesia.   The effect was observed even six hours after i.v. administration of the packaged enkephalin CDS, (5).   The method was recently extended to a centrally active thyrotropin releasing hormone (TRH) analog, 9 (pGlu-Leu-Pro-NH$_2$).   This compound presented a particularly difficult problem, as it does not have any free hydroxy or amino groups present.   The delivery was solved by incorporation of the progenitor sequence (QLPG), into the CDS.   The C-terminal glycine functions as an amide source for the proline by the enzyme peptide glycine amidating monoxygenase (PAM).   Glutamine is the precursor of the N-terminal pyroglutamyl residue. Cyclization on the N-terminal glutamine is catalyzed by a specific enzyme, glutaminal cyclase. The hypothetical "locked-in" precursor T+-AQLPG is, indeed processed to the prolinamide as validated *in vitro* (50).   Thus, the processing to the desired TRH analog 9 is only dependent on the slow release of 10, which proceeds similarly to that of the enkephalin analog from the "locked-in" peptide conjugate.   The CNS delivery of a pharmacologically significant amount of a TRH analog is evidenced by the profound decrease in barbiturate induced sleeping time, the measure of the activational effect on cholinergic neurons in mice (50).   Most recent studies (51) demonstrated that manipulation of the spacer, that is use of two proline to separate the targetor and the peptide provides even better release and activity.

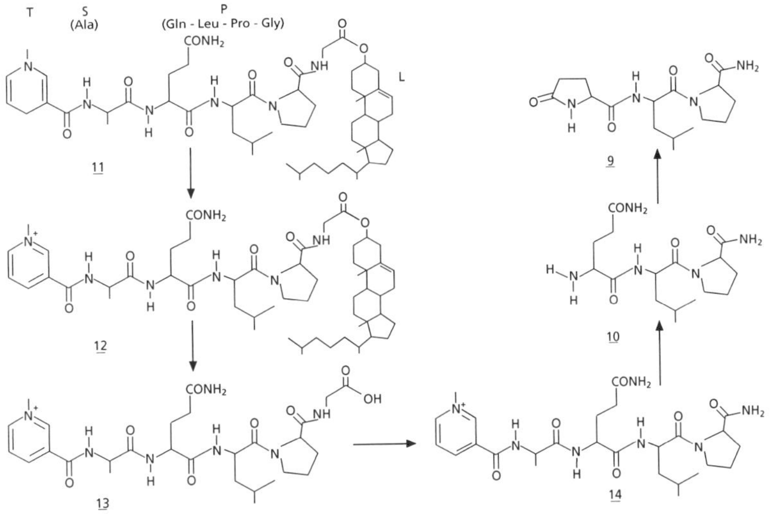

Scheme 3. CNS targeting of pGlu-Leu-Pro-NH$_2$, 9 by sequential metabolism.

REFERENCES

1.   N. Bodor, *Advances in Drug Research*, p. 255, Academic Press, London (1984).
2.   N. Bodor and H. Farag. *J. Med. Chem.* **26**, 313 (1983).
3.   A.A. Sinkula and S.H. Yalkowsky. *J. Pharm. Sci.* **64**, 181 (197).
4.   A. Albert, *Selective Toxicity*, 7th Edition, Chapman and Hall, New York (1985).
5.   N. Bodor, *Trends in Medicinal Chemistry '90*, p.35, Blackwell Scientific Publications (1992).
6.   N. Bodor. *J. Ocular Pharmacology* **10**, 3 (1994).
7.   S.T. Rapaport, *The Blood-Brain Barrier in Physiology and Medicine*, Raven Press, New York (1976).
8.   N. Bodor and M. Brewster. *Pharm. Ther.* **19**, 337 (1983).

9.   N. Bodor, R. Roller and S. Selk. *J. Pharm. Sci.* **67**, 685 (1978).
10.   N. Bodor, H. Farag and M. Brewster. *Science*, **214**, 1370 (1981).
11.   E. Palomino, D. Kessel and J.P. Horvitz. *J. Med. Chem.* **32**, 622 (1989).
12.   N. Bodor and H. Farag. *J. Med. Chem.* **26**, 528 (1983).
13.   N. Bodor and J. Simpkins. *Science* **221**, 65 (1983).
14.   J. Simpkins, N. Bodor and A. Enz. *J. Pharm. Sci.* **74**, 1033 (1985).
15.   N. Bodor, T. Nakamura and M. Brewster. *Drug Des. & Del.* **1**, 51 (1986).
16.   W. Anderson, J. Simpkins, P. Woodard, D. Winwood, W. Stern and N. Bodor. *Psychopharmacol.* **92**, 157 (1987).
17.   K. Rand, N. Bodor, A. El-Koussi, I. Raad, A. Miyake, H. Houck and N. Gildersleeve. *J. Med. Virol.* **20**, 1 (1986).
18.   R. Little, D. Bailey, M. Brewster, K. Estes, R. Clemmons, A. Saab and N. Bodor. *J. Biopharm. Sci.* **1**, 1 (1990).
19.   M. Brewster, E. Pop, A. Braunstein, A. Pop, P. Druzgala, W. Anderson, A. ElKoussi and N. Bodor. *Pharm. Res.* **10**, 1356 (1993).
20.   E. Pop, M. Brewster, W. Anderson and N. Bodor. *Med. Chem. Res.* **2**, 457 (1992).
21.   K. Raghavan, E. Shek and N. Bodor. *Anti-Cancer Drug Des.* **2**, 25 (1987).
22.   N. Bodor and M. Brewster. *Eur. J. Med. Chem.* **18**, 235 (1983).
23.   E. Shek, T. Murakami, C. Nath, E. Pop and N. Bodor. *J. Pharm. Sci.* **78**, 837 (1989).
24.   E. Pop, W. Anderson, K. Prokai-Tatrai, J. Vlasak, M. Brewster and N. Bodor. *Med. Chem. Res.* **2**, 16 (1991).
25.   E. Pop, W. Wu, E. Shek and N. Bodor. *J. Med. Chem.* **32**, 1774 (1989).
26.   M. Phelan and N. Bodor. *Pharm. Res.* **6**, 667 (1989).
27.   N. Bodor, V. Venkatraghavan, D. Winwood, K. Estes and M. Brewster. *Int. J. Pharm.* **53**, 195 (1989).
28.   W. Anderson, J. Simpkins, M. Brewster and N. Bodor. *Neuroendo.* **50**, 9 (1989).
29.   N. Bodor and H. Farag. *J. Pharm. Sci.* **73**, 385 (March 1984).
30.   N. Bodor and M. Abdelalim. *J. Pharm. Sci.* **75**, 29-35 (1986).
31.   M. Brewster, K. Estes and N. Bodor. *Pharm. Res.* **3**, 278 (1986).
32.   N. Bodor, J. McCornack and M. Brewster. *Int. J. Pharm.* **35**, 47 (1987).
33.   M. Brewster, J. Simpkins and N. Bodor. *Reviews in the Neurosciences* **2**, 241 (1990).
34.   K. Fotherby, *Contraception* **31**, 367 (1985).
35.   N.M. Kaplan, *Ann. Rev. Med.* **29**, 31 (1978).
36.   S.P. Kalra and P.S. Kalra, *Endocrine Rev.* **4**, 311 (1983).
37.   B.S. Hurst and J.A. Rock, *Obstet. Gynecol. Surv.* **44**, 297 (1989).
38.   L.W. Christensen and L.G. Clemens, *Endocrinol.* **95**, 984 (1974).
39.   D.P. Faff, *J. Comp. Physiol. Phsychol.* **73**, 349 (1970).
40.   L.C. Huppert, *Med. Clin. North. Am.* **71**, 23 (1987).
41.   B.K.S. McEwen, *Science* **211**, 1303 (1981).
42.   A. Maggi and J. Perez, *Life Sci.* **37**, 893 (1985).
43.   G. Mullersman, H. Derendorf, M. Brewster, K. Estes and N. Bodor, *Pharm. Res.* **5**, 172 (1988).
44.   D.K. Sarkar, S.J. Friedman, S.S.C. Yen, and S.A. Frautschy, *Neuroendocrinology* **50**, 204 (1989).
45.   W. Anderson, J. Simpkins, M. Brewster and N. Bodor. *Pharmacol. Biochem. & Behav.* **27**, 265 (1987).
46.   M. Brewster, K. Estes, R. Perchalski and N. Bodor. *Neurosci. Letters* **87**, 277 (1988).
47.   J. Howes, N. Bodor, M.E. Brewster, K. Estes and M. Eve, *J. Clin. Pharmacol.* **28**, 951 (1988).
48.   N. Bodor and L. Prokai, *Peptides: Chemistry, Structure, and Biology*, Edmonton, Alberta, Canada, in press (1993).
49.   N. Bodor, L. Prokai, W. Wu, H. Farag, S. Jonnalagadda, M. Kawamura and J. Simpkins. *Science* **257**, 1698 (1992).
50.   L. Prokai, X. Ouyang, W. Wu and N. Bodor. *J. Am. Chem. Soc.* **116**, 2643 (1994).
51.   X. Ouyang, N. Bodor, L. Prokai, and W. Wu. *J. Med. Chem.* submitted for publication (1995).

Rational design of peptides with enhanced membrane permeability

Ronald T. Borchardt

Department of Pharmaceutical Chemistry, The University of Kansas, Lawrence, KS 66045

Abstract: A major challenge confronting medicinal chemists in the future will be the design of drug candidates having structural characteristics adequate to circumvent the biological barriers (e.g., intestinal mucosa) that often prevent their clinical development. Through rational drug design and combinatorial library strategies, medicinal chemists are capable of synthesizing very potent and very specific drug candidates. These drug candidates are developed with molecular characteristics that permit optimal interaction with the specific macromolecules (e.g., receptors, enzymes) that mediate their pharmacological effects. However, these drug design strategies, as currently practiced in many pharmaceutical companies, do not necessarily insure optimal delivery of the drug to its site of action. This manuscript describes ongoing research in our laboratory designed to elucidate the general structural features of peptides and peptide mimetics that afford optimal permeation of these types of molecules through the intestinal mucosa. In addition, we describe a novel prodrug methodology used to prepare cyclic derivatives, which increases their ability to permeate membranes and stabilizes them to metabolism by peptidases.

Introduction

Recent advances in synthetic chemistry permit production of large quantities of various peptides and peptide mimetics possessing a diverse array of pharmacological effects. The clinical development of these peptide-based drugs, however, has been restricted due to their very low oral bioavailabilities and short *in vivo* half-lives (1). Successful design of such molecules as orally available drugs is a major challenge for pharmaceutical scientists. Designing a suitable structure necessitates a balance between optimal pharmacological (e.g., receptor binding) and optimal pharmaceutical (e.g., membrane permeability, metabolic stability) properties.

The epithelium lining of the gastrointestinal tract acts as a strategic interface between the external milieu (e.g., intestinal lumen) and internal milieu (e.g., blood) of the body. This interface is both a physical (Fig. 1A) and a biochemical barrier (Fig. 1B). Biochemically, the gastrointestinal tract is designed to break down dietary proteins into subunits (e.g., peptides, amino acids) sufficiently small to be absorbed (2). Digestive processes for peptides and proteins are catalyzed by a variety of proteases and peptidases both extracellularly and intracellularly (Fig. 1B). Due to the wide substrate specificity of these enzymes, it is not surprising that the metabolic barrier is considered to be important in limiting the absorption of peptide-based drugs. Another important aspect of the biochemical barrier is the existence of apically polarized efflux systems in the intestinal mucosa (3) (Fig. 1B).

Physiologically, the intestinal epithelium also represents an important physical barrier. The organization and architecture of the intestinal mucosa, which have been extensively reviewed elsewhere (2), limit peptides to traversing the cell barrier via the paracellular and/or the transcellular route (Fig. 1A). The paracellular pathway is an aqueous, extracellular route across the epithelia that is followed by molecules according to their hydrophilicity, size and charge. The main barrier to the paracellular diffusion of molecules is the region of the tight junctions or zonula occludens. Although the degree of permeability at the tight junctions varies significantly within different epithelia, tight junctions are generally reported to be impermeable to molecules with radii larger than 11–15 Å (5).

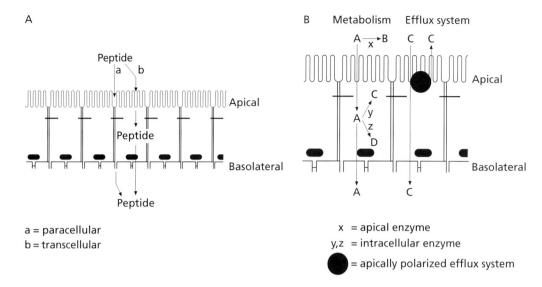

Fig. 1 Intestinal mucosa: A, physical barrier; B, biological barrier.

The transcellular pathway involves movement of the solute across the apical cell membrane, through the cytoplasm and across the basolateral membrane by either active or passive processes (Fig. 2). It is well known that di- and tripeptides are absorbed by both active, carrier-mediated processes and simple passive diffusion. Generally, active processes are fairly substrate-specific, although exceptions have been found (5). Although there is evidence that mucosal peptide/protein uptake is mediated by endocytotic processes (2), in most cases this does not lead to transcytosis. Transcellular permeation by passive diffusion requires a solute to have optimal physicochemical properties, including size, lipophilicity (hydrophobicity and hydrogen bonding potential) and conformation (6).

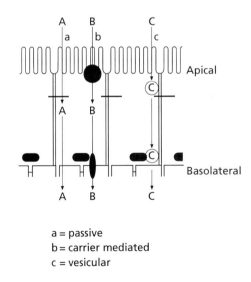

Fig. 2 Transcellular pathways for peptide transport across the intestinal mucosa.

Results and Discussion

Transcellular permeation by a passive route requires a distinct interaction of the solute with the membrane. Since the latter represents a very diverse chemical environment, the ability of a peptide to traverse this barrier by passive diffusion depends on its interaction not only with the lipid bilayer but also with many different integral and peripheral membrane proteins (6). Most of the studies performed to date have focused on the individual contributions of physicochemical properties, including size, lipophilicity (hydrophobicity and hydrogen bonding potential) and conformation, on the transcellular permeation of peptides. However, the controlling features for this route of transport of peptides are still not well understood.

Traditionally, hydrophobicity has been viewed as the most important molecular characteristic for organic molecules in determining passive diffusion through biological membranes, mainly because a membrane is simplistically considered a "lipophilic" barrier. Nevertheless, early *in vivo* data suggested that intestinal absorption may decline when hydrophobicity becomes too high (7). These results imply an "optimal" rather than a high hydrophobicity for improved transmucosal permeability of an organic molecule. In an attempt to discover whether the hydrogen bonding potential or the hydrophobicity of a peptide is the major factor in determining its permeation across the intestinal mucosa by passive diffusion, we have measured the permeability coefficients (P_e) for a series of peptide mimetics using an *in situ* perfused rat ileum model (8). The P_e values determined using the *in situ* model were compared to the transcellular permeability coefficients (Ptranscell) determined using an *in vitro* cell culture model (Caco-2) (9,10). The model peptide mimetics, which were all blocked on the N-terminal (acetyl, Ac) and the C-terminal (amide, NH_2) ends, consisted of D-phenylalanine (Phe) residues (*e.g.*, Ac-Phe-NH_2, Ac-Phe-Phe-NH_2, Ac-Phe-Phe-Phe-NH_2). To alter the degree of hydrogen bonding potential, the nitrogens of the amide bonds were sequentially methylated [*e.g.*, Ac-Phe-Phe(Me)-Phe-NH_2, Ac-Phe(Me)-Phe(Me)-Phe-NH_2, Ac(Me)-Phe(Me)-Phe(Me)-Phe-NH_2, Ac(Me)-Phe(Me)-Phe(Me)-Phe-NH(Me)]. These peptides were shown not to be metabolized in the *in situ* perfused rat ileum system. The results of the transport experiments showed a dramatic effect of incorporation of the *N*-methyl bioisostere into these peptides. For example, the P_e values for Ac-Phe-Phe-Phe-NH_2, Ac-Phe-Phe(Me)-Phe-NH_2, Ac-Phe(Me)-Phe(Me)-Phe-NH_2, Ac(Me)-Phe(Me)-Phe(Me)-Phe-NH_2 and Ac(Me)-Phe-(Me)-Phe(Me)-Phe-NH(Me) determined using the *in situ* model were 0.333, 0.569, 5.12, 9.42 and 13.3×10^{-6} cm/sec, respectively. A similar effect of incorporation of the *N*-methyl bioisostere in this series of peptides was observed when the transport studies were conducted in Caco-2 cells (9,10). The results of the transport experiments also showed that there were poor correlations between the P_e values determined in the *in situ* perfused rat ileum model and the octanol-buffer partition coefficients (r = 0.60). However, good correlations were observed between the P_e values determined *in situ* for these peptide mimetics and their partition coefficients in heptane-ethylene glycol (r = 0.96) and the differences in their partition coefficients between octanol-buffer and isooctane-buffer (r= 0.86), which are measures of hydrogen bonding potential. These results suggest that hydrophobicity may not be the major factor in determining the intestinal permeability of these peptide mimetics and that hydrogen bonding potential may be a major contributing factor. It should be noted that a good correlation (r =.940) was observed between the P_e values determined for these peptides in the *in situ* perfused ileum model and those Ptranscell values determined in the *in vitro* cell culture model (Caco-2). These results suggest that the permeability values determined in the Caco-2 cell culture model may be good predictors of the intestinal permeability of peptide mimetics. It should also be noted that Burton *et al.* (11) have recently shown that these same model peptides, which contain the *N*-methyl bioisostere, are substrates for the apically polarized efflux mechanism in Caco-2 cells (Fig. 1B). However, if the Ptranscell values for these model peptides are corrected for the effect of this efflux mechanism, a good correlation still exists between intrinsic Ptranscell (permeability value corrected for the effect of efflux mechanism) and heptane-ethylene glycol partition coefficients and the difference in partition coefficients between octanol-buffer and iso-octane-buffer (P.S. Burton, unpublished data).

The paracellular route has been of interest for the delivery of peptides and peptide mimetics because of the perception that it has limited proteolytic activity (1). Recently, our laboratory has become interested in elucidating the structural features that influence the diffusion of peptides via the paracellular route. In one study, our laboratory determined the effect of conformation flexibility on the permeation of peptides through Caco-2 cell monolayers, an *in vitro* model for the intestinal mucosa (F.W. Okumu, G.M. Pauletti, D.G. Vander Velde, T.J. Siahaan and R.T. Borchardt, unpublished data). We compared linear hexapeptides (H$_2$N-Trp-Ala-Gly-Gly-Asp-Ala-OH and Ac-Trp-Ala-Gly-Gly-Asp-Ala-NH$_2$) with a cyclic analog (cyclo(Trp-Ala-Gly-Gly-Asp-Ala)), which was covalently linked by the N-terminal and C-terminal ends. Solution structural analysis by means of 2-D NMR revealed that for the linear hexapeptides significant amounts of secondary structure (e.g., βI turns) exist in a dynamic equilibrium with unfolded solution structures. In comparison, the cyclic analog existed in well-defined conformations containing βII turns. When the permeation of the Asp-containing peptides was determined across Caco-2 cell monolayers, the cyclic peptide was shown to be *ca.* 3 times more able to permeate than the linear, protected hexapeptide H$_2$N-Trp-Ala-Gly-Gly-Asp-Ala-OH and *ca.* 55 times more able to permeate than the linear, unprotected hexapeptide Ac-Trp-Ala-Gly-Gly-Asp-Ala-NH$_2$, which was rapidly metabolized (Table 1). This implies that cyclization has stabilized the peptide to metabolism, and it may have decreased the average molecular size, accounting for the observed increase in flux.

TABLE 1. Comparison of the permeability coefficients of the linear (uncapped and capped) and cyclic DSIP turn 1 hexapeptides and the N- to C-terminal linked acyloxyalkoxycarbamate and phenyl propionic acid prodrugs of the DSIP turn 1 hexapeptide through Caco-2 cell monolayers.

Peptide	P_{app} x 10^8 cm/s	Relative Increase in Permeability
$^+$H$_3$NTrp-Ala-Gly-Gly-Asp-AlaCOO$_2^-$	< 0.17	1
AcNHTrp-Ala-Gly-Gly-Asp-AlaCONH$_2$	2.10 ± 0.50	12x
Trp-Ala-Gly-Gly-Asp-Ala HN———————C⩰O	9.27 ± 0.31	55x
Trp-Ala-Gly-Gly-Asp-Ala NH−C−O−CH$_2$−O−C⩰O (C=O)	11.65 ± 1.07 [a]	69x
Trp-Ala-Gly-Gly-Asp-Ala HN-C-CH$_2$-CR$_2$Aryl-O-C⩰O (C=O)	12.09 ± 1.24	71x

[a] corrected for chemical degradation

Proteins within the junctional complexes consist of polar amino acids with ionizable side chains. As a consequence, the junctional space exhibits an electrostatic field with a negative net charge that may affect the paracellular flux of molecules due to charge-charge interactions. Our laboratory recently demonstrated that the flux of several model peptides, Ac-Trp-Ala-Gly-Gly-X-Ala-NH$_2$ (X= Asp, Lys, Asn) and Ac-Tyr-Pro-X-Z-Val-NH$_2$ (X=Gly, Ile and Z=Asp, Asn), in Caco-2 cell monolayers did not show any discrimination based on the difference in charge. Similar effects were observed with a series of cyclic peptides cyclo(Trp-Ala-Gly-Gly-X-Ala) (X = Asp, Lys, Asn) (F.W. Okumu, G.K. Knipp, G.M. Pauletti, D.G. Vander Velde, T.J. Siahaan and R.T. Borchardt, unpublished data). It was concluded that molecular radius and not charge predominantly limits the permeation of these penta- and hexapeptides through cell monolayers. However, with smaller peptides (e.g., tripeptides), where size is not the predominant factor, charge could influence their flux via the paracellular pathway.

In an effort to alter the molecular radius of a peptide transiently and thus improve its permeation via the paracellular route, our laboratory has recently developed novel approaches for preparing cyclic prodrugs of the hexapeptide H$_2$N-Trp-Ala-Gly-Gly-Asp-Ala-OH (Fig. 3). These prodrug systems utilize either an acyloxyalkoxycarbamate pro-moiety (Fig. 3A) (S. Gangwar, G.M. Pauletti, T.J. Siahaan, D.G. Vander Velde, V.J. Stella and R.T. Borchardt, unpublished data) or a 3-(2'-hydroxy-4',6'-dimethylphenyl)-2,2-dimethyl propionic acid pro-moiety (Fig. 3B) (B. Wang, S. Gangwar, G.M. Pauletti, T.J. Siahaan, D.G. Vander Velde and R.T. Borchardt, unpublished data). The conversion of both prodrugs to the peptide was significantly more rapid in rat or human blood than in buffer, suggesting esterase-mediated hydrolysis. In addition, when permeabilities were determined in Caco-2 cell monolayers, the permeation of the prodrugs was *ca.* 5 times greater than that of the metabolically stabilized, linear hexapeptide Ac-Trp-Ala-Gly-Gly-Asp-Ala-NH$_2$ and *ca.* 70 times greater than the linear, unprotected hexapeptide H$_2$N-Trp-Ala-Gly-Gly-Asp-Ala-OH, which was rapidly metabolized and showed no transport (Table 1).

The multiplicity of barrier mechanisms in the gastrointestinal tract represents a challenge for successful oral delivery of peptides. However, intestinal absorption of biologically active peptides may be possible through an understanding of the many different mechanisms that regulate the mucosal barrier. Hence, peptides can be modified to achieve enhanced chemical and enzymatic stability and improved permeation properties. Nevertheless, each biologically active peptide must be treated individually in order to improve its permeation through the intestinal mucosa.

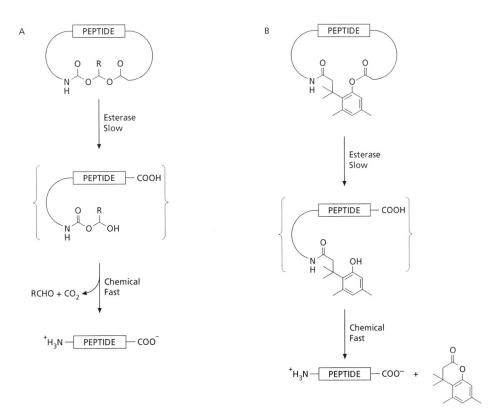

Fig. 3 Prodrug moieties for preparing cyclic peptides. A, acyloxyalkoxycarbamate pro-moiety; B, 3-(2'-hydroxy-4',6'-dimethylphenyl)-2,2-dimethyl propionic acid pro-moiety

Acknowledgments

The author's work in this area has been supported by grants from Glaxo, Inc. and The United States Public Health Service (GM-51633, GM-88539).

References

1. G.L. Amidon and H.J. Lee. *Ann. Rev. Pharmacol. Toxicol.* **34**, 321–341 (1994).
2. J.L. Madara and J.S. Trier. In L.R. Johnson (Ed.), *Physiology of the Gastrointestinal Tract*, Raven Press, New York, 1994, pp. 1577–1622.
3. G.K. Collington, J. Hunter, C.N. Allen, N.L. Simmons and B.H. Hirst. *Biochem. Pharmacol.* **44**, 417–424 (1992).
4. N.F.H. Ho, J.S. Day, C.L. Barsuhn, P.S. Burton and T.J. Raub. *J. Control. Rel.* **11**, 3–24 (1990).
5. J. Lowther, S.M. Hammond, K. Russel and P.D. Fairclough. *J. Antimicrob. Chemother.* **25**, 183–184 (1990).
6. P.S. Burton, R.A. Conradi and A.R. Hilgers, *Adv. Drug Delivery Rev.* **7**, 365–386 (1991).
7. C. Hansch and T. Fujita. *J. Amer. Chem. Soc.* **86**, 1616–1626 (1964).
8. D.C. Kim, P.S. Burton, and R.T. Borchardt. *Pharm. Res.* **10**, 1710–1714 (1993).
9. R.A. Conradi, A.R. Hilgers, N.F.H. Ho and P.S. Burton. *Pharm. Res.* **9**, 435–439 (1992).
10. R.A. Conradi, A.R. Hilgers, N.F.H. Ho and P.S. Burton. *Pharm. Res.* **8**, 1453–1460 (1991).
11. P.S. Burton, R.A. Conradi and N.F.H. Ho, *Biochem. Biophys. Res. Commun.* **190**, 760–766 (1993).

Review on the development of a polymer conjugate drug: SMANCS

Hiroshi Maeda

Department of Microbiology, Kumamoto University School of Medicine, Kumamoto 860 Japan

Abstract: SMANCS is the first approved anticancer polymer drug. It utilizes the unique character of tumor blood vessels for it tumor selective delivery. Namely, the enhanced permeability and retention (EPR) effect is a tumor selective targeting principle. SMANCS formulated in Lipiodol® (lipid contrast medium) is yielding currently unprecedented clinical efficacy against hepatoma. The rationales of the SMANCS therapy as well as side effects (which is virtually nill) are discussed. The most promising future outlook of SMANCS is against the renal cell carcinoma using Lipiodol formulation. Also a pronounced effect of SMANCS against carcinomatosis in the pleural or peritoneal cavity by intracavitary route is discussed. Intravenous administration of aqueous formulation of SMANCS under the angiotensin II induced hypertension further doubled macromolecular tumor targeting in addition to the EPR effect observed under normotensive state.

1. Introduction

Despite the emergence of numerous varieties of anticancer agents, very few are comparable to the pronounced effects and benefits of many antimicrobial agents. The reason for this can be attributed to the fact that all tumor is equipped with essentially identical biochemical and molecular metabolisms and anticancer drugs have almost no tumor-selective properties. Furthermore, we would emphasize that most of the antitumor drugs developed so far are within a limited molecular range, i.e. less than 1000 Da, not to mention of a rare capacity of protein-binding in vivo, and they are mostly only water soluble and used for intravenous injection.

SMANCS is the first polymer conjugate anticancer drug approved by Japanese Government in 1993 and it is being marketed in Japan since 1994. It is a conjugate of two chains of poly(styrene-co-maleyl-half-butylate)[SMA, Mr 1.6kDa] and a small proteinaceous antitumor agent neocarzinostatin[NCS, Mr 12k], and possesses a mass of about 16 kDa [Fig. 1] (1-5). Uniqueness of SMANCS is its high hydrophobicity thus we are able to develop an oily formulation using Lipiodol which can be best utilized as arterial injection to the tumor feeding artery. SMANCS with Lipiodol confers thus tumor selective imaging and provides accurate dose quantification depending upon the extent of SMANCS occupied area seen in X-ray CT-scan (6-10). Another unique aspect is that it can bind with albumin so that it will behave macromolecules of ~80kDa in vivo (11). This character, i.e. large size, makes the retention of SMANCS in the tumor interstitium at much higher concentration for longer period than low molecular weight drugs which readily traverse the vascular wall and reach equilibrium between blood

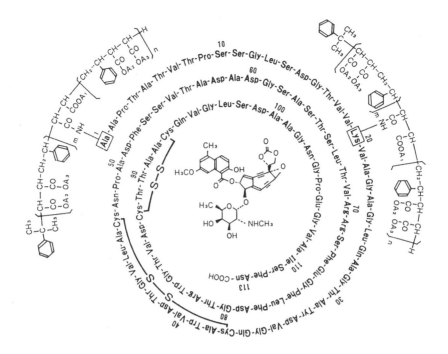

Fig. 1   Chemical structure of SMANCS

plasma and the substrates (4, 5, 12).   Thus, no tumor selective targeting is expected for these low molecular weight drugs.

2. Tumor Selective Targeting: at tumor vasculature

Most tumor vesseles are unique in numbers of aspects (2, 4, 5, 7, 12, 13):   They are mostly higher in vascular density than normal tissues and anatomical architecture is incomplete and irregular (14, 15), and mostly they do not respond to neuronal signals and missing those contractile smooth muscular layers.

During the course of the development of SMANCS, we have established that (a) arterial injection of the oily formulation is most preferable for tumor-targeted delivery of anticancer agents such as SMANCS, and (b) most biocompatible polymer conjugates or macromolecular drugs exhibit selective accumulation and long term retention (EPR effect) in tumor tissue, and their plasma half-live are usually very long (2, 4, 5, 7, 12, 16).   Both facts are found to be a universal phenomenon including soluble polymer poly(hydroxypropylmetaacrylate) with branched peptides (see ref. 16) and a term 'EPR' (enhanced permeability and retention) effect of macromolecules and lipids in the solid tumor has been introduced (2, 4, 5, 12, 13, 16) [Fig. 2].

Vascular properties exhibit a greater contrast than biochemical or molecular events among tumor or normal tissues.   However, they have not been fully utilized for clinical applications or drug targeting.   The basic mechanism for the EPR effect of macromolecules and lipids in a solid tumor is that tumor vasculature is uniquely different from that of normal tissue as shown in references (2-10, 12-16).

Namely, the following points clarify the mechanism of the EPR effect:

(i) Tumor angiogenesis results in hypervasculature particularly in tumor circumference where tumor growth is maximal, although many metastatic tumors tend

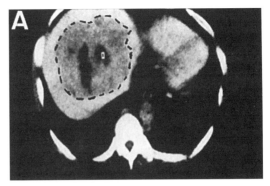

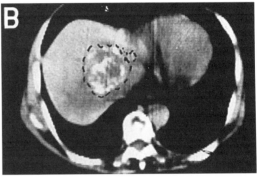

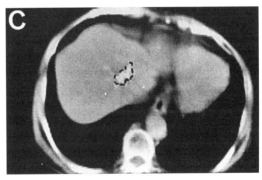

Fig. 2   CT-scan of hepatocellular
carcinoma and SMANCS treatment.
(A) Before injection of SMANCS/Lipiodol;
(B) 11 months after injection;
(C) 17 months after injection.
Tumor site are circled by dotted line.
Note white area of SMANCS/Lipiodol
which remains only in tumor.   The size
reduction in first 11 months is remarkable.

to be hypovascular towards their central areas.   High vascular density provides more
access for drugs from the circulating blood to the tumor tissue.

(ii) Tumor vasculatures become very permeable even for macromolecules and
plasma proteins (12) as well as synthetic polymer (16) due to numbers of
permeability enhancing factors such as tumor vascular permeability factor of protein
nature (16, 19), bradykinin (20-23), nitric oxide (NO) (23), prostaglandin (23), TNF
(tumor necrosis factor) and other cytokines derived from infiltrated leukocytes also
show this effect.

(iii) Ineffective functioning of the lymphatic drainage system is observed in
tumor tissue than normal tissue (2, 4, 7, 8-10), which resulting in longterm retention

of macromolecular and lipid drugs such as albumin, SMANCS, Lipiodol or HPMA polymers (2, 6, 7, 12, 13, 16, 17).

(iv) The structural incompleteness of tumor blood vessels is demonstrated (14, 15). Furthermore, there are numbers of physiological factors which facilitate extravasation or enhanced vascular permeability (EPR-effect), such as activation of type IV collagenase mediated by nitric oxide (24).

Permeability enhancing factors as described just above play important role in sustaining tumor growth and angiogenesis. NO is now known as angiogenic (25). The control of this factor will regulate tumor growth. For instance steroids are known angiostatic and inhibits induction of NOS thus suppress NO production. On the contrary, kininase inhibitors (angiotensin I convertury enzyme inhibitor) will enhance bradykinin level and thus facilitate vascular permeability by inhibiting kinase which degrade kinin (20, 23).

Other than these factors, physical elevation of blood pressure results in more enhanced tumor targeting which become possible by infusing angiotensin II (26-28).

Above points are not relevant to generally described tumor specific or associate antigens or alike but all common to vascular functions and properties of solid tumor. They are mostly forgotten or unnoticed in cancer chemotherapy.

3. Clinical Effect of SMANCS

SMANCS is approved for the clinical case in hepatoma or hepatocellular carcinoma, and its effect is most pronounced when given arterially using lipid contrast agent Lipiodol as a carrier solvent. Tumor targeting efficiency as revealed by tumor/blood ratio can be as high as 2,500 (7), thus, there is no bone marrow suppression, nor toxicity to the liver, the kidney, the intestine or the heart; nor hair loss or neuronal toxicity are observed (4, 5). Repeated use for more than ten times is possible. Usual dosage is about 4-6 mg SMANCS/4-6 ml of Lipiodol per injection for hepatoma. It is given initially once in 4 weeks, and a few~six months later, once in every two to five months (8). The details depend on each patient, susceptibility of each tumor, or initial size and spread of tumor etc. Adequate tumors filling with SMANCS/Lipiodol is essential for good response (8-10).

As revealed by tumor size reduction to less than 50% of the initial size, the response is observed in 90% of patients based on images in X-ray CT-scan at 1 year after treated, whereas it is about 50% when evaluated at 6 months although 90% show some degree of size reduction [Fig. 3]. Decrease of tumor marker such as AFP is observed in about 90% of patients (4, 5).

SMANCS treatment has advantage over other conventional methods such as surgical resection, ethanol injection, or chemoembolization. First, SMANCS/Lipiodol by arterial route can be applicable for multicentered tumor contrary to other methods (all other methods can treat only a limited tumor foci of 1~2, and restricted location (one, at most, two segmental spread) as well as limited size (<2~3 cm) range. The present method can be applicable even to the cirrhotic patients because SMANCS is the least hepatotoxic than all above methods. Embolization or chemoembolization method will frequently damages non-tumorous portion because they are not selective to tumor, whereas SMANCS/Lipiodol can be only retained in tumor and damages the tumor only. SMANCS can cover the entire liver but normal part of the liver will clear the SMANCS/Lipiodol. Thus, it will be retained only in tumor, and thus effect of SMANCS can be retain longer than 4 weeks in hepatocellular carcinoma (drug is released very slowly from Lipiodol during this period) accompanying good response rate.

As general side effects, low grade fever is most frequently (~50%) observed for a few days to a week after intraarterial (i.a.) injection but easily controlled. Bone

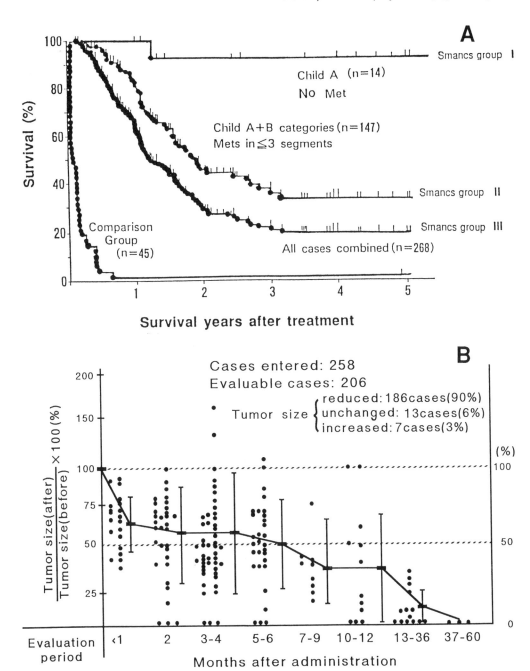

Fig. 3 (A) Survival of patients with unresectable hepatocellular carcinoma (HCC) treated with intra-arterial administration of SMANCS/Lipiodol. Comparison of survival of patients after conventional therapy and after SMANCS/Lipiodol. Survival of HCC patients classified according to the extent of liver cirrhosis (Child's classification A, B, and C). All SMANCS-treated patients are inoperable and/or at highly advanced stage. Mets in the figure means metastasis. (B) Tumor size reduction after SMANCS/Lipiodol treatment in unresectable primary hepatoma. Most patients received SMANCS/Lipiodol multiple times intrahepatic artery as observation period progress. Average patients received about 5-9 times in the first year.

marrow suppression, or toxicity to the liver, kidney, intestine, heart and other organ's are unremarkable. The shock is much less frequent than 10% which depends on the patients and also procedural skillfulness of the angiographists, but it is mostly attributed to iodine of the contrast media during the angiography.

4. Future outlook of SMANCS

Next to the hepatoma using Lipiodol formulation, I believe use of SMANCS for renal cell carcinoma (9, 29, 30), for which there is no effective remedy, will be one of the most preferred target because our pilot study have shown the efficacy of 90% response although numbers of renal cancer patients are small.   Renal cell carcinoma has high vascular density similar to hepatocellular carcinoma, thus good clinical response would result.   For this use I recommend more consideration in both higher viscosity and drug concentration of SMANCS/Lipiodol because the renal blood flow is very rapid, and it tends to be washed out.

Secondly, use of SMANCS for intracavitary application is highly recommendable; this notion is again based on the results of pilot study and pharmacokinetic data.   The results of pilot study provided >90% response as judged by Papanicolaou smear for those patients with pleural fluid or ascetic fluid originated from the carcinomatoses (31).   Ascitic or pleural fluid formation is blocked by SMANCS as well.   For other solid tumors, it may be similarly effective including that of the lung, gall bladder, bile duct, pancreas, stomach, etc., however, the use of SMANCS/Lipiodol for lung needs to be carried out cautiously because the lung cancer has usually its own feeding artery through the bronchial artery which is occasionally near to the branching of the spinal artery.   The latter is vulnerable to the ischemia and paralysis which may result from saturation of the artery with Lipiodol.   The use for colon cancer needs less viscous and lower drug concentration since damage to the intestinal wall will result in of perforation.   This is because the blood flow is very slow and the drug might cause necrotic damage.

In general intravenous injection of SMANCS became most effective when it is given under angiotensin II induced hypertension, e.g. 120 → 160 mmHg (26, 27, 28). This will not only enhance the drug targeting efficiency to tumor, but also drug delivery to the normal tissues or organs become suppressed to about 60% of normotension [Fig. 4].   Thus, this method enhancing drug delivery and therapeutic efficacy but also suppress side effect (27).   The ideal targets of this hypertension chemotherapy using aqueous formulation of SMANCS is the tumors of brain, lung and colon.   Since the drug retention of aqueous SMANCS is not as pronounced as lipid formulation, more frequent administration is required.   Recently, Suzuki et al. reported most remarkable effect of SMANCS against mouse B16 melanoma (32).

5. The Mode of Action

It has been long established for parental compound neocarzinostatin(NCS) that NCS affects on DNA metabolism, both degradation of preexisting DNA and inhibition of DNA synthesis in prokaryotes and eukaryotes as the primary target (33).   We found the same is true for SMANCS.   Non-protein chromophore [enediyene] of NCS is most responsible for its action, forming aducts with DNA (34).   More recently, superoxide radical generation is found to be catalyzed by NCS or SMANCS in the presence of NADPH and cytochrome P450 reductase (35).   In this reaction system this effect takes place at the drug concentration close to cell toxicity (i.e. $0.01 \mu g/ml$), whereas it is required to have 50 $\mu g/ml$ in cell free system).   It is also reported that it has a strong

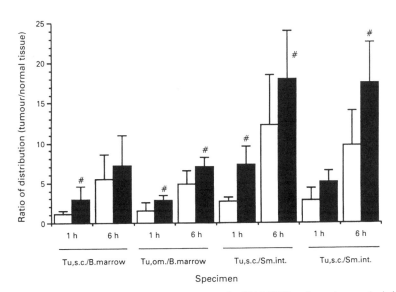

Fig. 4   Tumor-specific increase in accumulation of SMANCS after i.v. administration under angiotensin-II-induced hypertensive conditions in rats.   The concentration of SMANCS in subcutaneous tumor (Tu,s.c.) or omentum tumor (Tu,om.) was divided by that in the bone marrow (B.marrow) or the small intestine (Sm.int.) for each animal to obtain the distribution ratio of SMANCS.   SMANCS was administered i.v. under normotensive (□) or hypertensive (■) conditions. Each column shows the distribution ratio at 1 or 6 hrs after administration.   Bars indicate s.d. Every value obtained at 6 hrs was significantly higher than that at 1 hr in each condition (P<0.05, Student's *t*-test).   # indicates that the values under hypertensive conditions are significantly higher than those for the normotensive control (P<0.05, Student's *t*-test).   (From reference 23).

immunopotentiating activity such as activation of NK-, T-cells, together with induction of interferon (36-38).

References

1. H. Maeda, J. Takeshita and R. Kanamaru. *Int'l J. Pept. Protein Res.* **14**, 81 (1979).
2. H. Maeda, T. Matsumoto, T. Konno, K. Iwai and M. Ueda. *J. Prot. Chem.* **3**, 181 (1984).
3. H. Maeda, M. Ueda, T. Matsumoto and T. Morinaga. *J. Med. Chem.* **28**, 455 (1985).
4. H. Maeda, L. Seymour and Y. Miyamoto. *Bioconjugate Chem.* **3**, 351 (1992).
5. H. Maeda, *Polymeric Site-Specific Pharmacotherapy.* p. 96, John Wiley & Sons Ltd. (1994).
6. T. Konno, H. Maeda, K. Iwai, S. Tashiro, S. Maki, T. Morinaga, M. Mochinaga, T. Hiraoka and I. Yokoyama. *Eur. J. Cancer Clin. Oncol.* **19**, 1053 (1983).
7. K. Iwai, H. Maeda and T. Konno. *Cancer Res.* **44**, 2115 (1984).
8. T. Konno and H. Maeda, *Neoplasms of the Liver* p. 345, Springer-Verlag, Berlin (1987).
9. T. Konno, H. Maeda, K. Iwai, S. Maki, S. Tashiro, M. Uchida and Y. Miyauchi. *Cancer* **54**, 2367 (1984).
10. S. Maki, K. Konno and H. Maeda. *Cancer* **56**, 751 (1985).
11. K. Oka, Y. Miyamoto, Y. Matsumura, S. Tanaka, T. Oda, F. Suzuki and H. Maeda. *Pharmaceut. Res.* **7**, 852 (1990).
12. Y. Matsumura and H. Maeda. *Cancer Res.* **46**, 6387 (1986).
13. C.J. Li, Y. Miyamoto, Y. Kojima and H. Maeda. *Br. J. Cancer* **67**, 975 (1993).
14. S.A. Skinner, P.J.M. Tutton and P.E. O'Brein. *Cancer Res.* **50**, 2411 (1990).

15. M. Suzuki, T. Takahashi and T. Sato. *Cancer* **59**, 444 (1987).
16. L.W. Seymour, Y. Miyamoto, H. Maeda, M. Brereton, J. Strohalm, K. Ulbrich and R. Duncan. *Eur. J. Cancer* **31A**, 766 (1995).
17. H.F. Dvorak, J.A. Nogy, J.T. Dvorak and A.M. Dvorak. *Am. J. Pathol.* **133**, 95 (1988).
18. D.W. Leung, G. Cachiunes, W.J. Kuang, D.V. Goeddel and N. Ferrara. *Science* **246**, 1306 (1989).
19. P. Kech, S.D. Hausen, G. Krivl, K. Sanzo, T. Warren, J. Feder and D.T. Connolly. *Science* **246**, 1309 (1989).
20. Y. Matsumura, M. Kimura, T. Yamamoto and H. Maeda. *Jpn. J. Cancer Res.* **79**, 1327 (1988).
21. Y. Matsumura, K. Maruo, M. Kimura, T. Yamamoto, T. Konno and H. Maeda. *Jpn. J. Cancer Res.* **82**, 732 (1991).
22. H. Maeda, Y. Matsumura and H. Kato. *J. Biol. Chem.* **263**, 16051 (1988).
23. H. Maeda, Y. Noguchi, K. Sato and T. Akaike. *Jpn. J. Cancer Res.* **85**, 331 (1994).
24. D. Vussalli and M.S. Pepper. *Nature* **370**, 14 (1994).
25. M. Ziche, L. Morbidelli, E. Masini, S. Amerini, H.J. Granger, C.A. Maggi, P. Geppetti and F. Ledda. *J. Clin. Invest.* **94**, 2036 (1994).
26. M. Suzuki, K. Hori, I. Abe, S. Saito and H. Sato. *J. Nat'l Inst. Cancer* **67**, 663 (1981).
27. C.J. Li, Y. Miyamoto, Y. Kojima and H. Maeda. *Br. J. Cancer* **67**, 975 (1993).
28. A. Noguchi, T. Takahashi, T. Yamaguchi, K. Kitamura, A. Noguchi, H. Tsurumi, K. Takahashi and H. Maeda. *Jpn. J. Cancer Res.* **83**, 240 (1992).
29. M. Kobayashi, H. Maeda, K. Iwai, T. Konno, S. Sugihara and H. Yamanaka. *Urology* **37**, 288 (1991).
30. S. Noda, S. Konno, J. Tanaka, M. Yamada and N. Yoshitake. *Anticancer Res.* **10**, 709 (1990).
31. M. Kimura, T. Konno, T. Oda, H. Maeda and Y. Miyauchi. *Anticancer Res.* **13**, 1287 (1993).
32. F. Suzuki, M. Kobayashi, R.B. Pollard and H. Maeda. *Proceedings of the AACR* **39**, No. 2747, 461 (1995).
33. T, Ono, Y. Watanabe and N. Ishida. *Biochim. Biophys. Acta* **119**, 46 (1964).
34. I.H. Goldberg and L.S. Kappen, *Enediyne Antibiotics as Antitumor Agents*, p. 327, Marcel Dekker, Inc. New York-Basel-Hong Kong (1994).
35. K. Sato, T. Akaike, M. Suga, S. Ijiri, M. Ando and H. Maeda. *Biochem. Biophys. Res. Commun.* **205**, 1716 (1994).
36. F. Suzuki, R.B. Pollard, S. Uchimura, T. Munakata and H. Maeda. *Cancer Res.* **50**, 3897 (1990).
37. E. Masuda and H. Maeda. *J. Cancer Immunopharmacol. Immunother.* **40**, 329 (1995).
38. F. Suzuki, T. Munakata and H. Maeda. *Anticancer Res.* **8**, 97 (1988).

Modulation of the carbohydrate structures of recombinant human glycoprotein therapeutics

Harald S.Conradt, Martin Gawlitzek, Eckart Grabenhorst, Andrea Hoffmann, Manfred Nimtz, Freerk Oltmanns-Bleck and Susanne Pohl

Dept. of Gene Regulation and Differentiation, GBF - Gesellschaft für Biotechnologische Forschung, Mascheroder Weg 1; D-38124 Braunschweig, Germany

Abstract

The carbohydrate moiety of glycoproteins (*N- and O-linked glycans*) is involved in numerous molecular recognition phenomena such as host–pathogen interaction, cell–cell recognition, receptor–ligand interaction and targeting of proteins. The expression of recombinant human glycoprotein in mammalian host cells is now widely used for their production in large quantities for clinical applications. In this contribution, we summarize recent advances and discuss future prospects in the biotechnology of recombinant glycoproteins. First, we consider the genetic engineering of glycoproteins with novel or improved *in vivo* properties through site-directed mutagenesis of the corresponding genes. Next, we consider the metabolic engineering of mammalian host cells to permit the production of glycoproteins with different human tissue-type glycosylation characteristics. Lastly, we discuss the effect of modulating cell culture conditions on protein glycosylation.

The carbohydrate moieties of glycoproteins (*N- and O-linked glycans*) are involved in numerous biological phenomena (in-vivo clearance, antigenicity, tissue targeting, cellular differentiation, cell-cell recognition and inflammatory events) (1).

The expression of recombinant human glycoprotein pharmaceuticals destined for clinical application in mammalian host cells, e.g.: erythropoietin, colony stimulating factors, tissue-plasminogen activator, interferons or interleukins, therapeutic antibodies etc., is now widely used for their production in large quantities in biotechnological processes.

The use of mammalian cell lines for the production of human glycoprotein therapeutics (chinese hamster ovary cells [CHO], baby hamster kidney cells [BHK-21], mouse L cells and mouse C127 cells) is favourable since these hosts provide a polypeptide glycosylation machinery which is very similar to that of many human cells. Other expression systems e.g. bacteria, yeasts or insect cell lines infected with recombinant baculovirus vectors are disadvantageous for several reasons: bacteria lack the enzymic machinery for N- and O-glycosylation of polypeptides and in many cases the secretory human proteins are deposited in "inclusion bodies", thus requiring the development of labourious renaturation strategies for each individual protein. Yeasts and insect cells (25) do not synthesize complex-type oligosaccharides. When administered in-vivo the products derived from these latter sources are expected to be rapidly cleared from the blood-stream by recognition and internalization by specific hepatocytic cell surface receptor molecules since they lack the characteristic masking of lactosamine-type oligosaccharide chains with terminal N-acetylneuraminic acid (NeuAc).

1.   Natural Human Glycoproteins:

Glycoproteins from natural sources, e.g. human serum, cerebrospinal fluid or from primary cells exhibit cell-specific, tissue-specific and differentiation specific glycosylation characteristics (1-5). It is noteworthy to mention that all glycoprotein preparations from so-called "natural sources" comprise populations of a large number of different polypeptide "glycoforms" (1) bearing different oligosaccharide chains linked to the same polypeptide backbone and even linked to individual glycosylation domains within a single glycoprotein; this phenomenon is termed microheterogeneity (site specific microheterogeneity).

Differentiation specific glycosylation has been demonstrated by analysis of human interleukin-2 from stimulated natural peripheral T-lymphocytes (2). Immediately after induction of T cells with Ca-ionophore A 23187/phorbolester, T cells secrete a nonglycosylated form of interleukin-2. In later stages after induction (>30 hours) O-glycosylated forms of this cytokine are detectable and these forms increase up to 96 hours after stimulation until a 6:4 ratio of the unglycosylated and O-glycosylated forms is reached (2).

Tissue specific N-glycosylation of proteins: O u r laboratory has recently shown that intrathecally synthesized glycoproteins which are secreted into the human cerebrospinal fluid (CSF) contain <u>"brain-type" N-glycosylated</u> oligosaccharide side chains which differ strikingly from those generally seen in human serum glycoproteins. "Brain-type" glycosylation is characterized by high amounts of : *asialo-agalacto chains, presence of bisecting N-acetylglucosamine, quantitative proximal fucosylation and significant amounts of peripheral fucose (Lewis X and sialyl-Lewis X motifs*; see Ref 3-5, 28). In contrast to serum glycoproteins (see Fig. 1) with a high prevalence of α2-6 linked N-acetylneuraminic acid (NeuAc), brain tissue synthesizes glycans with α2-6 as well as *α2-3 linked NeuAc* in almost equal amounts (63-5). This hypothesis has been confirmed by the analysis of human ß-trace protein and the asialo-variant of transferrin present in human CSF. The asialo-variant of human CSF transferrin (also named ß$_2$-transferrin in the literature) has turned out to be in fact an asialo-agalacto-transferrin with bisecting GlcNAc and proximal fucose in contrast to the transferrin forms that are secreted by human liver cells into the serum : serotransferrin contains oligosaccharides with intact lactosamine side chains which are almost fully sialylated with α2-6 linked NeuAc, they lack proximal fucose as well as bisecting GlcNAc (4).

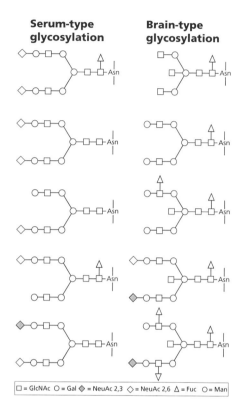

Serum-type glycosylation   Brain-type glycosylation

□ = GlcNAc ○ = Gal ◈ = NeuAc 2,3 ◇ = NeuAc 2,6 △ = Fuc ○ = Man

Fig. 1 *Serum-type versus "brain-type" glycosylation of human proteins*

major glycan structure of brain asialo-transferrin (ß$_2$-transferrin) :

GlcNAcß1-2 Manα1-6
      GlcNAc1-4 Manß1-4GlcNAcß1-4GlcNAc   αFuc1-6
GlcNAcß1-2 Manα1-3

major glycan structure of serotransferrin :

NeuAcα2-6Galß1-4GlcNAcß1-2 Manα1-6
                       Manß1-4GlcNAcß1-4GlcNAc
NeuAcα2-6Galß1-4GlcNAcß1-2 Manα1-3

2.   **Recombinant human glycoproteins:**   BHK-21 cells and CHO cells are the most frequently used host cell lines for the production of recombinant human pharmaceutical glycoproteins. Carbohydrate structural analysis of their N-linked glycans has revealed that NeuAc is exclusively attached in α2-3-linkage to Galß1-4GlcNAc-R (6-13, 16-18,28). Many human glycoproteins, however, contain NeuAc either in α2-6-linkage or as a mixture of α2-3 and α2-6-linkages. Therefore, recombinant human glycoproteins when produced from BHK-21 or CHO cells may differ from their

natural counterparts in their type of terminal sialylation. The structural characteristics of N-linked oligosaccharides that can be expected in recombinant glycoproteins from the most frequently used mammalian host cell lines are summarized in Table 1. Whilst the murine cell lines C127 and L-cells have the capability of α2-6 sialylation, these cells may be disadvantegeous for the expression of human glycoprotein therapeutics since they produce a vast array of terminal structural motifs on polypeptide-bound oligosaccharides (e.g.: Galα1-3Gal-GlcNAc-R, branched repeats, sulfated galactose, bisecting GlcNAc and considerable amounts of N-glycolylneuraminic acid against which circulating antibodies have been described in humans (14). BHK-21 cells and CHO cells exhibit very similar glycosylation characteristics (6-9,13).

Apart from the outer chain structural motifs of recombinant glycoproteins depicted in Table 1 (cell-type or host cell-specific terminal glycosylation) each given glycoprotein, even when expressed from different mammalian hosts shows high similarities in the antennarity of oligosaccharides.

Table 1   Structural characteristics of N-glycans of recombinant glycoproteins expressed from mammalian host cell lines:

Host cell line :	CHO	BHK-21	C127	Ltk⁻
proximal fucose	+	+	+	+
α2-6 NeuAc	-	-	+	+
α2-3 NeuAc	+	+	+	+
NeuGly	+	+/-	+	+
tri/tetra-antennarity	+	+	+	+
Galβ1-4GlcNAc repeats	+	+	+	+
Galβ1-3GlcNAc	-	+	-	-
sulfated glycans	+	?+	+	+
Galα1-3Gal	+	-	+	+
branched repeats	?	-	+	-
mannose-6-phosphate	+	+	?	?
bisecting GlcNAc	-	-	+	+
GalNAcβ1-4GlcNAc	-	+	-	-

a) based on structural analysis of recombinant proteins in our laboratory: human IFN-β, erythropoietin, antithrombin III, interleukin-6, "tissue plasminogen activator", β-trace protein and N-glycosylation muteins of interleukin-2 (6-8, 18, 19, 3-27)

Recombinant human erythropoietin (EPO), human tissue-plasminogen activator (t-PA) and interferon-ß (IFN-ß) are glycoproteins characterized in most detail from heterologous hosts. Human EPO produced from BHK-21 cells as well as from CHO cells contains preponderantly tetraantennary oligosaccharide chains without or with one, two or three lactosamine repeat units (8,12,13). The glycoprotein isolated from the urine of patients with aplastic anaemia contains also mostly tetraantennary glycans (12). Human IFN-ß from primary cultures of foreskin cells has 80% of biantennary carbohydrate chains. Published data on recombinant IFN-ß from several mammalian host cells (including CHO and BHK-21 cells) clearly demonstrate that biantennary chains predominate irrespective of the host cell line used for expression. These results indicate that glycosylation of proteins is polypeptide-specific. Furthermore individual glycosylation sites of glycoproteins are recognized specifically by the cellular glycosylation machinery (glycosylation site specific modification). This is examplified in the case of human t-PA expressed from several different mammalian host cells (7,9,16,17). t-PA contains four potential N-glycosylation sites. The Asn-218 is never glycosylated whereas Asn-117, Asn-184 and Asn-448 are recognized by the cellular glycosylation machinery; Asn-117 bears invariantly high-mannose type structures, Asn-184 and Asn-44 bear complex-type oligosaccharide chains when expressed from mammalian hosts. Interestingly, Asn-184 is not recognized quantitatively by the polypeptide oligosaccharide-dolichol-phosphate transferase, however, when this site has aquired a carbohydrate precursur chain it is trimmed and processed by the cellular enzymes.

O-glycosylation: In the case of O-glycosylation, we have shown that BHK-21 cells have a reduced O-glycosylation capacity when compared to CHO cells (8,25). The recognition of O-glycosylation domains in popypeptides, however, is identical and the final carbohydrate structures in all recombinant glycoproteins analysed so far have been shown to consist of either the classical NeuAα2-3Galβ1-3GalcNAc or the tetrasaccharide NeuAα2-3Galβ1-3(NeuAcα2-6)GalcNAc linked to Ser or Thr hydroxyl-groups.

3.   Modulation of recombinant protein glycosylation by cell culture conditions:

Recombinant cell lines expressing human therapeutic glycoproteins are grown routinely in large scale bioreactors e.g. in the presence or absence of fetal calf serum. Biotechnological processes have been designed which utilize cells grown either adherent to macroporous-carriers, in suspension or as suspended aggregates and production times of several weeks for a single process are routine.

It is necessary to fully characterize recombinant glycoprotein products that are obtained from controlled biotechnological processes with respect to their carbohydrate structures 8in order to ensure their product quality). Several bioprocess parameters have been shown to significantly affect glycosylation of recombinant polypeptides produced from mammalian host cells: e.g. serum/protein content of the medium, glucose levels, culture pH and growth of cells in suspension or on carriers (see 11,18 and 19 for review). NH$_4$-ions present in the culture medium have been reported to inhibit cell growth and productivity(20-22). Their effect on product quality has been studied. Ammonium ions in cell culture media are derived from thermal degradation of glutamine and have been shown to be present in 2-10 mM concentrations in standard cultivation media containing 7-15 mM glutamine. It has recently been shown that increased ammonium concentrations during cell culture are paralleled by an increase of intracellular UDP-GlcNAc and UDP-GalNAc. These nucleotide sugars are biosynthetic precursor molecules in both the cytoplasmic and the Golgi located biosynthetic pathways of polypeptide glycosylation (22).

In our work we have examined the effects of different glutamine and NH$_4$-ion concentrations, as well as glucosamine in controlled perfused BHK-21 cell cultures on the N-linked oligosaccharide structures of a recombinant human IL-2 N-glycosylation mutant protein (IL-Mu6), which was generated by site directed mutagenesis of a human IL-2 cDNA (resulting in a single AA exchange: Gln$_{100}$ to Asn).
Defined protein-free culture conditions have been used in the presence/absence of either Gln, NH$_4$Cl or glucosamine. The products were purified and then characterized by amino acid sequencing, carbohydrate structural analysis using HPAEC-PAD mapping, methylation analysis and MALDI-TOF/MS. In the absence of Gln, cells secreted glycoprotein forms with preponderantly biantennary carbohydrate chains (85%) with a higher NeuAc content. Under standard conditions in the presence of 7.5 mM glutamine N-glycans were found to be complex-type biantennary (68%) and 2,6-branched triantennary structures (33%) with 90% containing proximal α1-6-linked fucose; 37% of the Galβ1-4GlcNAc antennae were found to be substituted with α2-3 linked N-acetylneuraminic acid. In the presence of 15 mM exogenously added NH$_4$Cl, a significant and reproducible increase in tri- and tetraantennary oligosaccharides (45% of total) was detected in the secretion product. In Gln-free cultures supplemented with glucosamine, an intermediate amount of high antennary glycans was detected. The increase in complexity of N-linked oligosaccharides is considered to be brought about by the increased levels in the intracellular UDP-GlcNAc/GalNAc pool that were found both under high ammonium ion concentrations and glucosamine conditions present in the culture (22).

4.   Genetic engineering of glycoproteins: The concept of polypeptide-specific as well as glycosylation-domain-specific modification of glycoproteins with respect to antennarity and trimming/processing as described above, has led us to test this hypothesis (23+24) by using site-directed mutagenesis of model glycoproteins including single amino acid exchanges as well as creating potential novel glycosylation domains transfer of glycosylation domains from other glycoproteins and expression of these muteins in mammalian hosts (BHK-21 cells and Ltk⁻ cells).
O-glycosylation: The natural human interleukin-2 molecule contains a cluster of hydroxy-amino acids at the N-terminus (Ala-Pro-Thr-Ser-Ser-Ser-Thr-Lys-Lys-Thr$_{11}$....) where only Thr$_3$ serves as an acceptor for O-glycosylation (23-25.) We could show that O-glycosylation of human Il-2 can be prevented by a single amino acid substitution (Thr$_3$ > Ser). This Thr residue in position 3 of the polypeptide chain serves as a O-glycosylation recognition signal in the wild-type polypeptide, however, it is not recognized by the UDP-GalNAc:polypeptide α1-3GalNAC-transferase when the complete N-terminal hydroxy-amino acid cluster containing peptide is shifted to different locations of the IL-2 polypeptide chain (e.g. to position 80 or to the COOH-terminus of the IL-2 molecule) or when it is introduced at different locations of thuman IFN-ß

polypeptide chain. In contrast, when *IL-2 muteins* containing a novel artificial O-glycosylation peptide motif (Ala-Pro-Thr-Pro-Pro-Pro) are expressed from mammalian cells, they become readily O-glycosylated when introduced at locations where the wild-type O-glycosylation site of IL-2 is not recognized by the pertinent enzymes (23+24). This is also the case when the Ala-Pro-Thr-Pro-Pro-Pro.. sequence is inserted into different regions of the human IFN-ß protein. All of the muteins analyzed so far bear the above mentioned Galß1-3GalNAc core oligosaccharide with one or two NeuAc.

Thus, peptide motifs can be generated that are recognized by the O-glycosylation apparatus of mammalian cells and allow for the creation of muteins with O-glycosylated human-type carbohydrate chains.

N-glycosylation: N-glycosylation of polypeptides requires the consensus tripeptide recognition sequence Asn-Xxx-Thr/Ser. N-linked carbohydrate binding domains are generally located in loop-regions of polypeptides (Hecht and Conradt, unpublished observations). The Asn-GlcNAc bond prevents formation of e.g. α-helical structures of proteins. It is noteworthy, that some human glycoproteins are incompletely N-glycosylated when expressed from recombinant cells, e.g. human interferon-y and tissue-plasminogen activator. This can readily be explained by a competition between the kinetics of protein folding immediately after the growing polypeptide chain enters the rough endoplasmatic reticulum and the activity of ER resident oligosaccharidyl transferase which may be occupied by the synthesis of cell specific glycoproteins.

Individual glycosylation domains of recombinant glycoproteins are N-glycosylated with oligosaccharides of characteristical antennarity /trimming irrespective of the cell line used for their expression (see above).

As exemplified in Figure 2 for glycosylation sites of EPO, IFN-ß and antithrombin III, 9-17 amino acid peptides comprising individual glycosylation domains can be successfully transferred from one protein to another resulting in predictable modification of the acceptor protein when expressed in mammalian cells. The site specific glycosylation characteristics (antennarity) are conserved in the genetically engineered glycoprotein.

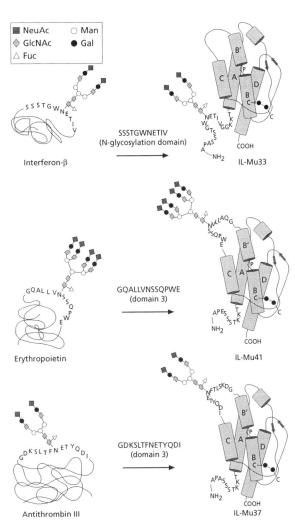

Fig. 2. *Transfer of N-glycosylation domains from one polypeptide to another. The schematic drawing of IL-2 was generated combining NMR-structural data (Bassan et al.) and an X-ray structural model published by Brandhuber et al.*

5. Metabolic engineering of mammalian host cell lines:

Construction and characterisation of mammalian cell lines with novel glycosylation properties for the expression of recombinant "human-type glycosylated" proteins:

Most of the mammalian host cell lines currently being used for the expression of secretory recombinant human pharmaceutical glycoproteins (CHO and BHK-21 cells) lack the capability of α2-6 sialylation as well as α1-3 fucosylation of Galß1-4GlcNAc-R motifs on N-glycans of polypeptides (see Table 1). However, α2-6-linked NeuAc has been shown to be the major linkage type of this monosaccharide in carbohydrates of a large number of "serum-type" glycoproteins isolated from natural sources (e.g. human interferon-γ, human antithrombin III, hemopexin, transferrin). Fucose attached in α1-3 linkage to N-acetyglucosamine in Galß1-4GlcNAc-R motifs constitutes the Lewis-X antigen and in its sialylated form, the sialyl-Lewis X motif. This structural motif when bound to cell surface glycoproteins is known to be implicated in many normal and pathological cell-cell recognition phenomena (1).

In a primary effort to obtain stable BHK-21 cell lines expressing human-type glycosylation characteristics we have cotransfected cells whith plasmids encoding the human secretory glycoprotein ß-trace (ß-TP) and the human CMP-NeuAc: Galß1-4GlcNAc-R α2-6 sialyltransferase (ST6N). The recombinant ß-TP was purified from supernatants of cells grown in spinner culture and was subjected to carbohydrate structural analysis of the N-glycans. The enzymatically liberated oligosaccharides were found to consist of 90% of biantennary chains as is the case for natural ß-TP from human cerebrospinal fluid. About 90% of the total oligosaccharides were recovered in the monosialo and disialo fractions in a ratio of 1:5. After HPLC-separation on NH$_2$-bonded phase and subsequent methylation analysis, the monosialylated oligosaccharides of ß-TP coexpressed with human ST6N were found to contain NeuAc in α2-6 and α2-3-linkage in the same ratio. From ^{1}H-NMR analysis as well as calculations of peak areas obtained by HPLC, 60% of the molecules of the disialo fraction were found to contain NeuAc in both α2-3- and α2-6- linkage to Galß1-4GlcNAc-R, whereas 40% of the molecules of this fraction contained NeuAc in only α2-3linkage to Galß1-4GlcNAc-R. The α2-6-linked NeuAc was shown to be attached exclusively to the Galß1-4GlcNAcß1-2Manα1-3 branch of the biantennary structure. Therefore the in vivo specificity of the newly introduced recombinant ST6N observed in this study supports the previously reported in-vitro branch specificity of bovine colostrum ST6N activity. Our studies demonstrate the sutability of genetically engineered mammalian host cell lines with <u>novel glycosylation properties</u> for the production of "human-typ glycosylated" recombinant polypeptides. The transfection of BHK or CHO cells with genes encoding human fucosyltransferase VI or GlcNAc-transferase III should result in host cell lines enabling the production of recombinant glycoproteins with the Lewis X or sialyl-Lewis X motif as well as with brain-type glycosylated structures (see chapter 1).

6.   Carbohydrate mapping and structural analysis of oligosaccharides: Since each individual host cell line exhibits its own intrinsic glycosylation characteristics for a given recombinant glycoprotein product, there is a demand for the detailed description of the producer cell line and a careful quality control of the final glycoprotein preparations to be used in patients. Considerable progress has been made during the past few years in the improvement of the biological/putative clinical properties of recombinant glycoprotein therapeutics from recombinant mammalian sources.

<u>H</u>igh-<u>p</u>H-<u>a</u>nion-<u>e</u>xchange <u>c</u>hromatography combined with <u>p</u>ulsed <u>a</u>mperometric <u>d</u>etection provides a versatile tool for the rapid mapping of N-linked oligosaccharides from proteins (see article by R.R.Townsend in Ref.26) and has simplified carbohydrate analysis of glycoproteins (26-27).

A further potent development in glycan analysis is <u>matrix assisted laser desorption/ionisation time of flight mass spectrometry</u> (MALDI-/TOF-MS; see Figure 3, A+B) of permethylated N-linked oligosaccharides.

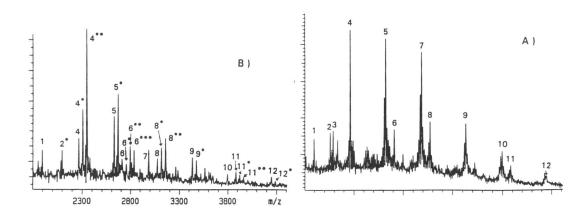

Fig, 3.   *MALD/TOF-MS of oligosaccharides from a genetically engineered model glycoprotein expressed in BHK-21 A and BHK-21B cells. Peak numbers are explained in the text.*

Positive ion MALDI mass spectra of the total reduced and permethylated N-glycans from a purified Interleukin-2 variant expressed in BHK-B cells are shown in Figure 2A. In panel B total N-glycans from the same Interleukin-2 variant expressed in BHK-A cells are shown. The structures (with the exception of structure **1** all glycans are "complex type" and contain proximal fucose, but not structures 1+3) are proposed based on the molecular weight (all oligosaccharides show predominantly Na^+ adducts with small amounts of K^+ adducts present) and methylation data. The number of asterisks indicates the number of GalNAcß1-4 versus Galß1-4 exchanges in the glycans from BHK-A cells. From the results it is obvious that there is only low sialyltransferase active in BHK-21 cells that sialylate GalNAcß1-4GlcNAc-R motifs ; **1**: high mannose type (Man_6, phosphorylated) = 1896 Da; **2**: diantennary asialo-agalacto (1Gal) = 2058 Da; **3**: diantennary asialo = 2088 Da; **4**: diantennary asialo = 2262 Da; **5**: diantennary monosialo = 2623 Da; **6**: triantennary asialo = 2711 Da; **7**: diantennary disialo = 2985 Da; **8**: triantennary monosialo = 3073 Da; **9**: triantennary disialo = 3434 Da; **10**: triantennary trisialo = 3796 Da; **11**: tetraantennary disialo = 3883 Da; **12**: tetraantennary trisialo = 4244 Da.

Conclusions:
Virtually every step in biochemistry of glycoconjugates is undergoing a rapid development at present.

References:
1. Rademacher, T.W., Parekh, R.B., Dwek, R.A. (1988) Ann.Rev. Biochem., **57**, 785-838
2. Conradt, H.S., Hauser,H.,Lorenz,C.,Mohr,H. and Plessing,A.(1988) Biochem.Biophys.Res.Commun.,**150**, 97-103
3. Hoffmann,A.,Nimtz, M.,Wurster,U. and Conradt, H.S.(1994). J.Neurochem.,**63**, 2185-2196
4. Hoffmann, A., Nimtz, M., Getzlaff,R. and Conradt, H.S. (1995) FEBS Lett., **359**, 164-168
5. Hoffmann, A., Conradt, H. S., Gross, G., Nimtz, M.& Wurster, U.(1993). J.Neurochem. 61,451-456
6. Zettlmeißl,G.,Conradt,H.S.,Nimtz,M. and Karges,P.(1989).J.Biol.Chem., **264**, 21153-21159
7. M.Nimtz, G.Noll, E.-P. Pâques and H.S.Conradt (1990). FEBS Lett. **271**, 14-18
8. Nimtz, M., Wray, V., Augustin,M., Klöppel,K.-D., Conradt, H.S. (1993). Eur.J.Biochem, **213**, 39-56
9. Spellman, M.W. et al.; 1989.J. Biol. Chem. **264**: 14100-14111
10. Cumming, D.A. 1991. Glycobiology **1**: 115-130
11. Goochee, C.F., Gramer, M.J., Andersen, D.C., Bahr, J.B., Rasmussen, J.R. 1991. Bio/Technology, **9**: 1347-1355
12. Sasaki, H., Bothner, B., Dell, A., Fukuda, M. 1987.J. Biol. Chem. **262**: 12059-12076
13. Tsuda, E., Goto, M., Akai, K., Ueda, M., Kawanishi, G., Takahashi, N.,1988.Biochemistry. **27**: 5646-5654
14. Hokke, C.H., Bergwerff,A.A.,van Oostrum,J.,Kamerling, J.P., Vliegenthart, J.F.G. 1990. FEBS Lett., **12**: 9-11
15. Conradt,H.S.,Egge,H.,Katalinic,J.and Siklosi,T.(1987) J.Biol.Chem,**262**,14600-14605
16. Parekh, R.B., Dwek, R.A., Thomas, J.R., Opdenakker, G., Rademacher, T.W., Wittwer, A.J., Howard, S.C., Nelson, R., Siegel, N.R., Jennings, M.G., Harakas, N.K., Feder, J. 1989. Biochemistry **28**: 7644-7662
17.Parekh, R.B., Dwek, R.A., Rudd, P.M., Thomas, J.R., Rademacher, T.W., Warren, T., Wun, T.-C., Herbert, B., Reitz, B., Palmier, M., Ramabhadran, T., Tiemeier, D.C. 1989. Biochemistry **28**: 7670-7679
18. Gawlitzek, M., Conradt, H.S. and Wagner, R. Biotechnol. Bioeng., **46**, 536-544 (1995)
19. Gawlitzek,M.,Valley,U.,Wagner,R.and Conradt,H.S.; J.Biotechnology, in press (1995);
20. McQueen, A., Bailey, J.E. 1991. Bioprocess Engineering **6**: 49-61.
21. Martinelle, K., Häggström, L. 1993. J. Biotechnol. **30**: 339-350.
22. Ryll, T., Valley, U., Wagner, R. 1994.Biotechnol. Bioeng. **44**: 184-193
23. Dittmar,K.E.J.,Conradt,H.S.,Hauser,H.,Hofer,B.,Lindenmaier,W.1988.In: Blöcker, H.,Collins, J.,Schmid,R.D.,Schomburg,D.(eds.)"Advances in Protein Design";VCH Publishers,Weinheim, Germany.,**12**: 145-156
24. H.S.Conradt, B.Hofer and H.Hauser (1990).Trends in Glycosience and Glycotechnology, **2**, 168 - 181
25. H.S.Conradt, K.E.J.Dittmar, W.Lindenmaier, M.Nimtz and H.Hauser (1989) J.Biol.Chem., **264**, 17368-17373
25. Grabenhorst,E.,Hofer,B.,Nimtz,M.,Jäger,V. &Conradt,H.S. (1993). Eur.J.Biochem., **215**, 189-197
26. Conradt, H.S.(editor)(1991). Protein Glycosylation:Cellular,Biotechnological and Analytical Aspects;GBF Monographs Vol.15,VCH publ. Weinheim,New York

27. Hermentin,P.,Nimtz,M.,Seidat,M.,Witzel,M.,Kammerling,H.,Vliegenthart,J.H.F.and Conradt,H.S.(1992).Anal.Biochem.,**203**, 281-289
28. Grabenhorst,E.,Hoffmann,A.,Nimtz,M.,Zettlmeißl,G.&Conradt,H.S.(1995) Eur.J. Biochem., in press
29. Maiorella, B.L. 1992. In vitro management of ammonia's effect on glycosylation of cell products through pH control. United States Patent Pat. Number 5,096,816

Acknowledgements: This work was supported by a grant from the EU (BIO 2-CT94-3069) to H.S.Conradt (Braunschweig) and Prof.Dr. André Verbert (University of Lille). We are grateful to Susanne Pohl, Christine Karger and Christiane Proppe for excellent technical assistance.

Role of sphingoglycolipids in signal transduction, and their medicinal application for control of tumor progression and inflammatory processes

Sen-itiroh Hakomori and Yasuyuki Igarashi

The Biomembrane Institute, 201 Elliott Ave W, Seattle, WA 98119; and Depts. of Pathobiology and Microbiology, University of Washington, Seattle, WA 98195.

Abstract

Sphingoglycolipids (SGLs), particularly GM3 and its primary degradation products (lyso-GM3 and de-N-acetyl-GM3), are capable of inhibiting or promoting tyrosine kinase-linked growth factor receptors. Sphingosine (Sph) derivatives, N,N-dimethyl- and N,N,N-trimethyl-Sph, inhibit protein kinase C in a stereospecific manner. Sph-1-phosphate inhibits cell motility through inhibition of actin nucleation. Fatty aldehyde conjugates of psychosine (galactosyl-Sph), collectively termed "plasmalopsychosine," activate tyrosine kinase associated with nerve growth factor (NGF) receptor, leading to neuronal differentiation. SGLs are therefore useful medicinal reagents for inhibition of tumor growth and metastasis, inflammatory responses, and perhaps for correction of neuronal disorders. We describe successful application of: (i) liposomes containing GM3 and Gg3, or TMS and Sph-1-P, for inhibition of metastasis; and (ii) TMS for prevention of myocardial reperfusion injury.

1. Introduction

Sphingoglycolipids (SGLs) are common membrane components in all types of animal and plant cells. Structures of carbohydrates linked to ceramide (Cer) vary extensively but can be classified into lacto-series type 1 and 2, globo-series, and ganglio-series. Recent studies indicate that SGLs are involved in cell-cell and cell-substratum interactions, and in control of transmembrane signaling. Here we will present an overview of SGL-dependent modulation of transmembrane signaling and growth control, and describe our recent studies of medicinal application of SGLs or sphingolipid derivatives for inhibition of tumor metastasis and inflammatory processes.

2. SGL-dependent modulation of transmembrane signaling

Function of SGLs in cell adhesion/recognition depends on the carbohydrate moiety. In contrast, function of SGLs in signal transduction depends on the entire SGL structure, including the Cer moiety. Some signal transducer molecules are highly susceptible to sphingosine (Sph) and its derivatives, including Cer. Various signaling pathways initiated by cell surface receptors and leading to activation of transducer molecules have been identified, but the interrelationships among them remain unclear. Three types of receptor system elicited by signaling molecules are known: (i) ion-channel linked; (ii) G-protein linked, and (iii) enzyme-linked (tyrosine kinase-dependent).

Ion-channel-linked receptors include Na^+/K^+ ATPase associated with Na^+/K^+ pump, synaptic signaling between electrically excitable cells, acetylcholine receptors at the neuromuscular junction, etc. The observed susceptibility of ion transport modulator in kidney cells to GM1 ganglioside (1), and susceptibility of Ca^{2+}-dependent kinase to GD1a (2) and polysialoganglioside-dependent myelin protein (3), suggest that ion-channel linked receptors in general are modulated by gangliosides. Further extensive studies along this line are needed.

Signaling via G-protein-linked cell surface receptor is the most frequently-observed system. Unfortunately, almost no clear studies on susceptibility of this type of receptor to SGL or ganglioside have been performed. One exception: β-adrenergic receptor, which is typical of this class, was reported to depend on unknown ganglioside(s) (4,5). Consequent to binding of signaling ligand to G-protein-linked receptor, activation of G_s through binding to GTP induces activation of various enzyme systems (e.g. adenyl cyclase and phospholipase C) which play crucial roles in a large variety of subsequent signaling events. Susceptibility to gangliosides or SGLs of these events subsequent to G-protein activation has not been carefully investigated, and obviously requires extensive future study. Sph-1-phosphate (Sph-1-P)-dependent activation of DNA synthesis in Swiss 3T3 cells was reported to be inhibitable by addition of pertussis toxin, which activates G-protein (6). This finding suggests that G-protein is involved in regulation of the mitogenic effect of Sph-1-P.

Activation of adenyl cyclase following G-protein activation leads to activation of protein kinase A (PKA). Activation of phospholipase Cγ leads to protein kinase C (PKC) activation via production of diacylglycerol (DAG). Many signaling pathways are opened through PKA and PKC. PKC is highly susceptible to not only DAG but also SGL and Sph derivatives.

Among these signaling system, that initiated by enzyme-linked (tyrosine kinase-dependent) receptor is the type most intensively studied in terms of susceptibility to gangliosides, their derivatives, SGLs, and other sphingolipids. In this system, several steps are modulated by these compounds: (i) All receptor-associated tyrosine kinases so far tested are susceptible to gangliosides. (ii) Some receptor kinases are inhibited or promoted by primary degradation products of gangliosides, or SGL derivatives. (iii) Sph and N,N-dimethyl-Sph (DMS) inhibit MAP kinase, probably through inhibition of either PKC or Raf kinase. (iv) *src, ras* and a few unidentified kinases are strongly stimulated by Sph and DMS.

In the following subsections, we will describe specific patterns of kinase modulation by specific SGLs.

Modulation of receptor-associated tyrosine kinases by gangliosides. GM3 inhibits epidermal growth factor (EGF) receptor kinase and formation of tyrosine phosphate (7). In this case, GM3 may directly inhibit tyrosine phosphate, rather than inhibiting receptor-receptor interaction (8). Exogenous addition of GM3 inhibits EGF-dependent growth of A431 cells and KB cells (7), even though GM3 already exists in these cells. Interpretation of the physiological role of GM3 in EGF-dependent growth of these cells is difficult, since GM3 is present in many cell types regardless of their susceptibility to EGF stimulation. One possibility is that EGF induces membrane perturbation, "flip-flop" reorganization of GM3 occurs, and cytoplasmic GM3 then directly interacts with EGF receptor. Alternatively, EGF receptor activation may stimulate phospholipase A, resulting in production of lyso-phosphatidylcholine (lyso-PC) which, together with GM3, shows a synergistic inhibitory effect on EGF receptor kinase (9).

Further evidence that GM3 modulates growth factor-dependent cell growth was provided by two different experiment, one by Weis and Davis (10) using an epimeraseless mutant of Chinese Hamster Ovary cells (LA1D), and another by us using GM3 mutant FUA169 derived from F28-7 cells (11). LA1D was incapable of converting UDP-Glc to UDP-Gal. Therefore, in the absence of Gal, the cells could not synthesize LacCer or GM3. On addition of Gal in culture medium, the cells became capable of synthesizing LacCer or GM3. This mutant did not have EGF receptor and did not show EGF-dependent cell growth. After transfection of EGF receptor gene, LA1D-EGFR+ cells were isolated. These cells grew well in the presence of EGF, and showed EGF receptor tyrosine phosphorylation. When the LA1D-EGFR+ cells were grown in the presence of Gal, whereby GM3 was synthesized, EGF-dependent cell growth and EGF receptor tyrosine phosphorylation were both inhibited.

GM3 mutant FUA169 was isolated from F28-7 cells. FUA169 did not show insulin-dependent cell growth. Mouse insulin receptor is highly sensitive to GM3, and insulin-dependent growth of F28-7 cells was strongly inhibited in the presence of GM3 (11). However, in human cells, insulin receptor is most strongly inhibited by sialosyl-paragloboside (SPG) (12).

Fibroblast growth factor (FGF)-dependent growth of BHK cells was observed to be inhibited by exogenous addition of GM3. GM3-enriched BHK cells became refractive to growth stimulation by FGF, and internalization of FGF was completely blocked (13). Further studies on the effect of GM3 on FGF receptor function are needed. It is clear that FGF does not interact directly with GM3.

Platelet-derived growth factor (PDGF)-dependent cell growth and receptor tyrosine phosphorylation are inhibited by GM1 (14), GD1a, GD1b, and GT1b (15). These gangliosides do not directly interact with PDGF receptor, and do not alter affinity of the receptor for PDGF binding (14). 2→3 SPG (IV^{3}NeuAcnLc$_4$Cer) inhibits tyrosine kinase of the 95 kDa subunit associated with insulin receptor. Other gangliosides including 2→6 SPG did not have this effect. Insulin-dependent growth of K562, HL60, and IM9 cells was inhibited by exogenous addition of 2→3 SPG, whereby phosphorylation of the 95 kDa subunit was also blocked (12). In contrast to human cells, insulin-dependent growth of mouse cells appears to be highly susceptible to GM3. For example, GM3 inhibits insulin-dependent growth of mouse mammary tumor cell line FM3A/F28-7 (11).

A synergistic enhancing effect of GM1 on nerve growth factor (NGF)-dependent NGF receptor tyrosine kinase was recently reported in association with the differentiation-inducing effect of GM1 in the presence of NGF on PC12 cells (16). GM1 alone has no effect, in contrast to plasmalopsychosine (see following section).

Enhancing effects of novel SGL derivatives on receptor kinase activity, cell proliferation, and differentiation. A primary degradation product of GM3 (by elimination of the N-acetyl group of sialic acid) is de-N-acetyl-GM3 (deNAcGM3), which is detectable in various types of cultured cells by application of its specific antibody DH5. DeNAcGM3 is present naturally in minimal quantity, but its exogenous addition promotes EGF receptor phosphorylation (of serine as well as tyrosine) and enhances cell proliferation (8,17). DeNAcGM3 is resistant to sialidase and is not further degraded. It may be re-N-acetylated into GM3. It is possible that a balance between de- and re-N-acetylation is involved in control of cell proliferation. De-N-acetylation may occur not only for GM3 but also other gangliosides. Recently, deNAcGD3 was detected in human melanoma cells. De-N-acetylation of GD3 to deNAcGD3 was inhibited by Genistein, suggesting that formation of deNAcGD3 is controlled by tyrosine kinase (18). Identification of target molecules that regulate cell proliferation and are controlled by deNAcGD3 has not been achieved.

A novel SGL termed plasmalopsychosine was isolated and characterized from human brain white matter (19). This compound is the conjugate of plasmal through 4,6 or 3,4 cyclic acetal linkage to β-Gal residue of psychosine. Plasmalopsychosine is found exclusively in white matter of brain, but not in gray matter. In contrast to all other complex lipids, plasmalopsychosine has a unique structure with two aliphatic tails oriented in opposite directions. It strongly induces differentiation in neuronal cells leading to neurite outgrowth, in the same way as NGF. Neuritogenic differentiation of most neuronal cells is induced by NGF, but rarely by SGLs alone. However, plasmalopsychosine by itself, without NGF, induces neuritogenic differentiation. In other words, it mimics NGF activity. When PC12 cells were incubated with plasmalo-psychosine, tyrosine kinase associated with NGF receptor (P140^{trk}) was strongly activated. No other SGL is capable of activating P140^{trk}. This effect is the same as when PC12 cells are incubated with NGF. MAP kinase (MAPK) is also strongly and immediately activated when PC12 cells are incubated with plasmalopsychosine, and the activity is sustained for a long period. This enhancement of MAPK occurred within a few minutes of incubation, similar to the effect of NGF. No other SGL or growth factor showed such activation of MAPK. The effect of plasmalopsychosine may not directly involve upon activation of NGF receptor and P140^{trk}, since binding of [^{125}I]NGF to its receptor was not inhibited by plasmalopsychosine. Yet plasmalo-psychosine enhanced P140^{trk} activity to the same extent as NGF, leading to strong sustained enhancement of MAPK. Thus, plasmalopsychosine may have an independent receptor which interacts with the lipid bilayer and causes membrane perturbation, activating tyrosine phosphorylation of P140^{trk}.

Effects of gangliosides and sphingolipids on protein kinase C. PKC is a ubiquitous molecule present in essentially all living cells, from bacteria through higher animals and plants. It plays a key role in signal transduction. The direct substrate of PKC appears to vary extensively from one cell type to another. Polysialogangliosides strongly inhibit PKC in HL60 cells (20), and the effect of gangliosides on PKC is probably a general phenomenon. "Lyso-gangliosides" were claimed to strongly inhibit PKC in brain (21). Lyso-GM3 inhibits PKC in A431 cells (22). In 1986, Hannun et al. observed that Sph inhibits brain PKC; stereoisomeric Sph specificity was not observed (23). Sph is converted to DMS by methyltransferase in brain. DMS is a more effective inhibitor of PKC compared to Sph, and displays stereospecific effect, i.e. D-erythro-DMS was a stronger inhibitor than L-threo, L-erythro, or D-threo isomers (24).

Enhancement of signal transduction by PKC is often observed in cancer cells, and in activated platelets or endothelial cells in association with acute or chronic inflammatory response. DMS was effectively applied for inhibition of tumor growth (25). *N,N,N*-trimethyl-Sph (TMS), an analog of DMS, was synthesized and found to be an even more effective PKC inhibitor and to give a stable aqueous solution, unlike DMS. TMS was applied for inhibition of tumor cell metastasis (26) and inflammatory response (27). TMS and DMS effectively inhibit P-selectin expression on platelets (28) and endothelial cells (27); this is the basis of the anti-inflammatory effect of these compounds.

3.   Medicinal application of SGLs for inhibition of tumor metastasis

Tumor cell (TC) metastasis is initiated by selective adhesion of TCs to target structures such as basement membrane and endothelial cells (ECs), followed by transvascular migration of TCs. Two adhesion systems have been considered for TC-EC interaction; one based on carbohydrate-carbohydrate interaction, and the other based on recognition of TC carbohydrates by selectins expressed on ECs (29). Signaling for selectin expression on ECs, mediated by PKC, is highly susceptible to sphingolipids, particularly TMS. Here we will describe two examples involving (i) use of GM3 or Gg3 liposomes for inhibition of adhesion of TCs to ECs, resulting in inhibition of metastasis; (ii) application of TMS and Sph-1-P to inhibit EC or platelet signaling and thereby prevent TC-EC or TC-platelet interaction, with consequent inhibition of metastasis.

GM3/Gg3 liposomes. Variants of murine B16 melanoma having different metastatic potentials (in the order BL6≥ F10> F1> WA4) have been characterized by the same decreasing order of cell surface GM3 expression level, relative adhesiveness to nonactivated ECs, and relative degree of GM3-dependent adhesion to Gg3Cer- or LacCer-coated plates. Degree of integrin-dependent cell adhesion and adhesion to IL-1-activated ECs was similar for BL6, F10, and F1. These results suggest that metastatic potential of these B16 variants is closely dependent on relative adhesion to nonactivated ECs, which is based on GM3-Gg3Cer or GM3-LacCer interaction. This possibility has been supported by further studies showing that blocking of GM3-dependent melanoma adhesion by μM-order concentrations of GM3 or Gg3Cer in liposomes, or by sialidase treatment of melanoma cells, strongly inhibited BL6 metastasis to lung. Paragloboside or SPG did not affect GM3-dependent BL6 cell adhesion and did not inhibit metastasis. Spontaneous metastasis from subcutaneously-grown tumors was significantly reduced if GM3- or Gg3Cer-liposomes were intravenously injected during tumor growth (30). Thus, blocking of TC adhesion to nonactivated ECs based on carbohydrate-carbohydrate interaction may provide effective anti-adhesion therapy against tumor progression, in analogy to the anti-metastatic effect produced by blocking of integrin-dependent cell adhesion.

TMS/Sph-1-P liposomes. DMS and TMS have been shown to stereospecifically inhibit activity of PKC and other kinases essential for active proliferation of TCs, as well as for activation of platelets and ECs. DMS and TMS thereby inhibit tumor growth *in vivo*, and TMS inhibits *in vivo* metastatic potential of B16 melanoma cells. When TMS was administered in liposomes, its drug efficacy was increased and its undesirable side-effects were greatly reduced (31). Sph-1-P, long known as the initial catabolite of Sph metabolism, has aroused considerable interest recently because of its inhibitory effect on cell motility (32). We found that liposomes containing both TMS and Sph-1-P, in comparison to liposomes containing TMS or Sph-1-P alone, exert a much stronger inhibitory effect on B16 melanoma cell metastasis (33). This is ascribable to their inhibitory effect on TC invasiveness through motility inhibition, in conjunction with the previously-observed inhibitory effect of TMS on activation of platelets and ECs. Furthermore, the liposomal formulation resulted in prolonged circulation time of both TMS and Sph-1-P in blood, and consequent higher concentration of these compounds in tumor tissues.

4.   Medicinal application of sphingolipid derivatives for inhibition of inflammatory processes

TMS, a synthetic analogue of naturally-occurring DMS, shows a stronger inhibitory effect on PKC, resulting in phenotypic changes in ECs, neutrophils, and platelets. The secretory response of ECs and platelets (*i.e.* translocation to the cell surface of Weibel-Pallade bodies in ECs and α-granules in platelets) is strongly inhibited by 1-5 μM TMS. Thus, P-selectin expression, a crucial step in initiation of inflammatory responses, is effectively inhibited by TMS (27,28).

In view of these findings, TMS was investigated in a feline model of myocardial ischemia (90 min) and reperfusion (270 min) injury. TMS (60 μg/kg), administered intravenously 10 min before reperfusion, significantly reduced myocardial necrosis (15 ± 3 vs. $31 \pm 4\%$ necrosis of area at risk; $p < 0.01$) and cardiac myeloperoxidase activities, a marker of neutrophil accumulation, compared with vehicle-treated cats. Endothelium-dependent relaxation to acetylcholine in ischemic-reperfused coronary artery rings treated with TMS was also significantly preserved compared with vehicle (73 ± 4 vs. $34 \pm 4\%$ vasorelaxation; $p < 0.01$). Polymorphonuclear neutrophil (PMN) adherence to coronary endothelium 270 min after reperfusion was greatly reduced in the TMS group compared with vehicle-treated cats (37 ± 5 vs. 76 ± 5 PMN/mm^2; $p < 0.01$). TMS also blocked up-regulation of P-selectin on coronary venular endothelium by immunohistochemistry (27). These *in vivo* results were consistent with *in vitro* findings that TMS reduces PMN adherence to thrombin-stimulated coronary endothelium and P-selectin up-regulation on thrombin-stimulated cat platelets. TMS at physiological concentrations exerts cardioprotective effects and preserves coronary endothelial function following myocardial ischemia and reperfusion *in vivo*. These effects are apparently mediated by inhibition of PMN-endothelial interaction and subsequent accumulation into the ischemic myocardium. TMS may therefore be a useful agent for reducing myocardial reperfusion injury.

References
1.   S. Spiegel, J.S. Handler, P.H. Fishman, J. Biol. Chem. **261**, 15755 (1986).
2.   J.R. Goldenring, L.C. Otis, R.K. Yu, et al, J. Neurochem. **44**, 1229 (1985).
3.   K.-F.J. Chan, J. Biol. Chem. **263**, 568 (1988).

4. M. Chorev, A. Feigenbaum, A.K. Keenan, et al, <u>Eur. J. Biochem.</u> **146**, 9 (1985).

5. A. Levitzki, <u>Science</u> **241**, 800 (1988).

6. K.A. Goodemote, M.E. Mattie, A. Berger, et al, <u>J. Biol. Chem.</u> **270**, 10272 (1995).

7. E.G. Bremer, J. Schlessinger, S. Hakomori, <u>J. Biol. Chem.</u> **261**, 2434 (1986).

8. Q. Zhou, S. Hakomori, K. Kitamura, et al, <u>J. Biol. Chem.</u> **269**, 1959 (1994).

9. Y. Igarashi, K. Kitamura, Q. Zhou, et al, <u>Biochem. Biophys. Res. Commun.</u> **172**, 77 (1990).

10. F.M.B. Weis, R.J. Davis, <u>J. Biol. Chem.</u> **265**, 12059 (1990).

11. T. Tsuruoka, T. Tsuji, H. Nojiri, et al, <u>J. Biol. Chem.</u> **268**, 2211 (1993).

12. H. Nojiri, M.R. Stroud, S. Hakomori, <u>J. Biol. Chem.</u> **266**, 4531 (1991).

13. E.G. Bremer, S. Hakomori, <u>Biochem. Biophys. Res. Commun.</u> **106**, 711 (1982).

14. E.G. Bremer, S. Hakomori, D.F. Bowen-Pope, et al, <u>J. Biol. Chem.</u> **259**, 6818 (1984).

15. A.J. Yates, J. VanBrocklyn, H.E. Saqr, et al, <u>Exp. Cell Res.</u> **204**, 38 (1993).

16. T. Mutoh, A. Tokuda, T. Miyada, et al, <u>Proc. Natl. Acad. Sci. USA</u> **92**, 5087 (1995).

17. N. Hanai, T. Dohi, G.A. Nores, et al, <u>J. Biol. Chem.</u> **263**, 6296 (1988).

18. E.R. Sjoberg, R. Chammas, H. Ozawa, et al, <u>J. Biol. Chem.</u> **270**, 2921 (1995).

19. E.D. Nudelman, S.B. Levery, Y. Igarashi, et al, <u>J. Biol. Chem.</u> **267**, 11007 (1992).

20. D. Kreutter, J.Y.H. Kim, J.R. Goldenring, et al, <u>J. Biol. Chem.</u> **262**, 1633 (1987).

21. Y.A. Hannun, R.M. Bell, <u>Science</u> **235**, 670 (1987).

22. Y. Igarashi, K. Kitamura, T. Toyokuni, et al, <u>J. Biol. Chem.</u> **265**, 5385 (1990).

23. A.H.Jr. Merrill, S. Nimkar, D. Menaldino, et al, <u>Biochemistry</u> **28**, 3138 (1989).

24. Y. Igarashi, S. Hakomori, T. Toyokuni, et al, <u>Biochemistry</u> **28**, 6796 (1989).

25. K. Endo, Y. Igarashi, M. Nisar, et al, <u>Cancer Res.</u> **51**, 1613 (1991).

26. H. Okoshi, S. Hakomori, M. Nisar, et al, <u>Cancer Res.</u> **51**, 6019 (1991).

27. T. Murohara, M. Buerke, J. Margiotta, et al, <u>Am. J. Physiol. (Heart Circ. Physiol.)</u> **269**, H504 (1995).

28. K. Handa, Y. Igarashi, M. Nisar, et al, <u>Biochemistry</u> **30**, 11682 (1991).

29. S. Hakomori, <u>Cancer Cells</u> **3**, 461 (1991).

30. E. Otsuji, Y.S. Park, K. Tashiro, et al, <u>Int. J. Oncol.</u> **6**, 319 (1995).

31. Y.S. Park, S. Hakomori, S. Kawa, et al, <u>Cancer Res.</u> **54**, 2213 (1994).

32. Y. Sadahira, F. Ruan, S. Hakomori, et al, <u>Proc. Natl. Acad. Sci. USA</u> **89**, 9686 (1992).

33. Y.S. Park, F. Ruan, S. Hakomori, et al, <u>Int. J. Oncol.</u> **7**, 487 (1995).

Recent aspects of glycoconjugates synthesis: orthogonal glycosylation strategy

Osamu Kanie* and Tomoya Ogawa*,**

*The Institute of Physical and Chemical Research (RIKEN) Wako-shi, Saitama 351-01 Japan
**Department of Cellular Biochemistry, University of Tokyo Yayoi, Bunkyo-ku, Tokyo 113 Japan

Abstract

An orthogonal glycosylation strategy was developed by combined use of phenylthioglycosides and glycosyl fluorides both as donors and acceptors. It was shown that both functional groups served as leaving group and protecting group under a set of distinct chemoselective glycosylation conditions. This strategy should provide flexibility in designing synthetic schemes for oligosaccharides. Further, it is expected that orthogonal glycosylation will ideally be suited for solid-phase oligosaccharide synthesis because only coupling reactions are necessary to extend a sugar chain.

Introduction

A great deal of effort has been defining the functions of the oligosaccharide chains found on the cell surface where they covalently attached to a variety of organic molecules. Because of the diversity of the oligosaccharide structures, they are often described as the "finger print" of a certain cell. The possibility that the complexity of the structures directly relates to their functions has attracted many synthetic chemists to synthesize glycoconjugates. During the last two decades the technology to access these structures have hence improved dramatically.

In 1951 the importance of sugars was first recognized when the relation between ABH blood type specificities and oligosaccharide structure was shown[1]. It was in 1953 that Lemieux et al. demonstrated the chemical synthesis of a disaccharide, sucrose[2]. In the 70's, it became possible to synthesize a variety of oligosaccharide structures by way of Helferich and Koenigs-Knorr methods which employ glycosyl halides as glycosylating reagents and heavy metal salts as activators[3]. We now have reliable glycosylation methods which enable us to access higher oligosaccharides in a very systematic manner[4]. Although this is mainly due to the development of a wide range of leaving groups and mild activating methods[5], it should be remembered that numerous protecting groups required, especially for the anomeric position, and the strategies to synthesize such molecules efficiently were developed and applied for the synthesis of oligosaccharides.

Glycosylation reaction

There are several ways to control stereochemistry of glycosidic bond formation (Fig. 1): e.g. (i) 1,2-trans glycoside formation by means of C-2 participating group[3, 6], (ii) axial glycoside formation resulting from the anomeric effect[7], (iii) equatorial glycoside formation using carbocation solvation by acetonitrile[8], (iv) axial glycoside formation by SN2 displacement of β-halides formed by halide equilibration in the presence of quaternary ammonium halide[9], (v) β-mannosylation by way of insoluble silver salts[10], and (vi) β-mannosylation by intramolecular aglycon delivery[11].

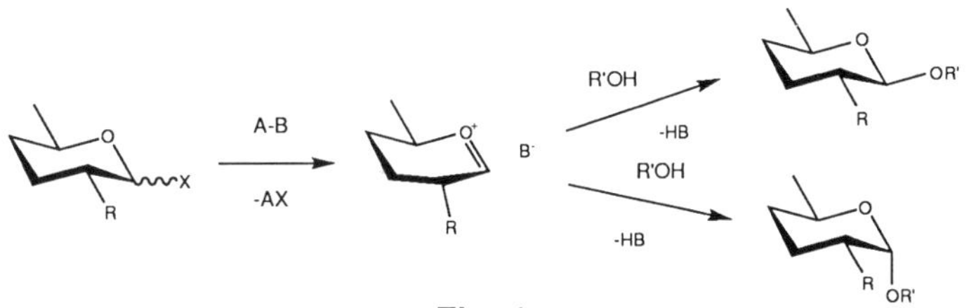

Fig. 1

The generalized scheme for the synthesis of oligosaccharides is shown in Fig. 2 as an example of complex oligosaccharide syntheses. A decision has to be made about which method (i-vi) will be applied for each coupling step. Also, a protecting group must be removed after each glycosylation step to release the hydroxyl function required for the next glycosylation reaction.

The strategy for oligosaccharide synthesis

There are two major strategies for oligosaccharide synthesis termed stepwise synthesis and block condensation. Although both strategies have to be employed to achieve an efficient synthesis, the stepwise method has classically been used for the oligosaccharide synthesis due to the difficulty of transforming anomeric protecting group into a leaving group. Recently, however, block synthesis utilizing temporary anomeric protecting groups (e.g. allyl, methoxyphenyl (MP), n-pentenyl, 2-trimethylsilylethyl (SE), t-butyldimethylsilyl or t-butyldiphenylsilyl group) has become more reliable (Fig. 2-1)[12]. These protecting groups are stable toward standard protection-deprotection schemes and glycosylation reactions and can easily be cleaved to yield the hemiacetal which can then be converted into donors.

One of the breakthroughs in new strategies is the idea of using lightly protected acceptor where there are several hydroxyl groups unprotected especially near the site where the glycosylation reaction takes place[13]. The lack of additional steric impediments caused by the protecting groups next to the hydroxyl group might be the reason of the success achieved using this strategy. This strategy was not only successfully applied to α-sialic acid containing oligosaccharide synthesis but also to block condensation leading to polymeric Le^X structures.

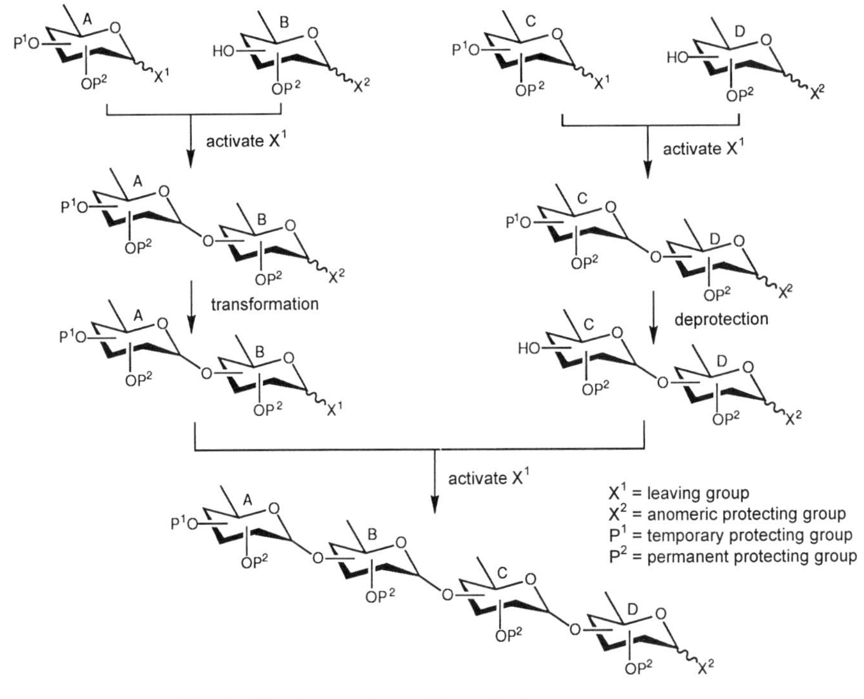

Fig. 2 Block condensation strategy

1 X^2 = OSE, OMP etc.   Transformation step have to go thorough hemiacetal.

2 X^2 = SR, OCH$_2$CMe=CH$_2$ etc.   Transformation step is usually one step. This method dealt with chemoselective glycosylation and latent-active method.

Another significant contribution to the block condensation strategy in oligosaccharide synthesis was the introduction of thioglycosides which could be used directly as glycosyl donors or transformed into halides or other leaving groups (Fig. 2-2)[14]. Thioglycosides can be activated under alkylating or oxidative conditions but are stable under the traditional glycosylation conditions such as Koenigs-Knorr method, so that halides can be chemoselectively activated in the presence of thioglycosides. For this reason, thioglycosides are ideal candidates as intermediates in a flexible synthetic strategy. The possibility of chemoselective activation of anomeric leaving groups has thus been emerged (Fig. 3-1)[14-16].

The armed and disarmed concept (Fig. 3-2), which employed n-pentenyl glycosides for both the donor and acceptor, has been developed on the basis of the observation that the reactivity of glycosyl donors is affected by the protecting groups (i.e. ether or ester) especially at O-2[3b, 17]. The utility of n-pentenyl glycosides is obvious since small fragments of oligosaccharides can be synthesized in a very efficient manner and the n-pentenyl group can be temporarily protected as the dibromide to be used as the acceptor for block condensations. However, changing the acyl protecting group into ether group in the middle of the synthesis, when one wants to extend the saccharide chain, presents a drawback of this methodology. The armed and disarmed concept has proved to be applicable also to thioglycosides, glycals, and glycosyl phosporoamidates.

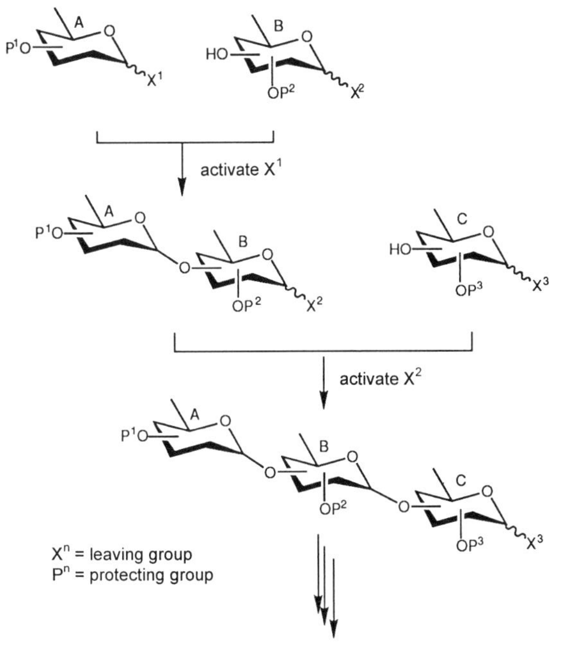

1 $X^1 \neq X^2 \neq X^3$, Require many leaving groups.
The mthod does not depend on Ps.

2 $X^1 = X^2$, $X^3 = OR$, $P^1 =$ ether, $P^2 =$ acyl
$P^3 = P^1$ or P^2; armed-disarmed concept.

3 $X^{2n+1} = X^1$, $X^{2(n+1)} = X^2$, $X^1 \neq X^2$
The orthogonal glycosylation does not depend on Ps.

$X^n =$ leaving group
$P^n =$ protecting group

Fig. 3 Chemoselective Sequencial Glycosylation

The orthogonal glycosylation strategy

The advantage of the chemoselective activation strategy lies in efficiency since the glycosyl acceptor has a potential leaving group which also temporarily protects the anomeric center. Therefore, in principle, there is no need for deprotection or protecting-group manipulations prior to coupling reactions. However, the number of available protecting groups and chemoselective glycosylation conditions pose a limitation on the chemoselective sequential glycosylation strategy.

The minimum requirement for a scheme to overcome the aforementioned limitations would be to combine two chemically distinct glycosylation reactions where one of the leaving groups is activated while the other behaves as a protecting group and *vice versa* (Fig. 3-3 and Fig. 4). To fulfill the requirement for this orthogonal system[18], each selected leaving group should be unaffected under the conditions used to activate

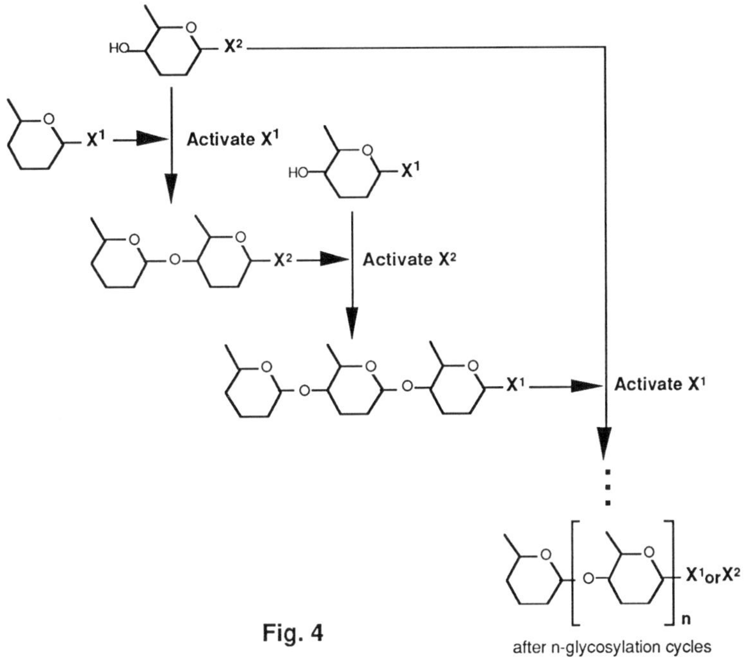

Fig. 4

after n-glycosylation cycles

the other. Also both leaving groups should remain compatible with routine manipulations of temporary protecting groups. For this orthogonal strategy, we selected the phenylthio group for X^1 and fluoride for X^2 as the leaving groups, and (a) NIS-TfOH (or AgOTf)[19] and (b) Cp_2HfCl_2-$AgClO_4$[20] as promoters, respectively[16s].

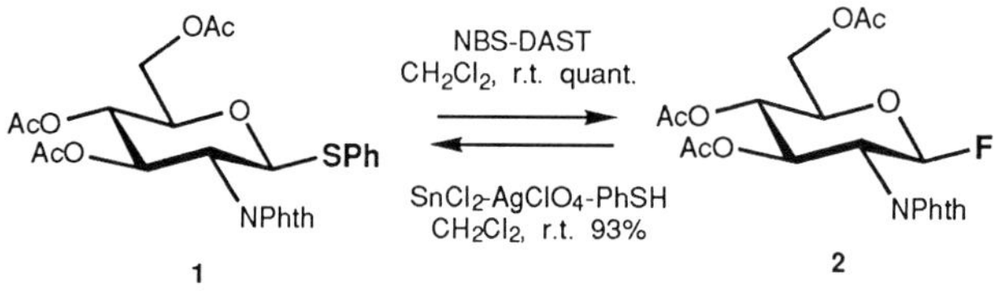

Fig. 5

In the initial attempt to demonstrate the feasibility of the orthogonal strategy, N-phthaloyl (Phth) protected glucosamine (GlcN) derivatives were chosen as the monosaccharide units. This decision was based solely on the assumption that any stereochemical ambiguity could be eliminated by the strong 1,2-trans-directing nature of the NPhth group. However, it is to be stressed that the basic principle should be applicable to a wide variety of oligosaccharide structures. In addition, the biological significance of β-1,4 linked oligomers of glucosamine (e.g. chitin) are well recognized. Also, the hydroxyl group at the C-4 position of GlcN is known to be relatively unreactive. Therefore, the construction of this type of oligosaccharide is a challenging task.

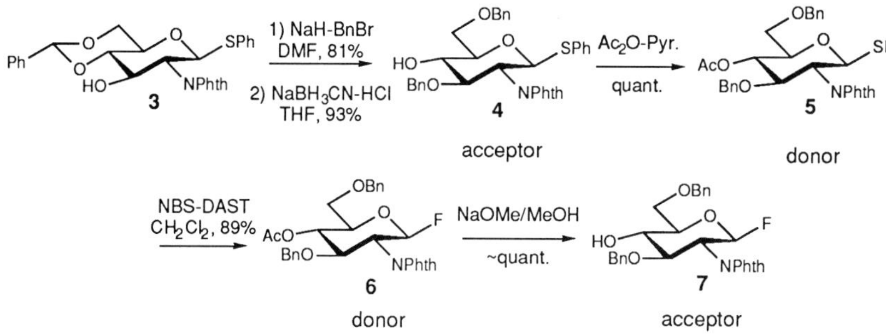

Fig. 6

In order to assess the orthogonality of the above-mentioned combination of reactions, the following reactions were performed. Thiophenyl glycoside 1 was treated under Nicolaou's conditions[14b] to yield the glycosyl fluoride 2 quantitatively. On the other hand, the fluoride 2 could be activated under Mukaiyama's conditions[21] in the presence of thiophenol to give the thioglycoside 1 in 93% yield (Fig. 5). This inter-conversion prompted us to investigate further. Next, to examine the glycosylation reactions, required GlcN derivatives 4, 5, 6 and 7 were synthesized according to the procedure described for closely related compounds (Fig. 6). Thus, a set of glycosylation reactions using 1 and 2 as donors and 7 and 4 as acceptors was examined. The thioglycoside 1 was reacted with the acceptor 7 under condition (a) to afford disaccharide 8 in 90% yield. The fluoride 2 was also successfully activated under condition (b) and reacted with 4, without affecting the thioglycosidic linkage, to give disaccharide 9 in 78% yield (Fig. 7). No α-isomer nor self-condensed product was detected in either sequence. These results indicate that less reactive acyl protected donors were activated preferentially compared to the potentially more reactive ether protected acceptors. The chosen set of reactions is therefore shown to be orthogonal.

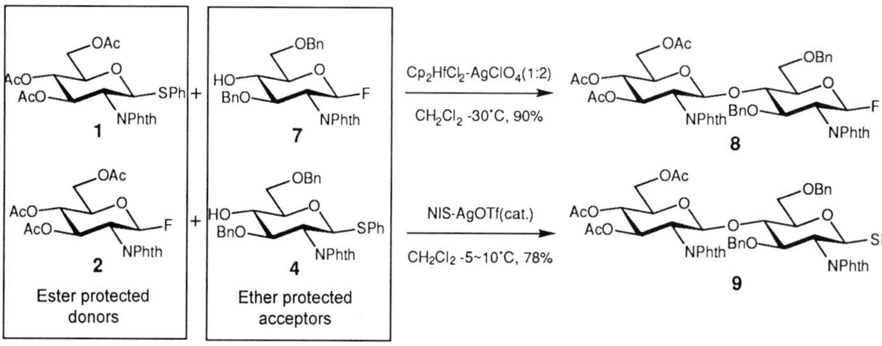

Fig. 7

The applicability of the present strategy to the synthesis of longer chain oligosaccharides was further examined by constructing the heptasaccharide **14** from two monosaccharide units **4** and **7** derived from **1** (Fig. 8). First, thioglycoside donor **5** was coupled with acceptor fluoride **7** under condition (a) to give disaccharide fluoride **10** (85%) which was then reacted with the acceptor **4** to produce **11** [condition (b), 72%]. Subsequent reaction of **11** with **7** [condition (a), 65%] gave tetrasaccharide **12**. Having accomplished the stepwise synthesis of a tetrasaccharide, we next examined a block condensation approach. Tetrasaccharide acceptor **13**, prepared by Zemplen deacetylation of **12**, was coupled with its precursor **11** to give compound **14** [condition (a), 67%], which is again ready for further use as a oligosaccharide donor.

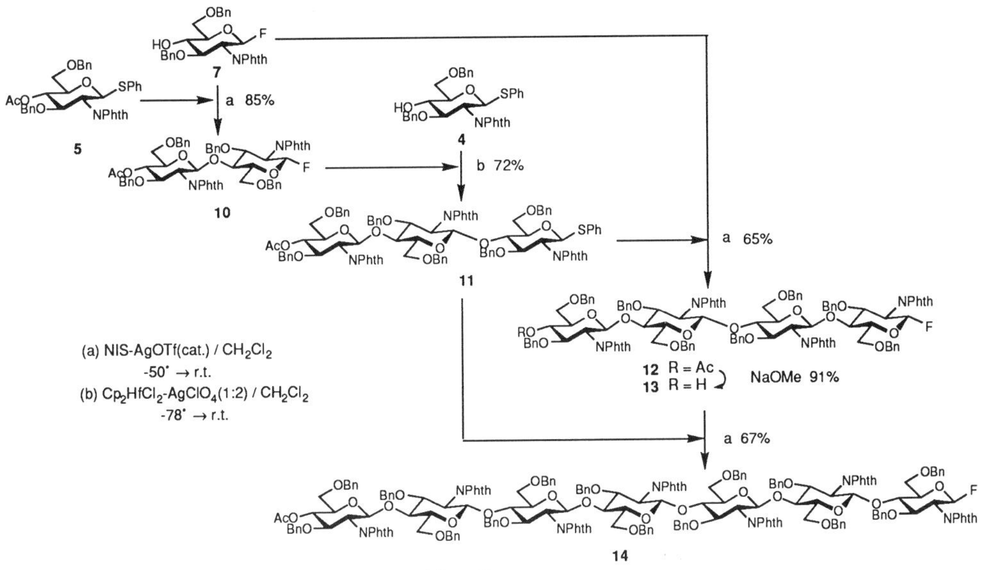

Fig. 8

Conclusion and future prospects

An orthogonal glycosylation strategy was developed by combined use of phenylthioglycosides and glycosyl fluorides both as donors and acceptors. It was shown that both functional groups served as leaving group and protecting group under a set of distinct chemoselective glycosylation conditions. This strategy should provide flexibility in designing synthetic schemes for oligosaccharides[22]. Further, it is expected that orthogonal glycosylation will ideally be suited for solid-phase oligosaccharide synthesis[23] because only coupling reactions are necessary to extend a sugar chain.

Acknowledgements

We thank Professor Ole Hindsgaul for technical advice. This work was supported by a Grant-in-Aid for Scientific Research from the Ministry of Education, Science, and Culture (T.O), and by a Special Researchers' Basic Science Program at RIKEN from the Science and Technology Agency of the Japanese Government (O.K).

References

1   (a) Watkins, W. M.; Morgan, W. T. J. *Nature*, **1952**, *169*, 825-826. (b) Watkins, W. M. *Science*, **1966**, *152*, 172-181. and references cited therein.

2   Lemieux, R. U.; Huber, G. *J. Am. Chem. Soc.*, **1953**, 4118.

3   (a) Igarashi, K. "The koenigs-knorr reactions" in *Adv. Carbohydr. Chem. Biochem.*, **1977**, *34*, pp243-283. (b) Paulsen, H. *Angew. Chem. Int. Ed. Engl.*, **1982**, *21*, 155-175.

4   (a) Nicolaou, K. C.; Caulfield, T. J.; Kataoka, H.; Stylianides, N. A. *J. Am. Chem. Soc.*, **1990**, *112*, 3693-3695. (b) Kameyama, A.; Ishida, H.; Kiso, M.; Hasegawa, A. *Carbohydr. Res.*, **1991**, *209*, C1-C4. (c) Matsuzaki, Y.; Ito, Y.; Ogawa, T. *Tetrahedron Lett.*, **1992**, *33*, 6343-6346.

5   (a) Schmidt, R. R. *Angew. Chem. Int. Ed. Engl.*, **1986**, *25*, 212-235. (b) Paulsen, H. *Angew. Chem. Int. Ed. Engl.*, 1990, *29*, 823-938. (c) Garegg, P. J. *Acc. Chem. Res.*, **1992**, *25*, 575-580. (d) Banoub, J.; Boullanger, P.; Lafont, D. *Chem. Rev.*, **1992**, *92*, 1167-1195. (e) Toshima, K.; Tatsuta, K. *Chem. Rev.*, **1993**, *93*, 1503-1531. (f) Khan, S. H.; Hindsgaul, O. in *Molecular Glycobiology-Frontiers in Molecular Biology*, Fukuda, M.; Hindsgaul, O. Eds., IRL Press, Oxford, **1994**, pp 206-229.

6   (a) Lemieux, R. U.; Takeda, T.; Chung, B. Y. *A. C. S. Symp. Ser.*, **1976**, *4*, 90-115. (b) Ito, Y; Ogawa, T. *Tetrahedron.* **1990**, *46*, 89-102. (c) Okamoto, K; Goto, T. *Tetrahedron.* **1990**, *46*, 5835-5857.

7   Lemieux, R. U. in *Molecular Rearrangements*, de Mayo, P. Ed., Interscience, New York, **1964**, p. 709.

8   Brccini, I.; Derouet, C.; Esnault, J.; Hervé du Penhoat, C.; Mallet, J.-M.; Michon, V.; Sinaÿ, P. *Carbohydr. Res.*, **1993**, *246*, 23-41. and referrences cited therein.

9   Lemieux, R. U.; Hendriks, K. B.; Stick, R. V.; James, K. *J. Am. Chem. Soc.*, **1975**, *97*, 4056-4062.

10   (a) Paulsen, H.; Lockhoff, O. *Chem. Ber.*, **1981**, *114*, 3102-3114. (b) Garegg, P. J.; Ossowski, P. *Acta Chem. Scand.*, **1983**, *B37*, 249-250. (c) Van Boeckel, C. A. A.; Beetz, t.; van Aelst, S. F. *Tetrahedron*, **1984**, *40*, 4097-4107.

11   (a) Barresi, F.; Hindsgaul, O. *J. Am. Chem. Soc.*, **1991**, *113*, 9376-9377. (b) Stork, G.; Kim, G. *J. Am. Chem. Soc.*, **1992**, *114*, 1087-1088. (c) Barresi, F.; Hindsgaul, O. *Synlett*, **1992**, 759-761. (d) Bols, M. *Tetrahedron*, **1993**, *49*, 10049-10060. (e) Barresi, F.; Hindsgaul, O. *Can. J. Chem.*, **1994**, *72*, 1447-1465. (f) Ito, Y.; Ogawa, T. *Angew. Chem. Int. Ed. Engl.*, **1994**, *33*, 1765-1767.

12   (a) Sugimoto, M.; Ogawa, T. *Glycoconjugate J.*, **1985**, *2*, 5-9. (b) Murakata, C.; Ogawa, T. *Carbohydr. Res.*, **1992**, *235*, 95-114. (c) Fraser-Reid, B.; Udodong, U. E.; Wu, Z.; Ottosson, H.; Merritt, J. R.; Rao, C. S.; Roberts, C.; Madsen, R. *Synlett.*, **1992**, 927-942. (d) Jansson, K.; Frejd, T.; Kihlberg, J.; Magnusson, G. *Tetrahedron Lett.*, **1988**, *29*, 361-362. (e) Jansson, K.; Ahlfors, S.; Frejd, T.; Kihlberg, J.; Magnusson, G. *J. Org. Chem.*, **1988**, *53*, 5629-5647. (f) Murase, T.; Kameyama, A.; Kartha, K. P. R.; Ishida, H.; Kiso, M.; Hasegawa, A. *J. Carbohydr. Chem.*, **1989**, *8*, 265-283. (g) Kanie, O.; Takeda, T.; Ogihara, Y. *Carbohydr. Res.*, **1989**, *190*, 53-64. (h) Kinzy, W.; Schmidt, R. R. *Liebigs Ann. Chem.*, **1985**, 1537-1545. (i) Nakahara, Y.; Ogawa, T. *Tetrahedron Lett.*, **1987**, *28*, 2731-2734.

13   Kanie, O.; Hindsgaul, O. *Curr. Opin.*, **1992**, *2*, 674-681. and references cited therein.

14   (a) Koto, S.; Uchida, T.; Zen, S. *Bull. Chem. Soc. Jpn.*, **1973**, *46*, 2520-2523. (b) Nicolaou, K. C.; Dolle, R. E.; Papahatjis, D. P.; Randall, J. L. *J. Am. Chem. Soc.*, **1984**, *106*, 4189-4192. (c) Lönn, H. *Carbohydr. Res.*, **1985**, *139*, 105-113. (d) Fügedi, P.; Garegg, P. J.; Lönn, H.; Norberg, T. *Glycoconjugate J.*, **1987**, *4*, 97-108. (e) Nilsson, M.; Norberg, E. *Carbohydr. Res.*, **1988**, *183*, 71-82.

15   (a) Roy, R.; Andersson, F. O.; Letellier, M. *Tetrahedron Lett.*, **1992**, *33*, 6053-6056. (b) Boons, G-J.; Isles, S. *Tetrahedron Lett.*, **1994**, *35* 3593-3596.

16   (a) Paulsen, H. ; Tietz, H. *Carbohydr. Res.*, **1985**, *144*, 205-229. (b) Friesen, R. W.; Danishefsky, S. J. *J. Am. Chem. Soc.*, **1989**, *111*, 6656-6660. (c) Trumtel, M.; Veyrières, A.; Sinaÿ, P. *Tetrahedron Lett.*, **1989**, *30*, 2529-2532. (d) Veeneman, G. H.; van Boom, J. H. *Tetrahedron Lett.*, **1990**, *31*, 275-278. (e) Veeneman, G. H.; van Leeuwen, S. H.; Zuurmond, H.; van Boom, J. H. *J. Carbohydr. Chem.*, **1990**, *9*, 783-796. (f) Mori, M.; Ito, Y.; Uzawa, J.; Ogawa, T. *Tetrahedron Lett.*, **1990**, *31*, 3191-3194. (g) Mehta, S.; Pinto, B. M. *Tetrahedron Lett.*, **1991**, *32*, 4435-4438. (h) Marra, A.; Gauffeny, F.; Sinaÿ, P. *Tetrahedron*, **1991**, *47*, 5149-5160. (i) Zegelaar-Jaarsveld, K.; van der Marel, G. A.; van Boom, J. H. *Tetrahedron*, **1992**, *48*, 10133-10148. (j) Jain, R. K.; Matta, K. L. *Carbohydr. Res.*, **1992**, *226*, 91-100. (k) Marra, A.; Esnault, J.; Veyrières, A.; Sinaÿ, P. *J. Am. Chem. Soc.*, **1992**, *114*, 6354-6360. (l) Hashimoto, S.; Yanagiya, Y.; Honda, T.; Kobayashi, H.; Ikegami, S. Presented at 18th Symposium on Progress in Organic Reactions and Syntheses-Applications in the Life Sciences, Sapporo, Japan, October **1992**, Abstr. pp 276-280. (m) Raghavan, S.; Kahne, D. *J. Am. Chem. Soc.*, **1993**, *115*, 1580-1581. (n) Mehta, S.; Pinto, B. M. *J. Org. Chem.*, **1993**, *58*, 3269-3276. (o) Sliedregt, L. A. J. M.; Zegelaar-Jaarsveld, K.; van der Marel, G. A.; van Boom, J. H. *Synlett*, **1993**, 335-337. (p) Yamada, H.; Harada, T.; Miyazaki, H.; Takahashi, T. *Tetrahedron Lett.*, **1994**, *35*, 3979-3982. (q) Yamada, H.; Harada, T.; Takahashi, T. *J. Am. Chem. Soc.*, **1994**, *116*, 7919-7920. (r) Chenault, H. K.; Castro, A. *Tetrahedron lett.*, **1994**, *35*, 9145-9148. (s) Kanie, O.; Ito, Y.; Ogawa, T. *J. Am. Chem. Soc.*, **1994**, *116*, 12073-12074.

17   (a) Mootoo, D. R.; Konradsson, P.; Udodong, U.; Fraser-Reid, B. *J. Am. Chem. Soc.*, **1988**, *110*, 5583-5584. (b) Friesen, R. W.; Danishefsky, S. J. *J. Am. Chem. Soc.*, **1989**, *111*, 6656-6666. (c) Veeneman, G. H.; van Leeuwen, S. H.; van Boom, J. H. *Tetrahedron Lett.*, **1990**, *31*, 1331-1334.

18   (a) Barany, G; Merrifield, R. B. *J. Am. Chem. Soc.*, **1977**, *99*, 7363-7365. (b) Barany, G; Merrifield, R. B. *J. Am. Chem. Soc.*, **1980**, *102*, 3084-3095.

19   (a) Konradsson, P.; Mootoo, D. R.; McDevitt, R. E.; Fraser-Reid, B. *J. Chem. Soc., Chem. Commun.*, **1990**, 270-272. (b) Veeneman, G. H.; van Leeuwen, S. H.; van Boom, J. H. *Tetrahedron Lett.*, **1990**, *31*, 1331-1334. (c) Konradsson, P.; Udodong, U. E.; Fraser-Reid, B. *Tetrahedron Lett.*, **1990**, *31*, 4313-4316.

20   (a) Suzuki, K.; Maeta,; H. Matsumoto, T.; Tsuchihashi, G. *Tetrahedron Lett.*, **1988**, *29*, 3571-3574. (b) Suzuki, K.; Maeta, H.; Matsumoto, T. *Tetrahedron Lett.*, **1989**, *30*, 4853-4856.

21   Mukaiyama, T.; Murai, Y.; Shoda, S. *Chem. Lett.*, **1981**, 431-432.

22   Paulsen, H. *Angew. Chem. Int. Ed. Engl.*, **1995**, *34*, 1432-1434.

23   (a) Danishefsky, S. J.; McClure, K. F.; Randolph, J. T.; Ruggeri, R. B. *Science*, **1993**, *260*, 1307-1309. See also references therein. (b) Carol, L. Y.; Taylor, M.; Goodnow, Jr. R.; Kahne, D. *J. Am. Chem. Soc.*, **1994**, *116*, 6953-6954. (c) Douglas, S. P.; Whitfield, D. M.; Krepinsky, J. J. *J. Am. Chem. Soc.*, **1991**, *113*, 5095-5096. (d) Douglas, S. P.; Whitfield, D. M.; Krepinsky, J. J. *J. Am. Chem. Soc.*, **1995**, *117*, 2116-2117. See also references cited therein.

Chemical aspects of
artificial gene regulatory molecules

Keisuke Makino

Department of Polymer Science and Engineering, Kyoto Institute of Technology, Matsugasaki, Sakyo-ku, Kyoto 606, Japan

Abstract: We have studied on artificial gene regulatory molecules composed of oligonucleotides or peptides. For oligonucleotides, we focus on oligonucleoside phosphorothioate (S-oligo) and studied influences of thiophosphate linkage chirality on its hybrid structure. Using diatereomeric S-oligos isolated by RPLC and based on the Tm of their duplexes, the duplex stability was found to be dependent on the thiophosphate configuration. NMR analysis indicated that the factor responsible for the decrease in the Tm is possibly the difference in the restricted pre-melting molecular motion dependent on the chirality: The introduction of thiophosphate does not disturb the B-form formation. For peptides, we built DNA-binding small peptides which recognize the specific palindromic and non-palindromic base sequences. For this trial, we chose specific basic region peptide such as that of GCN4 as a recognition element and a β-cyclodextrin-adamantane host-guest complex as model of leucine zipper to construct a noncovalent dimerization module.

In the last several decades, there were numbers of epoch-making findings in molecular biology. Also, in chemistry, there has been enormous growth in scientific and technological developments in organic synthesis. By combining accumulated knowledge in both fields, we are now able to design and synthesize model oligonucleotides and peptides of biological importance.

Recent studies have revealed the existence and basic nature of molecules playing essential roles in gene expression, such as antisense RNA/DNA and transcription factors. From a scientific and clinical point of view, *de novo* synthesis of such functional molecules that have minimum size for essential functions is extremely interesting. One of the typical example is oligonucleotide as an antisense oligo-DNAs, which has been investigated in numbers of clinical trials. Despite of such intensive studies for more than a decade, chemical nature of antisense oligo-DNAs has not been fully understood. In this short review paper, some of remaining problems and our trials to overcome them are described. Also here, our recent studies on building peptides as an artificial transcription factor which recognizes and binds to the specific base sequences of DNA are also shown.

Oligonucleotide - Effect of thiophosphate chirality on the duplex stability of oligonucleoside phosphorothioate (S-oligo) and trials for its stereoselective synthesis

As an antisense molecule, oligonucleoside phosphorothioate (S-oligo) has been investigated exclusively because of its established ability to suppress the translation processes[1]. For instance, our recent study has demonstrated that an S-oligo with a complementary sequence to the *tax* region of HTLV-1 transcript inhibits the proliferation of the virus effectively. We still have, however, a difficulty in optimizing this antisense molecule, because this S-oligo is a 20-mer consisting of enormous sets of stereoisomers arising from the Rp/Sp configuration at each thiophosphate linkage and each diastereomeric S-oligo should have different binding affinities to the target. To overcome this essential question, it is important to reveal the effect of the R/S configuration both on the antisense efficiency and the duplex structures. Because of the difficulties in developing stereoselective synthetic procedure, however, these essential problems still remain to be overcome so far. In the present study, we focused on the solution structure of the S-oligo duplexes and used a combination of spectroscopical measurements to look into it in the molecular level.

To see the effect of the configurational difference of the thiophosphate on the duplex stability, we prepared an S-oligo pentamer (5'-dCpsApsTpsCpsG-3', ps: thiophosphate) as a sample, which consists of 16 diastereomeric isomers. Each isomer was isolated in a single peak by RPLC[2]: The diastereomeric mixture, obtained after the solid-phase synthesis and subsequent cleavage from the CPG support, was first separated in four envelope peaks in the presence of 5'-DMTr, and subsequently each peak was separated into four peaks after the removal of the protecting group. Absolute configuration of the individual isomer was determined by means of the specific enzymic digestion of nuclease P1 (for Sp-configuration) and snake venom phosphodiesterase (for Rp-configuration). Using an oligo-DNA with the complementary base sequence as a target, Tm measurements were carried out in 10 mM phosphate buffer containing 1 M NaCl: The resultant values are Tm(SpSpSpSp)≈12°C and Tm(RpRpRpRp)≈0°C at the sample concentration of 0.2 mM. This result strongly indicates that the combination of the absolute thiophosphate configuration gives rise to the differed duplex stability with the target. Since this Tm difference indicates the different antisense effects for S-oligos with different combination of the chiralities, we should reveal the structural variation of the duplexes formed by the Rp/Sp isomers with their target.

We have applied conventional spectral measurements including CD and, however, they showed only similar spectra without any distinctive information for such duplexes. On the other hand, ^{1}H-NMR spectroscopy (500 MHz) gave some useful information. The sample we took was an oligo-DNA which contains a single thiophosphate linkage: The sequence is 5'-dGCpsATCG-3' and the target 5'-dCGATGC-3' (Tm(Sp)≈Tm(Rp)≈24°C for a duplex at 0.2 mM sample concentration in 10 mM phosphate buffer (pH 7.0) containing 100 mM NaCl). A pair of the isomers were separated individually and all of the proton signals of the duplex of each isomer in the duplex assigned by the combination of COSY, HOHAHA and NOESY NMR techniques (sample concentration being 2 mM in 10 mM phosphate buffer/D2O containing 100 mM NaCl). Based on these assignment, information on the distance of the atoms was collected by NOESY at 280 K: The mixing times applied were 100, 150, 200 and 300 msec. With 180 NOE constrains, the solution structure of each S-oligo duplex was estimated by restrained MD (Discover) set in Insight: The force field used was Amber. The simulated annealing procedure consisted of 9 steps: The temperature was lowered from 600 to 280 K for 4 psec. The result was that both the Rp and Sp thiophosphate linkage introduced in the oligo-DNA does not disturb the formation of B-form and, therefore, that the duplex structure is almost equivalent between the isomers: It should be noted that analogous structural coincidence has been observed previously for the duplex between Rp/Sp isomeric S-oligo and the target RNA. Only difference we observed in NMR spectra is characteristic line broadening for specific base protons in the vicinity of the thiophosphate, which depends on the chirality of the thiophosphate. Since this occurred just below the Tm (Tm≈315 K), the relaxation time of such protons in the duplex possibly seems to decrease specifically. It is conceivable that accumulation of such small local change produce the substantial difference in the Tm of larger S-oligos although detailed elucidation of the basic nature to generate Tm difference requires further investigations.

To make the antisense method with S-oligos practically useful in clinics, it is also a basic requirement that we should develop stereoselective synthesis of S-oligos. For this purpose, such a method utilizing a pentavalent phosphorus compound (oxathiaphospholane) has been proposed[3] and, however, have not been used widely, probably because of the difficulty in preparation of monomers and the insufficient reactivity. On the other hand, a method utilizing trivalent phosphorous compounds, which are regarded as more reactive than pentavalent ones, have not been developed yet. The chiral phosphite triesters can be converted into phosphorothioates with retention of the configuration at the phosphorous atom as shown in Figure 1. For this reason, we studied on synthesis of such chiral phosphite triesters, in particular a stereoselective transesterification reaction of phosphite triesters under basic conditions.[4] Conclusively, chiral phosphite triesters with modified phenoxy groups as leaving groups, especially 4-phenylphenoxy group and 3-chlorophenoxy group, can serve as good substrates of stereocontrolled phosphite backbone synthesis. Transesterification reactions of the phosphite triesters under basic conditions are fully stereoselective. However, overall stereoselectivity is affected by P-epimerization of the substrate occurring prior to the transesterification reaction. It is, therefore, essential

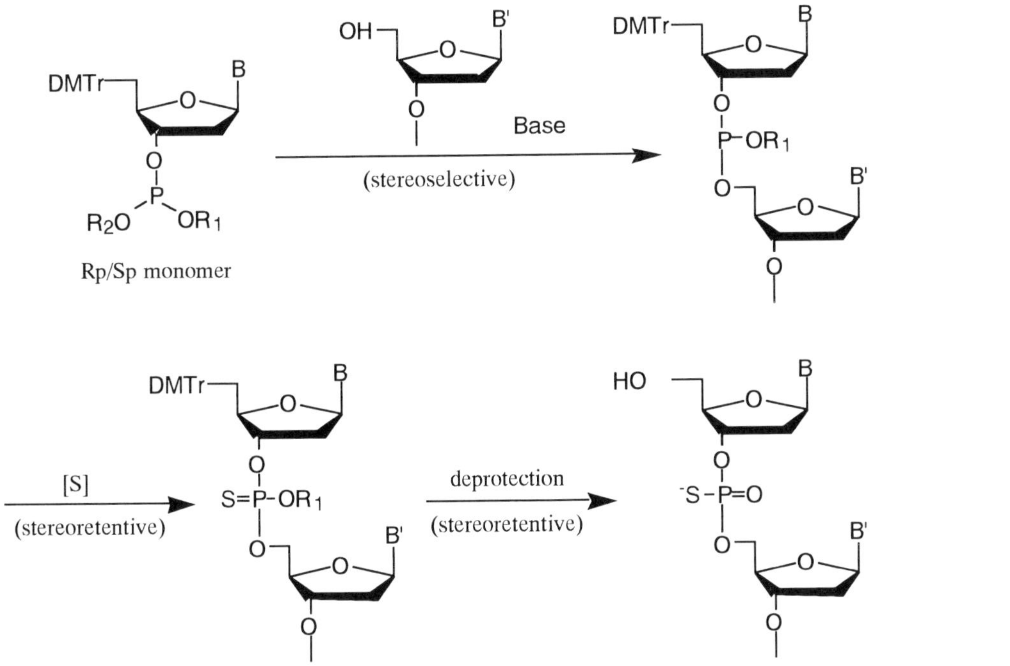

Figure 1. Proposed cheme for the stereocontrolled synthesis of oligonucleoside phosphorothioate.

to control the P-epimerization for reaction optimization.   This synthetic procedure is still under investigation.

As described above, S-oligos have a number of unsolved problems and each of them requires intensive work. We believe that the antisense method with this molecule will be easy to establish and much more useful than at present.

Peptide - *De novo* synthesis of DNA binding module

The other interesting gene regulatory molecule is a DNA-binding small peptide which recognizes and binds to the specific base sequence.

The participation of multiple proteins to control the transcription of eukaryotic genome clearly indicates that a recognition mechanism involving both the DNA-protein and protein-protein interactions is required for the high-precision interactions. A minimal example of such specific and non-covalent recognition mechanism is found in the sequence-specific DNA binding of dimeric proteins. The binding sites for many DNA-binding proteins have a palindromic nature, and x-ray crystallographic studies of many protein-DNA complexes have revealed mechanisms by which protein homodimers recognize palindromic DNA sequences. In most of the cases, relatively small portion of proteins contacts directly to the DNA surface, and the precise positioning of each recognition motif is reinforced by the three dimensional shape of the dimerization domain.

Although the mechanisms governing the specific heterodimer formation of proteins have began to be emerged recently, such as the mechanism of the specific heterodimer formation between Fos and Jun proteins, the rules by which a protein motif recognizes a specific protein partners are far from understood. Moreover, coupling of the protein folding to such specific protein-protein interactions further complicates the design of small peptides that could specifically recognize other peptides. It has been already reported by using covalently bonded dimeric peptides that steric constraints of the dimerization domain is important for the sequence-specific recognition of DNA.[5] However, this model will not allow us to study how the equilibrium governing the formation of dimeric species affects the sequence-specific DNA recognition by protein dimers. We describe here an artificial dimerization module that specifically effects a non-covalent dimer formation and cooperative DNA binding of short peptides.

Our concept to effect the non-covalent dimer formation is to utilize an artificial dimerization module consisting with β-cyclodextrin (β-Cd) and its guest compound.[6,7] Because formation of a specific host-guest inclusion complex will control the dimerization, attaching β-Cd or a guest molecule to a peptide would give a peptide with host and a peptide with guest, that could form a specific inclusion complex (Figure 2). G23 peptide, corresponding to the basic region of GCN4, possesses unique cysteine residue at the C-terminus. Modification of the peptide with 6-iodo-β-cyclodextrin and N-bromoacetyl-1-adamantanemethylamine (Ad), a guest molecule, gave G23Ad and G23Ad, respectively.

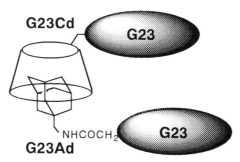

Figure 2. A schematic representation for the G23 peptide dimer with the β-cyclodextrin-adamantane inclusion complex.

GCN4 is known to recognize GRE or CRE sequences with the half-site sequence of 5'-ATGAC-3.' DNA binding of G23Ad/G23Cd was studied by gel mobility shift assay. A binding mixture containing G23Ad alone did not reveal a mobility-shifted band that indicates dimer binding. However, a mobility-shifted band was observed in a binding mixture containing both G23Ad and G23Cd. The binding of G23Ad/G23Cd to the target DNA sequence was inhibited by an addition of excess guest molecule. DNaseI footprinting experiments also confirmed the sequence-specific binding of G23Ad/G23Cd. Solution structures of G23Ad/G23Cd in the presence of target DNA sequence was analyzed by CD spectra. Although G23Ad and G23Cd are only partially helical in the solution, the structures changed into almost helical on addition of an oligonucleotide containing the GCN4 binding sequence. These results taken together, G23Ad and G23Cd bind the native GCN4 binding sequence in the same recognition mechanisms.

As already mentioned above, many transcription factors bind DNA as heterodimers that can have different binding and activation potential from that of the homodimers. However, a question of whether the protein heterodimers could recognize non-palindromic DNA sequences has yet to be established. When proteins with multiple subunits recognize the specific DNA sequence, the non covalent interaction between the proteins would play a predominant role in determining the sensitivity of specific DNA recognition and in the cooperative DNA binding. In order to explore these notions, we have used synthetic peptides containing two distinct domains: The DNA-protein interaction domain and the peptide-peptide interaction domain. We used the peptide corresponding to the basic region of the basic leucine zipper (bZIP) protein as the DNA-protein interaction domain, since the basic region peptides of bZIP proteins alone are sufficient for the sequence-specific DNA binding when covalently or noncovalently dimerized. A host-guest inclusion complex between β-cyclodextrin (β-Cd) and an adamantyl group (Ad) was used as the peptide-peptide interaction domain (Figure 1).

As described above, we has demonstrated a sequence-specific DNA binding of homodimers formed by the synthetic peptides at the palindromic DNA sequence.[6] We utilized peptides derived from the basic region of the yeast transcriptional activator GCN4 (G23) and that of an enhancer binding protein C/EBP (C23).

C23Ad/C23Cd

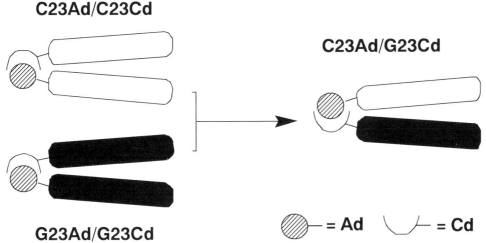

Figure 3. Schematic representation showing the formation of homodimers C23Ad/C23Cd and G23Ad/G23Cd, and a heterodimer C23Ad/G23Cd.

The artificial peptides containing β-Cd or Ad were synthesized for each basic region peptide (Figure 3). GCN4 and C/EBP are known to recognize palindromic sequences with a half-site of 5'-ATGAC-3' (CRE) and 5'-ATTGC-3' (CE), respectively. Combination of these half-sites gives a nonpalindromic sequence (CE/CR), 5'-ATGACGCAAT-3', for the target of the peptide heterodimer. Sequence specific binding to the nonpalindromic DNA sequence CE/CR by heterodimers G23Ad/C23Cd and G23Cd/C23Ad was evident from the DNase I footprinting assay. However, the homodimer C23Ad/C23Cd revealed partial protection at the nonpalindromic sequence CE/CR. DNA binding abilities of the homo- and heterodimer to the nonpalindromic sequence were comparable, as evident from gel shift assay. The heterodimer (C23Cd/G23Ad) preferentially bound to the non-palindromic CE/CR sequence over the palindromic CRE and CE sequence. The homodimer C23Ad/C23Cd showed comparable affinity to the CE/CR and CE sequences. Structures of the binding complexes were analyzed by circular dichroism spectroscopy. The heterodimers bound to the CE/CR site in the α-helical conformation as observed for the native bZIP protein. In contrast, the helicity of the homodimer C23Ad/C23Cd in the presence of the nonpalindromic sequence CE/CR was about a half of that in the presence of a palindromic CE sequence. Thus, the structure of the binding complex between the C23Ad/C23Cd and CE/CR sequence was partially helical, and most likely only the monomer at the matched half-site is in the α-helical conformation. These results indicate an existence of half-matched binding complexes for DNA binding of the homo- and heterodimers. Efficiency of the sequence discrimination by the peptide dimers depends on a stability of the half-matched binding complex.

As described above, by designing peptides according to the structure of naturally occurring DNA-binding proteins, we will be able to construct small peptide which can recognize and bind to specific base sequences to regulate its transcription.

References

1. J. S. Cohen and M. E. Hogan, *Scientific American*, 50-55, December, 1994.
2. A. Murakami, Y. Tamura, H. Wada, and K. Makino, *Anal. Biochem.*, **223**, 285 (1994).
3. W. J. Stec anf G. Zon, *Tetrahedron Lett.*, **25**, 5279 (1984).
4. M. Mizuguchi and K. Makino, *Nucleos. and Nucleot.*, in press.
5. T. Morii, M. Shimomura, S. Morimoto and I. Saito *J. Am. Chem. Soc.*, **115**, 1151 (1993).
6. M. Ueno, A. Murakami, K. Makino and T. Morii, *J. Am. Chem. Soc.*, **115**, 12575 (1993).
7. M. Ueno, M. Sawada, K. Makino and T. Morii, *J. Am. Chem. Soc.*, **116**, 11137 (1994).

Integrin adhesion receptors as targets for medicinal chemistry

Jeffrey W. Smith, Ph.D.

Program on Cell Adhesion and The Extracellular Matrix
La Jolla Cancer Research Foundation
10901 N. Torrey Pines Road
La Jolla, CA 92037

Abstract

Cell adhesion is a seminal event in many diseases and is an excellent therapeutic target. Integrin adhesion receptors mediate cell-cell contacts and the adhesion of cells to the extracellular matrix and are involved in inflammation, thrombosis, osteoporosis and tumor metastasis. Integrins are comprised of an α and β subunit, and many receptors in this family bind the arg-gly-asp (RGD) tripeptide motif. We have designed ways to create novel proteins that can bind the integrin ligand binding site. Human antibodies that bind the ligand binding site of β3-integrins have been engineered by using phage display. Antibodies that bind integrin with high affinity were selected by biopanning on integrin. This engineering effort has been expanded to develop a novel protein engineering technique called "protein loop grafting" in which the sequence of a protein surface loop is grafted to a protein of entirely different structure. This can be accomplished while maintaining the binding activity of the grafted loop.

A rationale for designing drugs to block cell adhesion

A large percentage of the biochemical reactions in the body either require cell-cell contact, or are greatly enhanced by cell adhesion. In fact, many diseases actually result from the improper movement of cells throughout the body. Theoretically, diseases including tumor metastasis, osteoporosis, arthritis and atherosclerosis all could be stopped by a blockade of cell adhesion. In each of these conditions sequential cycles of cell adhesion and release are involved in the movement of specific cell types. A precisely applied blockade of cell adhesion could prevent each disease. Therefore, a rational approach to treating disease is to use pharmaceutical means to block cell adhesion.

A strategy aimed at designing new drugs to interfere with cell adhesion is promising for three reasons. First, we now know the individual proteins that are responsible for mediating cell adhesion, and based upon the last five years of basic research, we know much about the mechanism by which these proteins operate. Secondly, we have developed a good understanding of the connection between specific diseases and the cell adhesion events that initiate these diseases. In many cases we can even pinpoint the genesis of a disease to a specific cell type and to a specific adhesion receptor. Finally, several initial studies, both in animals and now in man, prove the broad concept that a blockade of cell adhesion can be therapeutically beneficial. In essence, the biology of adhesion is largely understood, the protein targets are known, and the broad concept of anti-adhesion therapy has been proven. The challenges that remain are 1) to design drugs which exhibit a high degree of specificity for individual adhesion receptors, and 2) to devise new ways of exploiting adhesion receptors for drug design and application.

Integrin adhesion receptors: The machinery of cell adhesion

Cell-cell contact and cell adhesion are frequently mediated by a family of cell surface receptors called integrins. Integrins are αβ heterodimers that are expressed on virtually all cells with adhesive capacity (1,2). Thus, integrins are involved in almost all tissue remodeling events including organismal development, inflammatory responses and wound healing (1,3). More than twenty different integrins have been cloned and sequenced. Integrin α and β subunits can cross pair generating a structurally diverse protein family. Each subunit spans the plasma membrane (Figure 1).

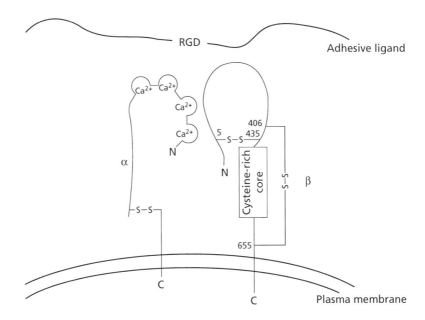

Figure 1. An Integrin

The α subunit is normally 140-150 kDa and the β subunits are usually 90 kDa. The integrins bind to a host of different ligands. A list of ligands includes extracellular matrix proteins, adhesive proteins that circulate in the blood, IgG-like proteins, snake venom proteins, viruses, and even other integrins. Clearly these ligands are structurally diverse, but there is a commonalty in the way in which each ligand binds to its integrin receptor. In most cases the integrin ligands contain a critical aspartic acid residue (4). At least seven integrins bind the tripeptide motif arg-gly-asp (RGD) with high affinity. Thus, the argument can be made that even though integrins bind numerous types of ligands, the ligand binding pocket of the integrins appears to accommodate related structural motifs. This paper will focus on a brief overview of the integrins that exhibit potential as drug targets and then describe novel ways to engineer lead proteins and peptides as integrin inhibitors. Finally, a novel protein engineering strategy called "loop-grafting" and its application in creating novel integrin binding proteins will be discussed.

Integrins associated with disease

Each of the twenty integrins should be considered potential drug targets. At present, information from basic research provides a strong rationale for developing antagonists of four prominent integrins. These are the platelet integrin αIIbβ3, integrin αvβ3, which is found on tumor cells and osteoclasts, integrin α4β1 which is expressed on monocytes, and integrin αLβ2 (LFA-1) which is a leukocyte adhesion receptor.

The platelet integrin αIIbβ3 is the best studied integrin and was also the first to be exploited for pharmaceutical benefit. Integrin αIIbβ3 is the platelet fibrinogen receptor and mediates the aggregation and adhesion of platelet at wound sites (5,6). On platelets that have been activated by thrombin, integrin binds to fibrinogen dimers which circulate in plasma. These fibrinogen dimers can bind to αIIbβ3 on two different platelets, thereby creating a platelet aggregate. The platelet aggregate is the body's first means of stopping blood loss and integrin αIIbβ3 is essential for the proper formation of a hemostatic plug at wounds. The function of αIIbβ3 was first recognized when it was discovered that patients with Glanzmanns thrombasthenia, who have persistent bleeding problems, lack expression or proper function of this integrin. Since many vascular diseases result from the improper adhesion and aggregation of platelets, it has been hypothesized that inhibitors of αIIbβ3 could be used as drugs. Two types of antagonists of αIIbβ3 have been tested in animal models and now in clinical trials. The first is monoclonal antibody (Mab) 7E3 which binds nearly irreversibly to αIIbβ3 and eliminates platelet adhesion and aggregation (7). This antibody, and its humanized counterpart have shown good efficacy in both animal models of thrombosis and in clinical trials (8,9). Others have developed small peptide antagonists of αIIbβ3. These peptide mimetics are based on the arg-gly-asp (RGD) sequence, the tripeptide motif is present in many adhesive ligands, including fibrinogen. Peptides and peptide mimetics based upon this motif have proven efficacious in animal models of thrombosis (10).

Integrin α4β1 is expressed on the surface of monocytes and can mediate the adhesion of these cells to injured vascular endothelium. It is this activity that accounts for the involvement of α4β1 in promoting monocyte adhesion during the formation of atherosclerotic plaques (11). It has been hypothesized that an oral inhibitor of α4β1 could be applied prophylatically as an inhibitor of atherogenesis. Integrin α4β1 has also been implicated in the infiltration of monocytes and eosinophils in a variety of inflammatory conditions (12-16). Thus, antagonists of this integrin could also be useful in preventing asthma. Although α4β1 does not bind to the prototypical RGD sequence, cyclic peptides containing this motif can be designed to have high affinity for this integrin (17). Consequently, lead compounds for drug design are available for this integrin.

Integrin αLβ2 was formerly known as the Leukocyte Function associated Antigen, or LFA-1. The initial clue to the biological function of this integrin came from the discovery that patients lacking the expression of αLβ2 and suffered from ongoing bacterial infections. Their condition was termed LAD, leukocyte adhesion deficiency (18,19). Conversely, several pathologies can also be initiated by leukocyte adhesion. These include arthritis, and transplant rejection. It has been hypothesized that antagonists of αLβ2 could be used as anti-inflammatory agents. This notion is supported by several studies showing that a blockade of the adhesion of leukocytes can prevent inflammatory responses (20-23).

Integrin αvβ3 is the fourth integrin to be linked to specific diseases. Interestingly, this integrin is associated with several diseases that involve tissue remodeling and cell migration. One example is osteoporosis. In bone αvβ3 mediates osteoclast adhesion to the bone surface (24). Since osteoclasts must adhere to the bone surface in order to initiate bone resorption, a blockade of this type of adhesion could be effective in preventing osteoporosis. Several recent studies in organ culture and in rats have proven this concept by showing that RGD peptide antagonists of αvβ3 block bone resorption (25). Integrin αvβ3 is also implicated angiogenesis, or the growth of new blood vessels. In fact, RGD-based antagonists of αvβ3 prevent vascular growth into tumors and thereby prevent tumor growth (26). Thus, antagonists of this integrin may be applicable in cancer. The application of αvβ3 antagonists in cancer could be efficacious for another reason because this integrin is also strongly implicated in tumor metastasis (27).

The RGD motif: A starting point for mimetic design and protein engineering

As stated above, many integrins bind to the RGD tripeptide motif that is present within their adhesive ligands. Although several thousand proteins within the known databases contain this sequence, not all of these proteins will bind to integrins. The conformation of the RGD sequence is key to its recognition. Within the last six years it has also become evident that not all integrins bind to RGD. However, all integrin ligands contain a critical aspartic acid residue (4) and this may be tied to the divalent cation dependence of ligand binding (28). Integrins that do bind RGD can often bind this sequence when it is present within small synthetic peptides, or when the motif is mimicked by organic compounds (10). Within natural adhesive proteins and in snake venom proteins that have evolved to bind integrins, the RGD sequence is present at the apex of a flexible loop. Even integrins that don't bind RGD, recognize a critical aspartate found at the apex of a flexible loop (4). These observations suggest unique approaches toward building integrin ligands and antagonists, but also lead to important technological challenges: Can we exploit our understanding of integrin biochemistry to devise ways of identifying novel lead compounds as integrin antagonists? Can our knowledge of the presentation of the RGD motif be used to test new methods of protein engineering?

As a first step in addressing these challenges we have designed a method of engineering highly selective human antibodies that interact with the integrin ligand binding site. The applications of human antibodies in treating disease are vast, but one major obstacle is the inability of host to recognize self. In other words, the human immune system will not generally produce antibodies against human proteins. As a partial solution to this problem many have turned to the "humanization" of biologically active rodent antibodies. This process involves substantial manipulations of the framework regions of the IgG. Another potential solution to this problem is the application of semi-synthetic antibodies in which the antigen binding site of a human IgG is manipulated by recombinant DNA technology so that it will bind a human antigen. Recent methods have enabled the screening of vast ($>10^7$) numbers of amino acid sequences displayed in the complimentarily determining region (CDR) of an antibody (29). This is accomplished by creating a library of antibodies all of which display different amino acid sequences in a CDR. The antibodies are expressed in a functional form on the surface of bacteriophage. Antibodies with high affinity for antigen can be recovered from this phage library based upon their binding activity. Then the amino acid sequence of the active CDR can be derived from the phage DNA sequence (29).

We recently applied this approach to engineer human antibodies against the two β3 integrins, platelet integrin αIIbβ3 and the osteoclast integrin αvβ3 (30,31). It is well established that residues which flank the RGD motif in natural proteins can influence the affinity and specificity of the protein for individual proteins. Therefore, the phage-display antibody library was constructed by inserting an RGD sequence at the apex of the antibodies heavy chain CDR3. The three residues that flank the RGD were randomized, thus creating 4×10^7 different version of the RGD motif (Figure 2).

Inserting a ligand motif into antibody

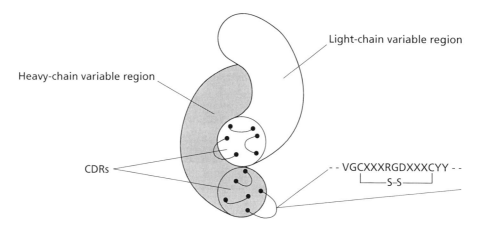

Figure 2. Semi-Synthetic Antibodies

The antibodies within this library that had highest affinity for αvβ3 were selected by biopanning on purified human αvβ3. This strategy yielded several antibodies that bound to αvβ3 with affinity of one nanomolar or greater. A biochemical characterization of these antibodies showed that they also bound to the highly homologous platelet integrin αIIbβ3 with high affinity. However, they did not bind to another homologous integrin, αvβ5. Thus, the antibody design strategy was successful in obtaining high affinity antibodies that recognize, and antagonize, a human antigen, integrin αvβ3. The initial step in this strategy also successfully obtained a high degree of specificity because the antibodies that bound αvβ3 did not react with the homologous αvβ5, even though this integrin binds to RGD peptides with very high affinity. These findings show that, in principal, the inability of the host immune system to recognize self can be overcome by re-engineering human antibodies via phage-display.

The first generation of "semi-synthetic" anti-integrin antibodies were selected for binding affinity for αvβ3, but they did exhibit cross-reactivity with the platelet integrin αIIbβ3. Therefore, a second step in the engineering strategy was to obtain antibodies that distinguish the ligand binding pockets of these highly homologous integrins. This change in selectivity was achieved by redesigning the phage library of antibodies. In the modified antibody library, the sequences within CDR3 that flank the RGD motif were maintained, but the RGDX motif was randomized. This modification was made with the intent of maintaining the conformation of the high affinity CDR, but to change the actual contact residues for the integrin. By using this strategy, we were able to obtain antibodies that could distinguish between αIIbβ3 and αvβ3 (31).

Collectively these results show that antibody engineering by phage display is an effective way of generating human antibodies that interact with high affinity and specificity with human antigens. Since the antibodies we have engineered block the activity of αvβ3 or αIIbβ3 they could be used as anti-thrombotic agents or as inhibitors of osteoporosis and tumor metastasis. These success in re-engineering antibodies also has broader implications. Antibodies are not the only proteins with an IgG-like fold. Other integrin ligands, like VCAM and ICAM, also have the IgG structure and also make contact with their integrin ligands through flexible loops. However, ICAM and VCAM have relatively low affinity for their ligands. The reported K_D's of these proteins for integrins range from several hundred nanomolar to even micromolar. The success in redesigning antibodies by phage display implies that other IgG-like proteins including ICAM and VCAM could be modified by phage display to have higher affinity for their integrin receptors.

Can semi-synthetic antibodies yield lead peptides?

Antibodies will be useful in treating acute conditions, but oral inhibitors are clearly better suited for chronic administration. Given the desire to identify small lead compounds for oral inhibitors, we tested the inhibitory activity of synthetic peptides derived from the CDR sequences of the anti-integrin antibodies. Do such peptides maintain the affinity of the parent antibody and do they maintain specificity for particular integrins? Both cyclic and linear peptides with sequences identical to CDR's of inhibitory antibodies were synthesized and tested for the ability to block ligand binding to integrins $\alpha IIb\beta3$, $\alpha v\beta3$ and $\alpha v\beta5$. Each of these integrins can bind the prototypical RGD sequence derived from fibronectin, GRGDSP. We found that the RGD-containing peptides derived from the CDR's of the phage-selected antibodies do indeed block integrin function. Whereas the parent antibodies exhibited affinities of near one nanomolar, the synthetic peptides had affinities that were roughly 10 to 100-fold lower. Thus, conversion of a CDR motif to a synthetic peptide results in a loss of binding affinity for the antigen. However, we did find that the synthetic peptides had the same integrin binding specificity as the parent antibody. Thus, although binding affinity is lost, binding specificity is maintained. These data show that the phage screening of antibodies can provide two potential drugs, 1) the human antibody itself for use in acute conditions, and 2) synthetic peptides derived from the antibodies CDR for use as a lead in developing small molecule inhibitors.

Protein loop grafting to develop novel protein-protein interactions

Although small molecule antagonists can be applied to combat many diseases, it is often beneficial to use recombinant proteins as drugs. Human growth hormone, insulin and tissue plasminogen activator are examples of such protein drugs. In such cases, it may be beneficial to target proteins to precise locations within the body by endowing them with novel binding activities. A typical approach for creating novel protein binding activities is to employ cumbersome and time consuming site-directed mutagenesis. Another approach is to link one protein to another. This is done either chemically or by recombinant means endowing the chimeric protein with the binding activities of both molecules. Unfortunately, this approach substantially increases the size of the potential protein drug. We recently investigated a new method for devising novel protein-protein interactions. We call this new technique "protein loop grafting." This method is based upon the observation that many protein-protein interactions are guided by protein surface loops.

To test the hypothesis that protein surface loops could be grafted from one protein to another, while maintaining the binding activity of the loop, we substituted the amino acid sequence of an antibodies CDR loop into the extended loop and an epidermal growth factor (EGF) module. It is important to emphasize that the protein loop is being taken from one protein backbone (IgG) and placed into a protein with a totally different structure (EGF). To test this concept, we used one of the RGD-containing antibodies selected from the phage library. This antibody is called Fab-9 and binds to both $\alpha v\beta3$ and $\alpha IIb\beta3$ with nonomolar affinity. When the amino acid sequence of CDR3 of Fab-9 is substituted in an EGF module, we find that the EGF module gains the integrin binding activity of Fab-9 (32). In fact, nearly all of the binding affinity is maintained in the loop graft. These data show that in principal, protein loop grafting could be applied as a relatively broad strategy to create novel protein binding activities. It combination, phage selection of active loop sequences, and then subsequent loop grafting between distinct protein modules may reduce protein engineering to a combinatorial process.

References

1. R. O. Hynes, *Cell* **48**, 549-550 (1987).

2. E. Ruoslahti, M. D. Pierschbacher, *Science* **238**, 491-497 (1987).

3. R. O. Hynes, *Cell* **69**, 11-25 (1992).

4. J. W. Smith, in *Integrins*, pp. 1-32, Academic Press, Inc., New York, 1994).

5. D. R. Phillips, I. F. Charo, L. V. Parise, L. A. Fitzgerald, *Blood* **71**, 831-843 (1988).

6. D. R. Phillips, I. F. Charo, R. M. Scarborough, *Cell* **65**, 359-362 (1991).

7. B. S. Coller, *J. Clin. Invest.* **76**, 101 (1985).

8. J. E. Tcheng, E. S. Topol, N. S. Kleinman, *Circulation* **88**, 1-506 (1993).

9. E. J. Topol, E. F. Plow, *Thrombosis & Haemostasis* **70**, 94-98 (1993).

10. J. A. Zablocki, M. Miyano, R. B. Garland, D. Pireh, L. Schretzman, S. N. Rao, R. J. Lindmark, S. G. Panzer-Knodle, N. S. Nicholson, B. B. Taite, A. K. Salyers, L. W. King, J. G. Campion, L. P. Feigen, *J. Med. Chem.* **36**, 1811-1819 (1993).

11. M. I. Cybulsky, M. A. Gimbrone,Jr., *Science* **251**, 788-791 (1991).

12. W. M. Abraham, M. W. Sielczak, A. Ahmed, A. Cortes, I. T. Lauredo, J. Kim, B. Pepinsky, C. D. Benjamin, D. R. Leone, R. R. Lobb, *J. Clin. Invest.* **93**, 776-787 (1994).

13. P. L. Chisholm, C. A. Williams, R. R. Lobb, *Eur. J. Immunol.* **23**, 682-688 (1993).

14. D. K. Podolsky, R. Lobb, N. King, C. D. Benjamin, B. Pepinsky, P. Sehgal, M. deBeaumont, *J. Clin. Invest.* **92**, 372-380 (1993).

15. V. B. Weg, T. J. Williams, R. R. Lobb, S. Nourshargh, *J. Exp. Med.* **177**, 561-566 (1993).

16. T. A. Yednock, C. Cannon, L. C. Fritz, F. Sanchez-Madrid, L. Steinman, N. Karin, *Nature* **356**, 63-66 (1992).

17. P. M. Cardarelli, R. R. Cobb, D. M. Nowlin, W. Scholz, F. Gorcsan, M. Moscinski, M. Yasuhara, S. L. Chiang, T. J. Lobl, *J. Biol. Chem.* **269**, 18668-18673 (1994).

18. T. K. Kishimoto, K. O'Conner, T. A. Springer, *J. Biol. Chem.* **264**, 3588-3595 (1989).

19. C. Nelson, H. Rabb, M. A. Arnaout, *J. Biol. Chem.* **267**, 3351-3357 (1992).

20. H. E. Jasin, E. Lightfoot, L. S. Davis, R. Rothlein, R. B. Faanes, P. E. Lipsky, *Arthritis Rheum.* **35**, 541-549 (1992).

21. L. S. Davis, A. F. Kavanaugh, L. A. Nichols, P. E. Lipsky, *J. Immunol.* **154**, 3525-3537 (1995).

22. A. F. Kavanaugh, L. S. Davis, L. A. Nichols, S. H. Norris, R. Rothlein, L. A. Scharschmidt, P. E. Lipsky, *Arthritis Rheum.* **37**, 992-999 (1994).

23. W. J. Mileski, R. Rothlien, P. Lipsky, *Eur. J. Pediatr. Surg.* **4**, 225-230 (1994).

24. T. J. Chambers, K. Fuller, J. A. Darby, J. A. S. Pringle, M. A. Horton, *Bone Miner.* **1**, 127-135 (1986).

25. M. Sato, M. K. Sardana, W. A. Grasser, V. M. Garsky, J. M. Murray, R. J. Gould, *J. Cell Biol.* **111**, 1713-1723 (1990).

26. P. C. Brooks, A. M. Montgomery, M. Rosenfeld, R. A. Reisfeld, T. Hu, G. Klier, D. A. Cheresh, *Cell* **79**, 1157-1164 (1994).

27. S. M. Albelda, S. A. Mette, D. E. Elder, R. M. Stewart, L. Damjanovich, m. Herlyn, C. A. Buck, *Cancer Res.* **50**, 675-676 (1990).

28. S. E. D'Souza, T. A. Haas, R. S. Piotrowicz, V. Byers-Ward, D. E. McGrath, H. R. Soule, C. Ciernieswki, E. F. Plow, J. W. Smith, *Cell* **79**:659-667 (1994).

29. C. F. III Barbas, J. D. Bain, D. M. Hoekstra, R. A. Lerner, *Proc. Natl. Acad. Sci. USA* **89**, 4457-4461 (1992).

30. C. F. Barbas, L. Languino, J. W. Smith, *Proc. Natl. Acad. Sci. USA* **90**:10003-10007 (1993).

31. J. W. Smith, D. Hu, A. Satterthwait, S. Pinz-Sweeny, C. F. III. Barbas, *J. Biol. Chem.* **269**, 32788-32795 (1994).

32. J. W. Smith, K. Tachias, E. L. Madison, *J. Biol. Chem.*, in press (1995).

Towards an orphan drug policy for the European Union

Patrick Deboyser

European Commission, Rue de la Loi 200, 1040 Brussels, Belgium

Abstract: So far, neither the European Union, nor its Member States have paid much attention to the issue of *Orphan Drugs*. Obviously, this issue would be best dealt with at Union level, and many voices have recently called upon the European Institutions to lay down the basis of an *Orphan Drug* policy. This would : an adequate legal framework, the setting up of an efficient telematic network, increased support of pre-normative research and, above all, incentives for the pharmaceutical industry, both from the European Union and the Member States.

Introduction

In Europe, very little has been done so far in respect of *Orphan Drugs*. On 2 June 1995, the Health Ministers of the European Union formally accepted the need of an *orphan drug* policy for the European Union, and invited the European Commission to set out the details of measures to be taken, whether at European or at national level. This paper discusses the prospect for establishing an *Orphan Drug* policy for the European Union.

What is an Orphan Drug ?

An Orphan Drug is a drug aimed at treating an Orphan Disease (see below). It can be an entirely new drug, offering a new preventive or curative treatment. But itcould also be the exploitation of subsidiary properties or uses of drugs which are already known. Sometimes, it may even entail maintaining old drugs on the market, or re-introducing drugs the use of which has been discontinued.

What is an Orphan Disease ?

In the European Union, it is generally considered that a rare disease is one that does not strike more than 500 people per million, i.e. 0,5 per thousand.

The WHO appears to have identified about 5,000 diseases that strike less than 650 to 1000 people. 4,000 of these seem to be linked to genetic factors.

From a more general point of view, one could consider as orphan diseases not only these rare diseases but also all diseases for which no treatment is available, because finding, developping and marketing it is considered un-economical under market-economy conditions.

Similarly, one might also have a topic-related approach, thus considering as orphan diseases, tropical or parasitic diseases which are prevalent in developing countries where they concern a significant number of patients, but are nearly inexistant in industrialized countries, and therefore less likely to be considered for research and development by the industry.

What is the situation in the Member States of the European Union ?

None of the Member States of the European Union has yet adopted the like of an Orphan Drug policy, although they are some elements in place in some of the Member States.

In France :
A law of December 1992 provides for the exceptional and limited *use* of certain drugs, when they are intended :

* either for the treatment of severe pathologies, whilst no alternative therapy exists and their efficacy is strongly presumed in view of the results of therapeutic trials (compassionate use),

* or for patients affected by rare diseases and they are no drug already authorised d likely to act as a substitute (orphan drugs),

In Spain :

A law of December 1990 provides that the Government can adopt special measures in relation to the manufacture, the distribution and the supply of "drugs without any commercial interest".

In the United Kingdom :

In the field of research, there is a definite collaboration between academic authorities and industry; in particular, the Medical Research Council and some industries work on programmes (diseases in developing countries).Additionnally, a marketing authorisation can be granted although not all requirements have been met, subject to commercialisation for a limited number of patients and availability of further data.

In Germany :

In 1990, the national Parliament proposed to the Federal Government that the possibilities for promoting R & D of orphan drugs be studied and that adequate measures be implemented without delay. However, this was not followed by action In 1993, the issues was reconsidered when the marketing of antidotes in case of intoxication, were withdrawn from the market for lack of economic return.

In Denmark :

The Health Minister has founded a research Centre for Rare Disorders and Disabilities in December 1994

In Sweden :

The status of orphan drug does not exist but a particular procedure is open for drugs of great medical value for which the sales forecast is low.

An organisation administering some 800 pharmacies is in charge of the manufacture and the distribution of non-commercialised orphan drugs. It establishes the galenic formulation, takes care of documentation and organises R & D programmes in collaboration with the doctors.

In 1990 the Swedish Medical Products Agency suggested a policy for orphan drugs to the Government and to the Pharmaceutical Industry which to date remains without concrete follow-up.

Furthermore, an interesting initiative has been launched in that country, in 1988, by a number of pharmaceutical companies. It is known as the *"Swedish Orphan AB"*.

Why an Orphan Drug policy at European Union level ?

Taking the issue at the European level offers several advantages.

First, since the number of patients in any one Member State is insufficient to support research, investment by the industry is more likely to be productive in the larger market of the European Union, the total population of which is now more than 370 million,

Secondly, the European framework offers many advantages, in terms of : epidemiology and observation of rare illnesses, involvement of a large number of pharmaceutical companies (about 2,000,

Thirdly, financial resources available for the public support of research is much larger, notably through European research programmes, and management of this support at Union level avoids distortion of competition,

Fourthly, since January 1995, companies can now file a central application to the European Medicines Evaluation Agency, and obtain a marketing authorization valid throughout the Union.

What is the situation at European Union level ?

Directive 65/65/EEC, which is the funding piece of the European pharmaceutical legislation, provides that under certain circumstances a drug can be maide available to patients before a marketing authorization is actually granted, thus allowing the so-called "compassionate use" of the drug.

Directive 75/318/EEC laying down the requirements to be fulfilled in support to an application for marketing authorization allows for a relaxation of the efficacy and safety requirements whenever the therapeutic indications requested is such that the applicant cannot reasonably be requested to supply complete information.

Regulation N° 2309/93 establishing the new European centralized procedure grants a 10-years protection of the data supplied in suppport of the marketing authorization. Admitedly, the benefit of this clause is not limited to Orphan Drugs, but when the product concerned is not a new substance, the clause will ensure some sort of market exclusivity for the holder of the marketing authorization in respect of the orphan indication.

Regulation N° 297/95 (not 267/95) establishing the fees to be levied by the European Medicines Evaluation Agency in respect of new drug applications provides that the fee can be waived by the Agency when the drug is likely to be used only in a limited number of applications. This may not be a very fortunate wording, but there is no doubt that the waiver is meant for Orphan Drugs.

What would be the main components of such a policy ?

Legal framework
The first thing to be done would be to amend Directive 65/65/EEC or Directive 75/18/EEC to include a definiition of an "orphan drug". For the purposes of European legislation, this definition could be linked to a prevalence of the corresponding disease within the Union of 0,05 % of population, or approximately 180,000 patients.

Pre-normative research
The 4th European framewok programme for action in the field of public health has identified rare diseases as an action area. Co-ordination with patient groups for the purposes of prioritisation of diseases and access to and communication with patients databases would need to be considered in this context. The BIOMED programme has allocated a small budget to Orphan Drugs (1.2m ECU - the equivalent of 1.6m ECU). This would be enhanced and made more focused, both to bring research closer to industry, and to achieve results i.e. products in the medium term.

Telematic networks
Given the dispersed location of patients with rare diseases, a telematic network of specialised hospitals/researchers would be established through telematic applications to Health programmmes. This would help in the identification of researchers in rare diseases and would provide a mechanism of exchange of information and as necessary of scientific knowledge.

Clinical Trials
It is important that patients with rare diseases benefit from the existence of medicines which are of the required standard of quality, safety and efficacy as is normally applied to medicines. However there are two other aspects to this.
Firstly, recruitement of subjects for clinical trials is complicated by the small numbers of patients in any one locality. Including patients from multiple regions or Member States introduces the complication of differing administrative systems for the comencement of a clinical trial and mechanisms for the approval of the trial protocol. This issue will have to be addressed specifically in a new frame-work Directive on Clinical trials that the Commission intends to propose next year.
Secondly, even with the facilitation of the commencement of clinical trials, the small number of patients available may mean that large scale phase III trials are not possible. As I have already indicated, a clause in Directive 75/318/EEC addresses this isue, but it would need to be tailored to the needs of Orphan Drugs.

Availability to patients
The same can be said in respect of the clause in Directive 65/65/EEC governing the availability to patients for compassionate use. This would have to be clarified, in order to address more specifically the needs of patients with rare diseases.

Marketing authorisation

The procedure for obtaining a marketing authorisation for an Orphan Drug should be reviewed. This would not only mean that the conditions for waiving the fee should be fixed, but also the length of the procedure should be considered.

Incentives from Member States

In some Member States, incentives for the investment in either research or development of rare diseases already exists. Under a co-ordinated European policy, the opportunity presents itself for Member States to act in harmony and to arrange incentives in an optimal fashion. This would notably cover fiscal incentives, an issue which so far remains under Member States' competence.

Incentives from the European Union

Even with such natonal incentivs and with public support for research, the return on investment for a pharmaceutical company in an orphan disease is unlikely to be attractive unless further incentives are offered. The Commission is currently reviewing various possibilities in this respect. One of them, would be to guarantee to the company concerned that no marketing authorization will be granted to anther company in a given time period.

A European Orphan Drug Bureau ?

There are already discussions about the opportunity to establish a European Orphan Drug Bureau. This would be encharged with following tasks :   the establishment of a European catalog of rare diseases, the creation and maintenance of an Orphan Drug data bank, the granting of Orphan Drug status, the allocation of research funds,   more generally, to encourage public and private research in Orphan Drugs.

Is there a need for such a Bureau ? It is to early to discuss this issue, as long as we are not sure that there is a real political will to develop a comprehensive Orphan Drug policy in the European Union. And if this indeed the case, one would then have to be convinced that the tasks just outlined could not be undertaken by the European Medicines Evaluation Agency.

Conclusion

Patients suffering from rare diseases deserve the same quality, safety and efficacy in medicinal products as other patients. For the first time, there seems to be a general consensus that the European Union should encourage research and development into *Orphan Drugs*. The European framework provides many advantages in this respect. If a genuine policy was developed over the next years in the European Union , this would in turn allow for international cooperation in this matter, perhaps within the framework of ICH - the International Conference on Harmonisation.

Support for orphan drug development: legislation in the United States, Japan and Europe

Marlene E. Haffner, M.D., M.P.H.
Office of Orphan Products Development, FDA
5600 Fishers Lane
Rockville, Maryland 20857
USA

ABSTRACT

Between 10 and 20 million Americans suffer from one of the estimated 5,000 rare diseases. Most of these rare diseases are orphan diseases; they have no parent organization, investigator, or agency dedicated to research on the prevention, diagnosis, or treatment of their victims. Orphan Products are those to treat diseases or conditions characterized by low prevalence, or special subsegment patient populations that often lack economic viability.

Congress enacted the U.S. Orphan Drug Act in 1983, after a long struggle by rare disease patient groups in the United States to attract attention to their hopeless plight. Incentives of the Act encourage pharmaceutical manufacturers to develop products that would otherwise provide little return on invested capital due to small patient populations. The Orphan Drug Act has brought hope to victims of rare, life-threatening diseases and has had a major impact on the discovery of ways to use already developed products to save lives and improve patient quality of life. The Orphan Drug Act amended the U.S. Federal Food, Drug, and Cosmetic Act, and has been amended several times to adjust to changing trends.

BACKGROUND

One of the earliest **major** focuses on issues surrounding orphan products in the United States was stimulated in June 1979, when a Department of Health, Education, and Welfare (HEW) [now the Department of Health and Human Services] interagency task force reported its assessment of issues surrounding significant drugs of limited commercial value in the United States. This report, directed to the Secretary for HEW, stated:

> " . . . whenever a drug has been identified as potentially life saving, or otherwise of unique major benefit to some patient, it is the obligation of society, as represented by Government, to seek to make that drug available to that patient."(1)

Conclusions reached by the task force centered attention on treatment for diseases affecting very small numbers of patients in the same way that Japan's 1972 "Outline of Measures for Intractable Diseases" directed attention to the political issues surrounding orphan diseases in Japan.(2)

Following the HEW task force report, congressional hearings were held in 1981 between representatives of the Government [Food and Drug Administration (FDA)], the pharmaceutical industry, voluntary disease groups, and affected individual patients who testified on their experiences in the area of rare diseases. Rare disease groups in the United States spawned a grass roots movement that exerted considerable influence on congressional legislators. Similar campaigns occurred in Japan in the very early 1990s.

Due to the significant concerns expressed, and because of the FDA's own desire to find ways to bring orphan drugs to the marketplace, FDA established its Office of Orphan Products Development (OPD) in 1982 -- nearly a year before the passage of the U.S. Orphan Drug Act. At that time, the outcome of legislation being considered by Congress was unclear. FDA's orphan product program relied then, as it does now, upon the private sector -- with the U.S. Government acting as catalyst in bringing drugs for rare diseases to the marketplace. In 1982, the OPD functioned mainly through communication and persuasion. There were no incentives that could be offered toward the development of such needed drugs.

Voluntary organizations worked aggressively to promote legislation to encourage the development of products necessary and often life-saving for patients with rare diseases. Despite their potential usefulness, these drugs and biologics -- known as orphan products -- often remained inadequately tested and/or unavailable to patients, due to the small number of patients with a particular disease. The National Organization for Rare Disorders (NORD) -- founded in 1983 by Abbey Meyers -- has been one of the most effective of these voluntary organizations. NORD evolved from an affiliation of national health agencies and support groups that formed together into a national foundation after succeeding in their unrelenting efforts for orphan drug legislation.

The Orphan Drug Act(3) was passed by the U.S. Congress with rare unanimity, and was signed into law by President Ronald Reagan on January 4, 1983. This legislation amended the U.S. Federal Food, Drug, and Cosmetic Act(4) and provided incentives for the pharmaceutical industry to develop drugs that otherwise have limited commercial value and consequently little industry support. NORD and its affiliates continue to be extremely helpful legislatively for the FDA in its endeavor to correct the deficiencies and neglect suffered by victims of orphan diseases.

The Orphan Drug Act established for the first time in the United States a Federal Government policy to cooperate and assist in a program to facilitate the development of drugs for rare diseases. Among the provisions included in the U.S. Orphan Drug Act legislation is seven years of exclusive marketing for manufacturers of drugs and biological products who have a product (1) designated as an orphan, and (2) approved by the FDA for that indication. Only drug and biological products may be designated as orphans in the United States -- not medical devices, which Japan's orphan drug law includes -- or medical foods. Orphan diseases are usually chronic, without definite treatment; however, "intractable and incurable" -- terms used to describe Japanese orphan products -- are not found in the definition of the term for purposes of the American Orphan Drug Act.

Initially, the law did not include criteria specifying a maximum number of patients affected by a disease or condition in order for it to be considered rare. **In 1984**, the Orphan Drug Act was amended(5) to base orphan designation on prevalence, rather than on the difficult calculation of profitability. Potential sponsors were no longer required to substantiate lack of profit. The amendment stipulates that the target disease or condition must affect fewer than 200,000 people in the United States. Currently, designation as an orphan -- achieving "orphan status" in the United States -- is based upon the product's use to treat or diagnose a rare disease or condition defined by the size of the population. The 200,000 population figure was established to serve as a surrogate for profit, and amounts to less than 1 percent of the U.S. population. This figure is roughly equivalent to two times the 50,000 total patient population required to qualify for orphan designation in Japan. It is possible for a drug with an indication affecting more than

200,000 in the United States to be designated as an orphan drug if the costs of developing the drug and making it available in the United States could not be recovered from sales in the United States of such drug in 7 years. This lack of profitability provision is used infrequently.

The prevalence figure stipulated by the U.S. Orphan Drug Act is determined at the time the application is made for orphan status. It is a "frozen snapshot" of that interval; orphan designation is maintained even if the original population grows. Sadly, this has occurred in frank AIDS and HIV infection with low CD_4 counts. Unlike Japan's orphan drug law, orphan designation by the FDA may not be rescinded when a population increases for a designated indication.

INTERNATIONAL IMPACT

Passage of the U.S. Orphan Drug Act triggered several important international initiatives. In 1985 the Nordic Council sponsored an analysis of rare diseases and rare disease consumer groups in Scandinavia. In 1987, I participated in a European "Health Orphans" meeting organized in Brussels to address the response of the pharmaceutical industry to the concerns of those with serious and rare illnesses for which no treatments exist and/or effective treatments are not available on the market. This meeting included representatives from France, Germany, the United Kingdom, and Switzerland. U.S. Senator Edward Kennedy, (then) Chairman of the Committee on Labor and Human Resources remarked " . . . I would hope that our Orphan Drug Act could serve as a model for other countries. Perhaps discussions held at your international meeting may provide alternatives for other countries so that all patients with orphan disease may benefit."(6) In 1989, the French National Academy of Pharmacy indicated support for European coordination on orphan drug development, and in 1990 the Swedish Medical Product Agency established an initiative on orphan drug policy. Thus, it became apparent that orphan drugs were to be a part of future programs affecting Europe and European Union (EU) member states.

In February 1995, I had the honor and pleasure to be part of an expert group convened in Brussels to develop recommendations on priorities for research and regulatory action in the field of orphan drugs within the EU. A representative from the Japanese Ministry of Health and Welfare participated in this assembly as well. The group determined the urgent need for European collaboration as well as industry involvement in efforts to develop orphan drugs.(7) It is anticipated that an orphan drug office for the EU -- modeled on the U.S. FDA's Office of Orphan Products Development -- will soon be established in Brussels.(8)

TABLE 1. Time Line of Progress Toward Orphan Drug Legislation

International Orphan Product Regulatory Actions

As a result of the 1995 meeting in Brussels, a European Orphan Drug Proposal, supported by the French Ministry of Health, was developed in June by the French drug industry (SNIP)[9]. France recently held the Presidency of the EU, and established an office of orphan drugs working to create a data bank for rare diseases as well as incentives for developing rare disease treatment and diagnoses. EU legislation on orphan drugs could be ready by the end of 1995. The following table outlines and compares the organization of orphan drug development in the United States, Japan, and Europe:

TABLE 2.   Comparison of Orphan Drug Legislation.

Policy	USA	JAPAN	EUROPE
Legislation Enacted	1983	1993	1995 Commission of European Communities proposals. Orphan development offices in some member states
Administered	FDA	MHW	EMEA
Definition: Prevalence	200,000 of 250 million population	50,000 of 120 million population	Proposals recommend profit limits/prevalence limit of 180,000
Marketing Approval	Same approval requirements for all drugs	Accelerated review for Orphans	—
Incentives: Exclusivity	7 years	Registration validity extended from 6 to 10 years	Exclusivity clauses proposed
Tax credit	For R&D costs	For R&D costs	Proposed
Research Grants	Yes	Yes	Proposed for European Office for Orphan Drugs

IMPLEMENTING THE U.S. ORPHAN DRUG ACT

In the United States, the primary organizational focus for activities within the FDA relating to products intended for rare diseases is vested in the OPD. The OPD exists to promote the development of safe and effective products for the diagnosis or treatment of rare diseases or conditions, and implements the Orphan Drug Act provisions of the U.S. Food, Drug, and Cosmetic Act using the following mechanisms:

1) Coordinating protocol assistance to sponsors,
2) Reviewing and evaluating sponsors' designation requests,
3) Managing the exclusivity program,
4) Encouraging sponsors to allow treatment use of investigational drugs for rare diseases or conditions, and
5) Managing an extramural grants program.

The original version of the U.S. Orphan Drug Act constituted a commitment to millions of Americans that they would not be ignored because treatment of their diseases is not profitable. The incentives established by the Orphan Drug Act have stimulated considerable interest in the development of products for treating rare diseases. The following explains these areas and how they are implemented:

Protocol Assistance: In the United States, the Orphan Drug Act requires the FDA to provide the sponsor of a drug for a disease or condition which is rare in the United States with written recommendations for the nonclinical and clinical investigations that must be conducted with the drug to receive approval. Even before passage of the U.S. Orphan Drug Act, it became evident that some rare diseases may affect as few as 30 patients throughout the United States; despite the uncommon occurrence of the disease, the parameters for approval of the drugs for these small populations could not be compromised. Therefore, it was apparent that there was need to provide sponsor assistance in order that studies could be devised that are adequately controlled to show safety and efficacy, even with small numbers of patients. Major changes in FDA's regulations and philosophy governing investigational new drug (IND) and new drug application (NDA) processes have occurred concurrently with orphan drug legislation during the past decade. Largely due to these changes, specific protocol assistance has been rarely used. Pre-IND meetings are common today; end of phase II sessions are routine; interaction between sponsors and reviewers as part of an ongoing process is now commonplace. Consequently, written protocol assistance has not been needed.

The Designation Process: Orphan designation is based upon a product's use in the treatment or diagnosis of a rare disease or condition and is conferred on a **specific product for a specific indication**. In order for a sponsor to obtain orphan designation for a drug or biological product, an application must be submitted to OPD, and approved by OPD staff for designation for the indicated use.

Oversight of the development and approval of the product is the responsibility of the appropriate reviewing division in the FDA. Frequently, OPD reviewers work with the FDA's Center for Drug Evaluation and Research (CDER) and the Center for Biologics Evaluation and Research (CBER) -- those areas of FDA that review and grant drug approval -- to fulfill the FDA mission to facilitate the availability of new therapies for serious and life-threatening conditions. The approval of an orphan designation request does not alter the standard regulatory requirements and process for obtaining FDA marketing approval. Safety and efficacy of a compound must be established through adequate and well-controlled studies.

Because of the life-threatening nature of many orphan products and the fact that many of these diseases have no other therapy, FDA often accomplishes its review within an accelerated timeframe. Accelerated evaluation of *all* orphan drug marketing applications, such as that provided by the Ministry of Health and Welfare in Japan, is not guaranteed by the FDA. While OPD does not have a role in the actual review of a product for approval, OPD staff are active ombudsmen, working together with the sponsors and assisting them in meeting agency requirements in order to achieve product approval.

Treatment Use of Investigational Drugs: Through treatment IND guidelines, FDA allows patients suffering from life-threatening diseases with no other treatment to obtain much-needed drugs still in investigational stages. As a result of these guidelines, a patient whose survival depends on obtaining an unapproved drug can be provided treatment in situations where studies are not yet complete, but where diligent clinical investigation has shown efficacy of the product. A large number of treatment INDs granted by FDA are for orphan products because of the serious nature of these rare disease indications.

Measures have also been proposed within the EU to make certain drugs available to patients prior to marketing approval. Such proposed availability would be based on incidence of the disease, the drug sponsor's ability to guarantee quality of the drug, and the risks associated with the drug's use.

The Orphan Products Grants Program: The Orphan Products Grants Program was established as part of the Orphan Drug Act legislation for the purpose of supporting clinical studies to determine whether products for rare diseases are safe and effective for premarket approval under the U.S. Federal Food, Drug, and Cosmetic Act. Each year, FDA issues in the *Federal Register* a Request for Applications (RFA) for orphan product grant funding. Orphan product grant clinical studies must be designed to assist in the approval of unapproved products or approval of unapproved new uses for already marketed products.

Applications are carefully reviewed by category experts in ad hoc panel meetings. Approximately 15-25 review group meetings are held each grant cycle, with FDA grants management officers and OPD grants administration staff. The nature of grant proposals submitted to FDA frequently involves highly complex and specialized medical, pharmaceutical, or bioengineering processes. The selection of reviewers, therefore, who are expert in the technologies involved is extremely important to achieving a fair evaluation and ranking of each proposal.

After panel reviews, OPD staff project officers prepare summary statements of the application evaluations which are provided to applicants after the second review level has been accomplished. Funding, from $100,000 to $200,000 per year, is provided primarily to academic researchers in the United States. Grants are also available for devices and medical foods.

Shortly after awards are made, OPD project officers contact the principal investigator for each grant to assist the researcher with patient accrual, possible study design changes, and sponsor acquisition. OPD project officers keep in touch with the reviewing division in FDA that maintains the IND for the grantee's study, and help maintain understanding and coordination between FDA and the principal investigator. This assistance is especially helpful for researchers facing special problems as a result of the fact that the disease or condition being studied is a rare one. Rare disease organizations also provide assistance to FDA researchers in locating patients willing to participate in orphan drug clinical trials.

The OPD grants program has been extremely successful. Orphan product grant studies have resulted in eleven approved products in the U.S. marketplace to date:

PEGADEMASE (Adagen)	Enzyme replacement for adenosine deaminase (ADA) deficiency in patients with Severe Combined Immunodeficiency Syndrome (SCIDS).
SUCCIMER (Chemet)	To treat lead poisoning in children.
HISTRELIN ACETATE (Supprelin injection)	To treat central precocious puberty.
NAFARELIN ACETATE (Synarel Nasal Solution)	To treat central precocious puberty.
BACLOFEN (Lioresal Intrathecal)	To treat spasticity associated with cerebral palsy.
IN-EXSUFFLATOR	To assist ventilator-dependent patients.
PULMONARY ANGIOSCOPE	For visual examination of life-threatening pulmonary emboli.
CLADRIBINE (Leustatin Injection)	To treat hairy cell leukemia
CYSTEAMINE (Cystagon)	To treat nephropathic cystinosis
LEVOCARNITINE (Carnitor)	To treat primary and secondary carnitine deficiency of genetic origin.
IOBENGUANE SULFATE	For localization of pheochromocytoma.

Tax Credit: The U.S. Orphan Drug Act exempts an orphan drug sponsor from tax liability on a portion of clinical research costs for a designated orphan product -- up to 50 percent of the qualified clinical testing expenses for the taxable year. Tax provisions are administered by the U.S. Internal Revenue Service.(10) Although there is no specific legislation concerning orphan drugs pertaining to the EU as a whole, most member states provide a research tax credit for a portion of research and development costs. Proposals for orphan drug legislation being developed in the EU stipulate exemption from required fees and also repayment of government subsidies at such time as an orphan drug sponsor realizes profits from the sale of the product.

CURRENT ISSUES

Biotech Products and Patentability: The advances in science that permit us to replicate natural, often scarce, compounds bring the issue of patentability to the forefront. Patent status and patentability of orphan products -- especially the growing proportion that are generated from new biotechnology processes -- are controversial issues. Initially, orphan status excluded any product eligible for patent; that limitation was removed by the Congress in 1985.(11) Patent protection in the United States for new biotechnology-derived products is limited, difficult to obtain, and requires years to resolve. Consequently, there are those who see the Orphan Drug Act's exclusivity as a surrogate for patent law, especially for biotech products.

Because of significant uncertainty in U.S. patent law, biotechnology companies have used Orphan Drug Act exclusivity to establish successful markets. Many such companies have become major pharmaceutical enterprises. Individuals recognizing the marketing potential offered by the benefits of the Orphan Drug Act have been responsible for the formation of an increasing number of small pharmaceutical firms designed specifically to capitalize on the possibilities of developing treatments for rare diseases.

Due to concerns about cost and profit of some orphan drugs, legislative proposals for amending the Orphan Drug Act have included changes in the exclusivity provision. Under one proposal, an initial marketing exclusivity period of 4 years for the manufacturer of a designated orphan product would replace the existing 7-year exclusivity provision; after 4 years, the sponsor of an orphan drug would be required to demonstrate continued limited commercial potential for the drug in order to obtain the remaining 3 years of exclusivity. Another proposed change would have permitted approval of a second sponsor's orphan drug to share exclusivity if development of the two products occurred simultaneously. Bills introduced in the U.S. Congress to enact these changes were not approved.(12)

Companies making profits on sales of orphan drugs in Japan must return a portion of the subsidy provided to them through the Japanese Orphan Drug Law. The proposed definition of orphan drugs within the EU will take into account the lack of expected profitability and may include a review of profitability after a period of time.

CONCLUSION

The FDA's Office of Orphan Products Development works to fulfill the needs of patients and physicians for whom information concerning treatments for rare diseases are essential. We maintain electronic databases of all orphan product designations as well as grant studies being funded. Lists of associations of patients with rare diseases, and complete information regarding our program are provided upon request. A toll-free telephone number to reach the OPD office is available for health care workers and patients seeking facts regarding designated orphan drugs and clinical research that is being conducted on diseases affecting very small numbers of patients.

Although small numbers of people are affected by each of these illnesses, the approximately 5,000 known orphan diseases affect tens of millions of people throughout the world and represent a notable health problem in the United States, Japan, and the countries of Europe. From its inception, the purpose of the FDA's orphan drug program has been to effect and implement benefits that would encourage firms to develop products for needy, previously ignored populations. The success and impact of the Orphan Drug Act in the United States have been far greater than anticipated, as major public health benefits accrue to many, many people. It is working! Our record shows more than 112 products developed and approved for rare diseases for which little or no therapy previously existed.

REFERENCES

1.   *Significant Drugs of Limited Commercial Value*; Interagency Task Force Report to the Secretary of Health, Education & Welfare (HEW); June 29, 1979.

2.   Y. Minameda; *Orphan Disease Recognition and Treatment in Asia*; World Congress of Pharmacy and Pharmaceutical Sciences (F.I.P.) Tokyo, Japan (1993).

3.   Public Law No. 97-414 (1983).

4.   Public Law No. 75-717 (1938).

5.   Public Law No. 98-551 (1984).

6.   Excerpt of message from Senator Edward Kennedy, Chairman of the U.S. Senate Committee on Labor and Human Resources, to Dr. Michel Salomon, Chairman of the European Symposium on Orphan Drugs, Brussels (1987).

7.   *European Union Priorities for Research & Regulatory Actions*; Meeting Proceedings; Brussels (1995).

8.   *Scrip*; July 14, 1995; p 3.

9.   *Orphan Drugs*: Report of the National Association of the Pharmaceutical Industry in France (1995).

10.   Title 26, Code of Federal Regulations (CFR) Section 1.28.1. *Federal Register*; October 3, 1988 (53 FR 38708).

11.   Public Law No. 99-91 (1985).

12.   U.S. Senate Proposal S.1981 (1994); U.S. House of Representatives proposal H.R. 4160 (1994).

System for the orphan drug development in Japan

Toshiki HIRAI

Director, Research and Development Division Pharmaceutical Affairs Bureau, Ministry of Health and Welfare, 1-2-2 Kasumigasei Chiyoda-ku, Tokyo 100-45, Japan

Abstract: In order to promote the development of orphan drugs, the Pharmaceutical Affairs Law and the Law Concerning the Drug Fund for ADR Relief and R&D Promotion were partially amended in April 1993 and a new system was established on 1st October 1993. In the new system, drug companies which wish to develop drugs for use of rare and severe diseases are encouraged to apply for orphan drug status to the Ministry of Health and Welfare (MHW). Once designated as an orphan drug by MHW, the following preferential measures are given: (a) grants for R&D expenses; (b) tax reduction on R&D costs; (c) guidance and advice for the development of the drug; (d) acceleration of NDA review; and (e) extension of the term for re-examination to 10 years.

MHW has been collaborating with Organization for Drug ADR Relief, R&D Promotion and Product Review to carry out the promotion. At present, 80 drugs have been designated as orphan drugs since 1993 and 12 drugs/usages have been approved. Grant for R&D expenses for orphan drugs amounts to 400 million Japanese Yen in fiscal 1995.

Japanese Orphan Drug Development system was started in 1993. The criteria of orphan drugs are described below.

> The number of patients is estimated to be less than 50,000 in Japan
> Drugs are highly demanded because of the absence of substitutional drugs or medical treatment
> Probability of development is high

Total 80 drugs/indications have been designated. Among them 12 drugs/indications are already approved by MHW and 14 drugs/indications are applying for NDA.

Development of scientific technology has produced many effective and useful drugs. However, we still have urgent needs for drugs and medical devices for treatment of rare and severe diseases. It is difficult for a private company to develop such drugs alone. Ministry of Health and Welfare (MHW) has taken some measures to improve such a situation. For example, human growth hormone was approved in 1974 with the clinical data tested on approx. 30 patients.

In 1979, MHW started to give financial support for basic research on some new drugs. The name of the grant was changed to "Research Grants for the Development of Orphan Drug" in 1993. It means, the project focuses exclusively on researches of orphan drugs rather than those of other new drugs. Subsidies are given directly to selected research groups which are composed of mainly researchers from such institutions as universities and hospitals. Moreover, MHW announced in 1985 the Administrative Guidance on the Drugs for rare disease.

However, it was supposed that the effect of those measure was limited. Hence, MHW considered to establish a comprehensive system for the orphan drug development which could deal with the issue effectively. For the establishment of the system, MHW proposed amendment of the Pharmaceutical

Affairs Law. The Amendment of the Law was put into force on 1st October 1993. The important point of the amendment of the Law is to change the purpose of the Law. The change means that the purpose article contains the phrase that aims promoting research and development of drugs and medical devices.

The criteria to be designated as orphan drugs are described below.

The number of patients is estimated to be less than 50,000 in Japan.
Drugs are highly demanded because of the absence of substitutional drugs or medical treatment.
Probability of development is high

Minister of Health and Welfare designates orphan drugs taking into account the Central Pharmaceutical Affairs Council's opinion. Pharmaceutical companies which seek designation of orphan drug are required to submit the data to MHW showing that their drugs could meet the criteria. The details of these criteria are following.

Considering the number of the patients, reasonable estimates could be accepted as data for the deliberation of the Central Pharmaceutical Affairs Council when established statistics do not exist. About the needs for the drugs, usually the data on early clinical trial which shows some possibilities of effectiveness of the drug or foreign clinical trial data could be accepted for the deliberation. With the probability of the development of the drug, the company should show the plan to develop a drug after the drug would be designated as an orphan drug.

If a drug is designated as an orphan drug by the Minister for Health and Welfare, following preferential measures are given.

Grants for R&D expenses of the orphan drug are given to drug companies
The companies tax could be reduced as to the costs of the orphan drug development (6%)
Guidance and advice are given to the companies in order to facilitate the development of the orphan drug by the Drug Organization
MHW accelerates the review procedure on New Drug Approval Application of the orphan drugs
MHW gives ten years' term for re-examination when MHW approves the orphan drug, A second company which seek an approval of the same drug during these periods have to repeat all kinds of studies for New Drug Application. Thus the system gives an advantage to the first company
The annual schedule of designation of orphan drugs is usually done as follows:

Pre-hearing	September/October
Apply for orphan drug designation	November
Consult with Central Pharmaceutical Affairs Council	January/February
Public announcement of designation	1 April
	(Japanese fiscal year starts at 1st April)
Application for Grants	May

Table 1 shows the number of orphan drugs designated, approved and so in each fiscal year.

Fiscal Year	'93		'94	'95	
	grant	no grant	grant	grant	total
Designated drugs	19	21	29	11	80
Companies	17	15	21	10	52
Applied NDA	6	8	0	0	14
Approved	1	11	0	0	12
Suspended	0	0	6	0	6

This system started on 1st October 1993. In 1993, drugs of which studies were already completed or were under the review of NDA in the Central Pharmaceutical Affairs Council could be designated as orphan drugs. In such case, it was no need to provide grants for pharmaceutical companies.

At present, 80 drugs/indications have been designated as orphan drugs. Among them 12 drugs are already approved by MHW and 14 drugs are applying for NDA.

The framework of the new system for development of orphan drug has been established. The system has functioned well, then the pharmaceutical companies which have been developing orphan drugs have been encouraged to promote their development.

Design and pharmacological evaluation of a series of non-peptide endothelin ET$_A$ selective and ET$_A$/ET$_B$ receptor antagonists

A. Doherty, W. Patt, B. Reisdorph, J. Repine, D. Walker, M. Flynn, K. Welch, E. Reynolds and S. Haleen

Parke-Davis Pharmaceutical Research, Division of Warner-Lambert Company, Ann Arbor, MI, 48105 .

Abstract: This report will describe the design and pharmacological evaluation of both ET$_A$ selective and ET$_A$/ET$_B$ antagonists from the PD 155080 and PD 156707 series of orally active non-peptide ET$_A$ selective antagonists. Modification of the substituents around the butenolide ring has lead to compounds with differing selectivity for human ET$_A$ and ET$_B$ receptors. For example, several analogs of the subnamolar affinity ET$_A$ selective antagonist PD 156707 have been designed as either potent ET$_A$ or balanced ET$_A$/ET$_B$ antagonists. In this series the di-allyloxy analog (PD 161867) of PD 156707 is 7500-fold selective for the human ET$_A$ receptor. ET$_A$/ET$_B$ antagonists from this series include PD 160874, 162073 and 160672. For example, PD 160874 is a competitive inhibitor of [125I]ET-1 and [125I]ET-3 binding to human cloned ET$_A$ and ET$_B$ receptors with IC$_{50}$'s of 3.5nM (ET$_A$) and 8.9 nM (ET$_B$) respectively while PD 162073 exhibits and pharmacological evaluation of the non-peptide orally active PD 156707 series of ET antagonists where the selectivity ratios for ET$_A$ and ET$_B$ receptors have been varied from >2000 to 20-fold will be described.

INTRODUCTION

The potent vasoconstrictor endothelin (ET) is implicated in several human disease states including hypertension, congestive heart failure, atherosclerosis and other vascular disorders, renal failure, pulmonary hypertension, renal, cerebral and myocardial ischemia and cerebral vasospasm (1-9). Over the last few years research has advanced significantly in both the biology and pathophysiology of the endothelin system, partly due to the discovery of several important ET$_A$ and ET$_B$ antagonists.

The first reported antagonists were largely peptidic in nature including BQ-123 (10,11), and FR 139317 (12), which are potent ET$_A$-selective antagonists; balanced ET$_A$/ET$_B$ antagonists including PD 142893 (13) and PD 145065 (14) and the recently reported ET$_B$ selective antagonist BQ-788 (15). These compounds were useful agents to study the basic physiology of the endothelins however they, for the most part, suffered from short half-lives and lack of oral bioavailability, making their use difficult in chronic disease models. However in the last year or so a number of non-peptide endothelin antagonists have been reported. These include the Shionogi steroid analog 97-139 (16), and several balanced ET$_A$/ET$_B$ non-peptide antagonists, including Ro 46-2005 (9), Ro 47-0203 (bosentan) (17), SKF 209670 (18), CGS 27830 (19) and L-749,329 (20). In addition, several ET$_A$-selective antagonists have been described including BMS 182874 (21), PD 155080 (Figure 1) and PD 156707 (Figure 1) (22-25).

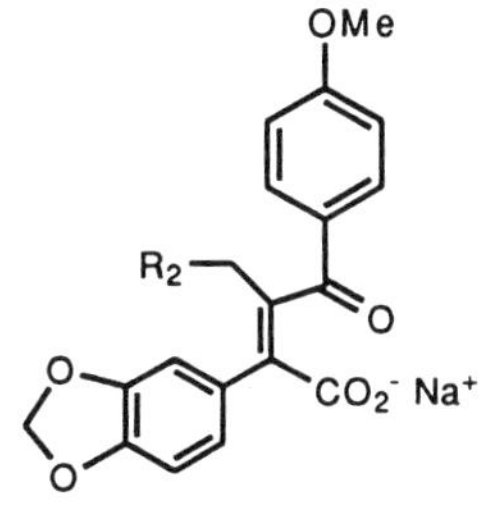

R$_2$ = Phenyl PD 155080

R$_2$ = 3,4,5-OMe-Phenyl PD 156707

Figure 1: Structures of PD 155080 and PD 156707

As has been reported previously PD 156707 is a highly selective ET_A antagonist (>1000 fold in several species) and has been reported to show efficacy in a cat model of cerebral ischemia with post-treatment administration (26,27).

In this report we will describe how structural modifications affect selectivity for the human ET_A and ET_B receptors ranging from over 1000-fold to less than 10-fold selectivity. Thus we have developed several analogs from the ET_A selective compounds, PD 155080 and PD 156707, that are ET_A/ET_B balanced agents. For example, PD 160874 is a competitive inhibitor of [125I]ET-1 and [125I]ET-3 binding to human cloned ET_A and ET_B receptors with IC_{50}'s of 3.5nM (ET_A) and 8.9 nM (ET_B) respectively (24). Another compound of interest in this series is the PD 156707 analog, PD 162073 with IC_{50}'s of 2.4nM (ET_A) and 50 nM (ET_B) respectively. The butenolide series of antagonists should prove very useful in elucidating the role of endothelin and the type of receptor selectivity required in a range of important human diseases.

MATERIALS AND METHODS

Chemistry

The synthesis of compounds in the butenolide series has been previously described (22, 24, 28). The butenolides were synthesized as shown in Scheme 1, via a modification of the method developed by Allen (29). Illustrated is the preparation of PD 156707: this route is efficient, high yielding and amenable to scale-up. The first step is the base-catalyzed condensation of 4-methoxyacetophenone 3 with piperonal 4 yielding the chalcone 5 in near quantitative yield. Reaction of this intermediate with potassium cyanide in acidic medium gave the corresponding nitrile 6 in excellent yield. Hydrolysis of the nitrile with p-toluenesulfonic acid gave the corresponding ketoester 7 in 80% yield. This key intermediate permitted easy access to the compounds of Tables 1-4. Reaction of the ester 7 and an appropriately substituted aldehyde with sodium metal in methanol, followed by acidification with acetic acid, generated the product butenolides 8-35 shown in Tables 1-4. Preparation of the sodium salt was achieved by reaction of the butenolide with sodium hydroxide followed by lyophilization. The aldehydes used were obtained commercially, or synthesized by alkylation of the corresponding phenol with alkyl halides in the presence of base. Certain commercially unavailable phenolic aldehydes were synthesized via an intermediate Schiff base by the method of Shulfin *et al.* (30).

SCHEME 1

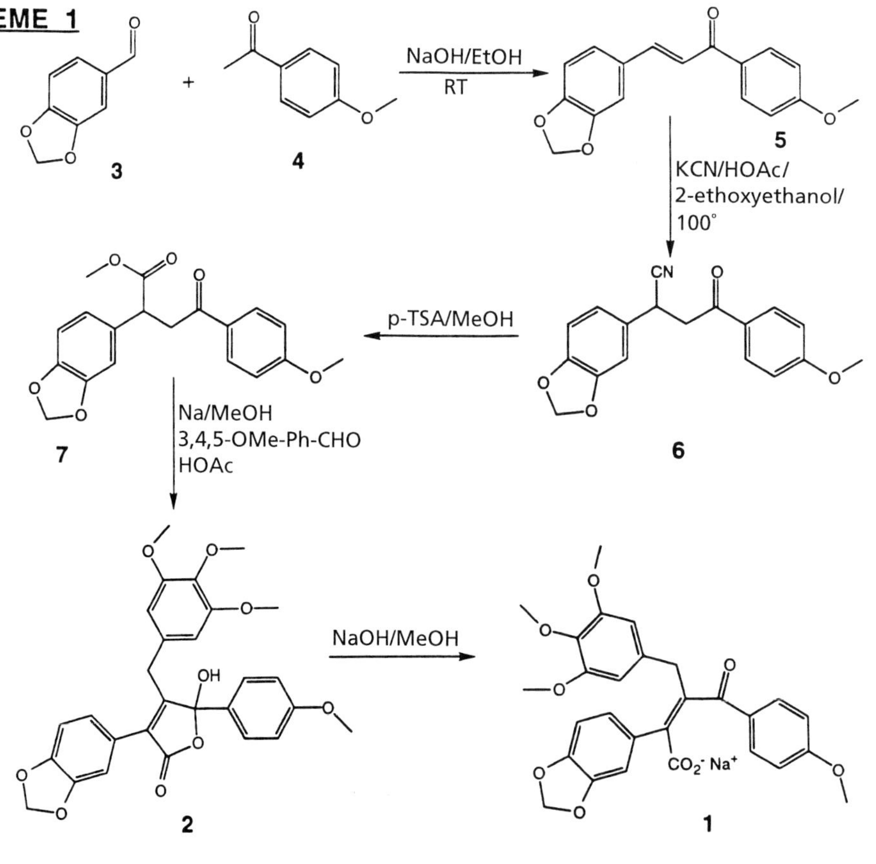

TABLE 1   Variation of the R_1 group

Compound	R_1	hET$_A$ IC$_{50}$(nM)	hET$_B$ IC$_{50}$(nM)	B/A
2	Methyl	0.3	780	2600
8	Ethyl	0.4	360	900
9	n-Propyl	0.3	170	566
10	n-Butyl	0.2	130	650
11	n-Pentyl	0.2	170	850
12	n-Octyl	1.3	700	538
13	**i-Propyl**	**0.7**	**70**	**100**
14	Allyl	0.3	82	273
15	n-(3)-Butenyl	0.4	140	350
16	Benzyl	0.6	114	184

TABLE 2   Variation of the R_2 group

Compound	R_2	hET$_A$ IC$_{50}$(nM)	hET$_B$ IC$_{50}$(nM)	B/A
2	Methyl	0.3	780	2600
17	Ethyl	0.2	220	1100
18	n-Propyl	0.6	640	1066
19	n-Butyl	2.6	148	57
20	**n-Pentyl**	**0.8**	**44**	**55**
21	**n-Hexyl**	**0.6**	**35**	**58**
22	**n-Octyl**	**2.4**	**120**	**50**
23	i-Propyl	0.5	120	240
24	Allyl	0.3	260	866
25	Benzyl	30	740	25

TABLE 3   Variation of the R_1 and R_2 groups

Compound	R_1	R_2	hET_A $IC_{50}(nM)$	hET_B $IC_{50}(nM)$	B/A
2	Methyl	Methyl	0.3	780	2600
26	i-Propyl	i-Propyl	0.2	34	170
27	n-Butyl	n-Butyl	2.7	1200	444
28	n-Pentyl	n-Pentyl	3.3	4200	1275
29	**Allyl**	**Allyl**	**0.2**	**1500**	**7500**
30	**i-Propyl**	**n-Pentyl**	**2.4**	**50**	**21**
31	Allyl	n-Pentyl	0.4	300	750

TABLE 4   Variation of the R_1, R_2 and R_3 groups

Compound	R_1	R_2	R_3	hET_A $IC_{50}(nM)$	hET_B $IC_{50}(nM)$	B/A
2	Methyl	Methyl	Methyl	0.3	780	2600
32	Ethyl	Ethyl	Ethyl	0.11	630	5727
33	n-Propyl	n-Propyl	n-Propyl	1.0	220	220
34	**i-Propyl**	**i-Propyl**	**i-Propyl**	**5.0**	**140**	**28**
35	Allyl	Methyl	Allyl	0.9	400	444

Biological Assays

The binding assay procedures using human cloned receptors have been reported previously (22,23). Cultured Ltk- cells expressing human ET_A receptors and CHO-K1 cells expressing human ET_B receptors were utilized. Specific binding was computer-analyzed by non-linear least squares curve fitting giving the best fit for a one-site model. IC_{50} values were derived from single competition experiments in which data points were measured in triplicate. The in vitro functional assays including inhibition of ET-1 or ET-3 induced arachidonic acid release and the contractility bioassays in ET_A and ET_B containing tissues have been previously described in detail (31-33).

RESULTS AND DISCUSSION

In Table 1 the R$_1$ group was varied while keeping the 4- and 5-methoxy substituents constant in PD 156707 (2). Increasing the chain length or chain branching of R$_1$ are well tolerated at the human ET$_A$ (hET$_A$) receptor. At the human ET$_B$ (hET$_B$) receptor an increase in activity is observed as the length of R$_1$ increases to butyl, while activity drops off as R$_1$ increases in length further. Compound 13, where R$_1$ is an isopropyl group, shows the most balanced hET$_A$/hET$_B$ receptor activities as well as the highest potency at the hET$_B$ receptor. In this series we have shown that hET$_A$ activity can be maintained below an IC$_{50}$ of 1.0 nM while hET$_B$ activity may be increased over 10-fold by varying the substitution at the 3-alkoxy position.

In Table 2 the R$_2$ group was varied while keeping the 3- and 5-methoxy substituents in PD 156707 constant. Increasing the chain length or chain branching of R$_2$ are well tolerated at the hET$_A$ receptor. At the hET$_B$ receptor an increase in activity is observed as the length of R$_2$ increases to n-hexyl, while activity is reduced as R$_1$ is increased further in length. Compounds 20, 21 and 22 show the best balanced profiles in this series, exhibiting about 50-fold selectivity for hET$_A$ over hET$_B$ receptors.

The compounds in Table 3 are analogs where both the R$_1$ and R$_2$ substituents have been varied. Increasing the chain length of R$_1$ and R$_2$ simultaneously from methyl to pentyl leads to decreased activity at both hET$_A$ and hET$_B$ receptors, except in compounds 26 and 29. In compound 26, hET$_A$ activity is maintained while a large increase in hET$_B$ potency is observed. ET$_A$ activity is maintained in compound 29 while ET$_B$ activity decreases leading to a highly hET$_A$ selective compound (7500-fold). Substitution with non-identical groups led to compound 30 which shows a 15-fold increase in ET$_B$ affinity, in addition the best balance of hET$_B$/hET$_A$ affinities. In this series we have shown that hET$_A$ activity can be maintained below 1.0 nM while improving the hET$_B$ potency by about 20-fold. Compound 30 where R$_1$ = i-propyl and R$_2$ = pentyl exhibits a 120-fold improvement in the ET$_B$/ET$_A$ affinity ratio. Compound 29 (PD 161867) is the most ET$_A$ selective compound discovered in this set of analogs.

The compounds in Table 4 combine modifications at all three methoxy groups in PD 156707 (2). ET$_A$ activity was maximized in compound 32, while ET$_B$ activity was only modestly improved, and selectivity towards the ET$_B$ receptor was decreased more than 2-fold. hET$_A$ activity was maintained at or below an IC$_{50}$ of 1.0 nM in compounds 33 and 35, with only small improvements in hET$_B$ affinity. The best hET$_B$ activity and the compound with the most balanced profile was 34 with IC$_{50}$'s of 5.0 and 140 nM for the hET$_A$ and hET$_B$ receptors respectively. In this series the triethoxy analog 32 is a highly selective ET$_A$ antagonist.

In conclusion, modification of the substituents around the butenolide ring has lead to compounds with differing selectivity for human ET$_A$ and ET$_B$ receptors. Several of these compounds have been evaluated for their functional effects at ET$_A$ and ET$_B$ receptors (Table 5). PD 155080, PD 156707 (2) and PD 161867 (29) are highly ET$_A$ selective antagonists exhibiting antagonism of ET-1 induced vasoconstriction in rabbit femoral artery with pA$_2$s of 6.3, 7.5 and 7.8 respectively. Compounds with a more balanced ET$_A$/ET$_B$ profile are exemplified by PD 160672 (20), the 4-O-n-pentyl analog of PD 156707. PD 160672 is potent competitive inhibitor of [^{125}I]ET-1 and [^{125}I]ET-3 binding to human cloned ET$_A$ and ET$_B$ receptors with IC$_{50}$'s of 0.8 nM and 44 nM respectively. PD 160672 inhibits ET-1 and SRTX-6c induced vasoconstriction in rabbit femoral (ET$_A$) and pulmonary (ET$_B$) arteries with pA$_2$ values of 7.2 and 7.1 respectively. Other analogs with ET$_A$/ET$_B$ balanced profiles include compounds 13, 20-22, 26 and 30. The functional activities of compounds 13, 20 and 26 are shown in Table 5. In addition to different substituents on the aromatic ring, completely saturated analogs also exhibit a balanced ET$_A$/ET$_B$ profile, exemplified by the previously reported compound PD 160874 (Figure 2) (24). This compound is potent competitive inhibitor of [^{125}I]ET-1 and [^{125}I]ET-3 binding to human cloned ET$_A$ and ET$_B$ receptors with IC$_{50}$'s of 3.5 nM and 8.9 nM respectively. The ability to vary the selectivity by simple modification of the substituents at the central phenyl ring make this an interesting and useful series of endothelin antagonists to study in various disease states.

Table 5. In vitro binding and functional data

Compound	Binding hET$_A$ IC$_{50}$/nM	hET$_B$	Functional AAR$_A$	AAA$_B$	Vasoconstriction ET$_A$ pA$_2$	ET$_B$ pA$_2$
PD 155080 sodium salt	7.4	4500	49.0	>1,000	6.3	4.8
PD 156707 (**2**) sodium salt	0.31	780	1.1	120	7.5	5.4 (4.5*)
PD 161867 (**29**)	0.20	1500	5.0	>1000	7.8	<5.0
PD 160874 (ref 24)	3.5	8.9	39.0	36.0	6.9	6.6
PD 160773 (**13**)	0.70	70	1.4	67	7.6	6.1
PD 160672 (**20**)	0.8	44	11.0	69.0	7.2	7.1
PD 162344 (**26**)	0.2	34	2.8	58	7.5	-

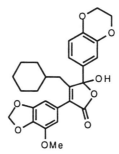

Figure 2: Structure of PD 160874

REFERENCES

1.   M. Yanagisawa *et al. Nature (Lond.)* 332, 411-415 (1988).
2.   A. Inoue *et al. Proc. Natl. Acad. Sci. USA* , 86, 2863-2867 (1989).
3.   A.M. Doherty. *J. Med. Chem.*. 35, 1493-1508 (1992) .
4.   T. Watanabe *et al. Nature* 344, 114 (1990).
5.   Y. Saito *et al. New Engl. J. Med.* 322, 205 (1989).
6.   A. Giaid *et al. New Engl. J. Med.* 328, 1732-1739 (1993).
7.   K. Takahashi *et al. Nephron* 66, 373-379 (1994).
8.   F. Cosentino and Z.S. Katusic. *Stroke* 25, 904-908 (1994).
9.   M. Clozel *et al. Nature* 365, 759-761 (1993).
10.   K. Ishikawa *et al. J. Med. Chem.*. 35, 2139-2142 (1992).
11.   M. Ihara *et al. Life Sci.*. 50, 257-255 (1992).
12.   K. Sogabe *et al.J . Pharmacol. Exp. Ther.* 264, 1040-1046 (1993).
13.   W.L. Cody *et al. J. Med. Chem.*. 35, 3301-3303 (1992).
14.   W.L. Cody *et al. Med. Chem. Res.* 3, 154-162 (1993).
15.   K. Ishikawa *et al. Proc. Natl. Acad. Sci.*.91, 4892-4896 (1994).
16.   S. Mihara *et al. J. Pharmacol. Exp. Ther.* 268, 1122-1128 (1994).
17.   S.P. Roux *et al. Circulation* 88, I-170 (1993).
18.   J.D. Elliott *et al. J. Med. Chem.* 37, 1553-1557 (1994).
19.   B. Mugrage *et al. Bioorg. Med. Chem. Lett.* 3, 2099-2104 (1993).
20.   T.F. Walsh *et al. ACS National meeting, Washington, August, MEDI 145 (1994).
21.   P.D. Stein *et al. J. Med. Chem.* 37, 329-331 (1994).
22.   A.M. Doherty *et al. J. Med. Chem.* 38(8), 1259-1263 (1995).

23. E.E. Reynolds *et al. J. Pharmacol. Exp. Ther.* 273(3), 1410-1417 (1995).
24. A.M. Doherty *et al.* (1995) 4th International Conference on Endothelin. April 23-26, London, UK, C40. J. Cardiovasc. Pharmacol. *in press* (1995).
25. J.J. Maguire *et al.* 4th International Conference on Endothelin. April 23-26, London, UK, C42 (1995).
26. T. Patel *et al.* 4th International Conference on Endothelin, April 23-26 London, UK, C43 (1995).
27. T. Patel *et al.* 4th International Conference on Endothelin, April 23-26, London, UK, C63 (1995).
28. B.R. Reisdorph *et al.* 210th ACS National meeting, Chicago, Illinois, August 20-24, MEDI # 39 (1995).
29. C.F.H. Allen and G.F. Frame. *Can. J. Res.* 605-613 (1932).
30. A.T.Shulgin and P.Jacob III. *Syn. Commun.* 11, 969-977 (1981).
31. A.M. Doherty *et al. Bioorg. Med. Chem. Lett.* 3(4), 497-502 (1993).
32. A.M. Doherty *et al. J. Cardiovasc. Pharmacol.* 22(Suppl. 8), S98-S102 (1993).
33. R.L. Panek *et al. Biochem. Biophys. Res. Commun.* 183, 566-571 (1992).

The DNA gyrase inhibitor cyclothialidine: progenitor of a new class of antibacterial agents

E. Goetschi,* P. Angehrn, H. Gmuender, P. Hebeisen, H. Link, R. Masciadri, J. Nielsen, P. Reindl, and F. Ricklin

Pharma Division, Preclinical Research, F. Hoffmann-La Roche Ltd., CH-4002 Basel, Switzerland

Abstract: Cyclothialidine (Ro 09-1437) is a new DNA gyrase inhibitor isolated from *Streptomyces filipinensis* NR0484. It represents a new class of natural products and was found to act by competitively inhibiting the ATPase activity exerted by the B subunit of DNA gyrase. However, cyclothialidine hardly exhibits any growth inhibitory activity against intact bacterial cells, probably due to insufficient penetration into the cytoplasm. To exploit the antibacterial potential of this lead compound, a general synthetic route was developed allowing for the systematic modification of its unique structure. From the total synthesis and the biological examination of a variety of its congeners, the minimal structural requirements for DNA gyrase inhibitory activity were found to be contained in a rather small partial structure of cyclothialidine. Based on this "minimal structure", several subclasses of new DNA gyrase inhibitors were found comprising bicyclic compounds containing 11- to 16-membered lactone rings as well as "*seco*-compounds" lacking the lactone moiety. Some of them exhibit a potent and broad activity *in vitro* against Gram-positive bacteria including *Staphylococcus aureus, Streptococcus pyogenes, Streptococcus pneumoniae* and *Enterococcus faecalis*, and overcome resistance against antibacterial agents clinically used today.

INTRODUCTION

The need to continually provide new agents for antibacterial chemotherapy has ensured a high interest in this research area. Besides the desire to develop drugs of higher potency, expanded spectrum of activity and improved safety profile, it is the emergence and spread of bacterial resistance to most of the antibacterial classes used clinically that necessitates a continuous search for new structural entities and novel targets for antibacterial attack. The bacterial type II topoisomerase, DNA gyrase, is targeted by two major classes of antibacterials: The fully synthetic quinolones, e.g. ciprofloxacin, are subunit A inhibitors interfering directly with the interactions between DNA and the A subunit of DNA gyrase (1), whereas the coumarin antibiotics, e.g. novobiocin and coumermycin A1, are competitive inhibitors of the ATPase activity conferred by the B subunit of DNA gyrase (2). Both drug families were discovered by the traditional method used for the identification of antibacterial principles that relies on the inhibition of the growth of bacterial cultures. By screening microbial broths for the *in vitro* inhibition of the supercoiling activity of DNA gyrase, a team at the Nippon Roche Research Center in Kamakura, Japan, discovered cyclothialidine 1 (Ro 09-1437) as a first representative of a new class of DNA gyrase inhibitors (3). The

Ciprofloxacin

Novobiocin

1

Cyclothialidine (Ro 09-1437)

compound was isolated from the fermentation broth of *Streptomyces filipinensis* NR0484 as an ampho-
teric solid. Its structure was elucidated by a combination of spectroscopic methods and amino acid analyses and turned out to comprise a twelve-membered lactone ring, partly integrated into a pentapeptide
chain containing a cis-3-hydroxy-L-proline, and fused to a highly substituted benzene ring (4). Cyclothialidine was found to be a potent and broad-spectrum inhibitor (5) that acts by abolishing the ATPase
activity of the B subunit of DNA gyrase (6). However, it exhibits only minimal activity against intact
bacterial cells - presumably due to poor penetration of the cytoplasmic membrane - and therefore had not
been identified in the traditional antibiotic screening. Nevertheless, the fact that growth of highly
susceptible bacterial species such as *Eubacterium moniliforme* was inhibited supported the view of
cyclothialidine being a promising lead structure for the development of a new class of antibacterials.

CHEMISTRY

In order to explore and exploit the potential of this natural product with respect to the development of a
new type antibacterial, we started a chemical modification programme. A classical strategy was applied
consisting of (i) investigating the structural requirements for DNA gyrase inhibition, and (ii) searching for
congeners of cyclothialidine which do exert growth inhibitory activity against intact bacterial cells. To
this end, we set out to develop an efficient and general synthetic route to bicyclic lactones IX that should
allow the preparation of a great variety of cyclothialidine analogues (Scheme 1) and indeed has proven its
flexibility so far in many different syntheses (7). A suitably substituted o-toluate I is converted into a
benzylating agent such as a bromide II. The peptidic moiety is then elaborated either by consecutive
attachment of a cysteamine derivative III followed by a peptide coupling with a 3-hydroxy-propionic acid
derivative V or, alternatively, by the alkylation of a thiol VII by the benzyl bromide II. Deprotection of the
carboxylic acid function provides a ω-hydroxy-acid VIII which is then cyclized to a lactone IX using
classical lactonization procedures.

Before illustrating this general synthetic route by discussing the preparation of the first analogue
exhibiting DNA gyrase inhibiting properties, an important result of the biological testing shall be
anticipated. Part of our initial synthetic objectives were simply substituted bicyclic lactones IX mainly
prepared to establish a feasible methodology. With one of these lactones IX (R^1=Me, R^2=R^3=H,
R^4=COOMe, R^5=NHBoc) in hand, we continued the modification work and completed stepwise the
pentapeptide moiety of cyclothialidine. The finding that 12,14-dideoxy-cyclothialidine was devoid of
DNA gyrase inhibitory activity - and the same was noted for all its precursors - clearly revealed that the
phenolic hydroxy groups were essential for the biological activity.

Scheme 1. General Synthetic Scheme

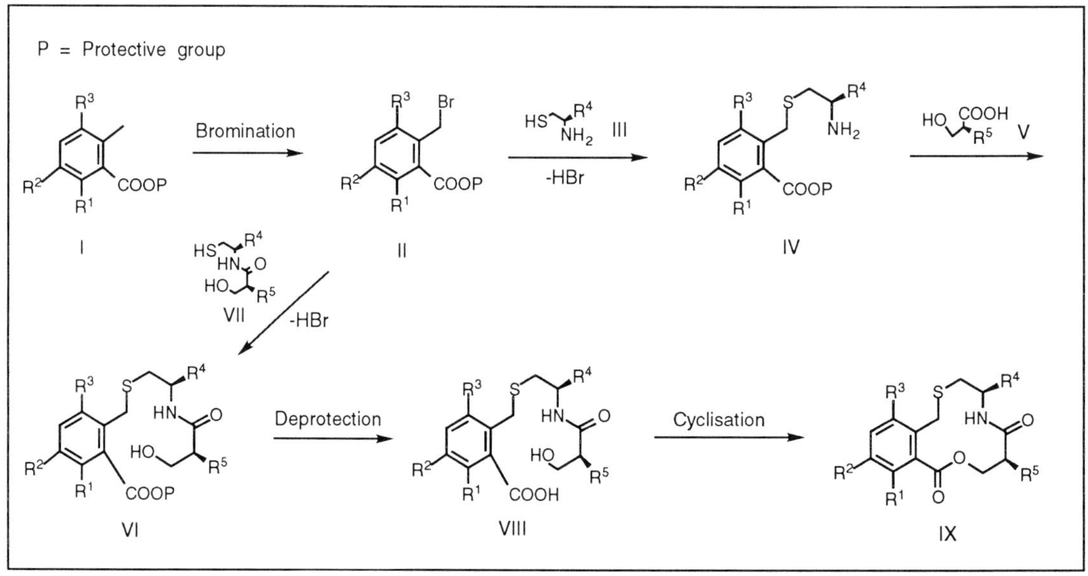

As a consequence of this, we focused our synthetic efforts on the bicyclic lactone featuring the original aromatic substitution pattern of cyclothialidine. Scheme 2 summarises the preparation of hydroxy-substituted lactones, such as 7 (Ro 42-9416) and of cyclothialidine (1) itself. In this synthesis, the tetra-substituted benzoic acid 2 was first protected at its oxygen functions by esterification and silylation, respectively, and then subjected to bromination with N-bromosuccinimide. The crude benzyl bromide derivative 3 contaminated with the o,o' bis-bromomethyl by-product was used directly for the coupling with the protected serinyl-cysteine dipeptide 4. The p-nitrobenzyl ester of the purified product was cleaved by hydrogenolysis to afford the hydroxy-acid 5. Lactonisation under Mitsunobu conditions afforded the valuable key intermediate 6. By cleavage of the silyl (TBDMS) protective groups, the first biologically active congener (7, Ro 42-9416) of cyclothialidine was obtained. By selective deprotection using standard procedures, the intermediates 8 and 9, suitable for further side-chain modification, were prepared.

Starting from the amine 9, we completed the peptidic side-chains of cyclothialidine. The missing amino acids were stepwise attached by a sequence of coupling and deprotection reactions, and the synthesis of cyclothialidine was finally accomplished by cleavage of the protective groups. The synthetic material thus obtained could not be distinguished from the natural one in its chromatographical, spectroscopic and biological properties.

Scheme 2. Synthesis of Ro 42-9416 and of Cyclothialidine

PNB = —CH$_2$—⟨⟩—NO$_2$

TBDMS = Si(Me)$_2$Bu^t

3Hyp = cis-3-hydroxy-L-proline

Cyclothialidine

$[\alpha]_D^{20}$ = -14.6° (c 1.0, H$_2$O)

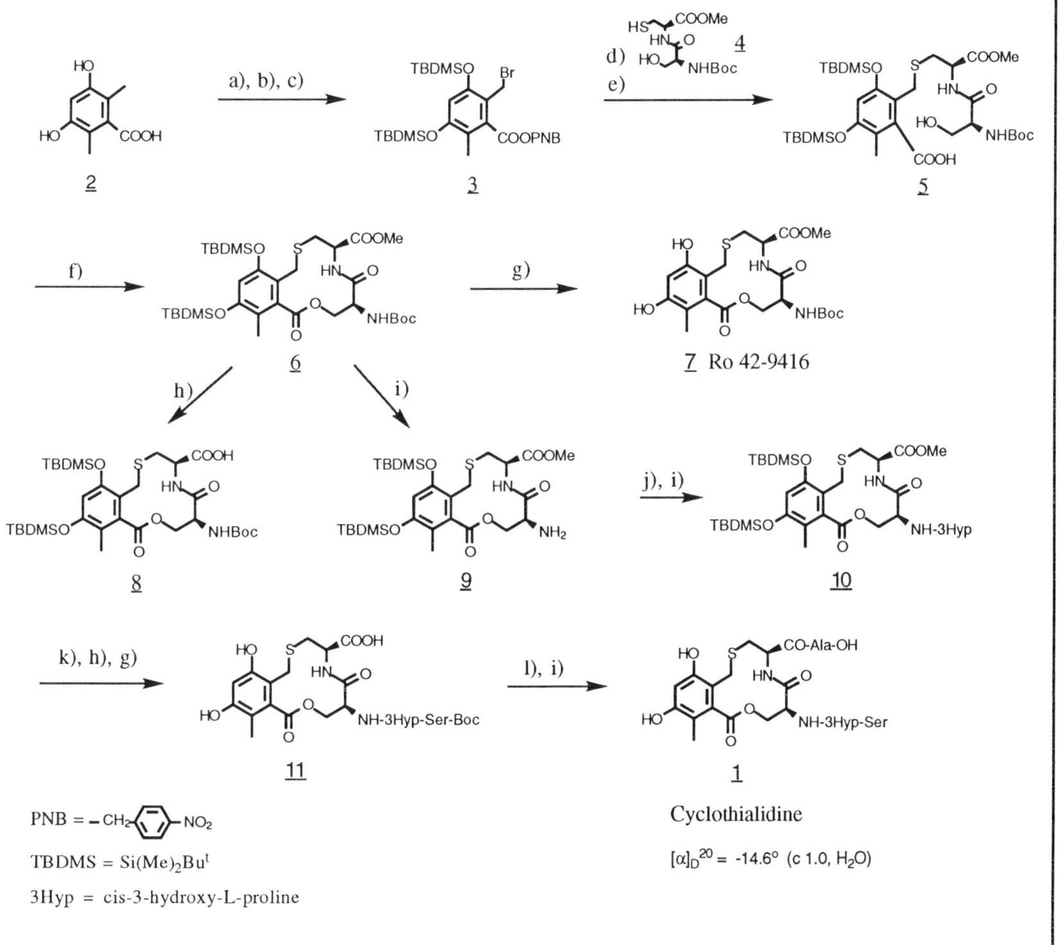

a) p-Nitrobenzyl bromide /1,1,3,3-tetramethylguanidine / DMF; b) TBDMSCl / NEt$_3$ / DMF; c) NBS / CCl$_4$ / hv; d) 4 / NEt$_3$ / CH$_2$Cl$_2$; e) H$_2$ / Pd-C / AcOEt; f) DEAD / TPP / toluene; g) NH$_4$F / MeOH; h) NaOH / H$_2$O / THF; i) TFA; j) Boc-3Hyp-OH / EDC / MeCN; k) Boc-Ser-OH / EDC / MeCN; l) i SuOH / EDC / MeCN, ii Ala-OBu^t.

BIOLOGICAL ACTIVITY OF CYCLOTHIALIDINE ANALOGUES

The synthetic route described in Scheme 1 has been used for the preparation of a large variety of cyclo-thialidine analogues. In the following, typical features of the structure-activity relationships with regard to the enzyme inhibitory activity and the *in vitro* antibacterial activity will be discussed. The DNA gyrase inhibitory activity of new compounds was determined in an *in vitro* supercoiling assay. In this assay (8), the introduction of supercoils into a relaxed plasmid is analysed by gel electrophoresis. For practical reasons, a MNEC value (Maximum Non Effective Concentration) was determined. It was found that in general this MNEC value of a compound was 3-5 times lower than the IC_{50}, and 10-20 times lower than the concentration needed to for complete inhibition of the supercoiling reaction. In our assay, cyclothia-lidine was found to be slightly more active than novobiocin. *In vitro* antibacterial activities were deter-mined using standard agar dilution methods. The discussion will follow more or less the development in our modification work. A first section comprises the conservative variations of the cyclothialidine molecule investigating the importance of the various substituents and exemplifying the emergence of *in vitro* antibacterial activity. In a second part, results from more extensive structural modifications affecting also the bicyclic core structure of cyclothialidine will be described.

1.Variation of the side chains of cyclothialidine

The lack of DNA gyrase inhibitory activity of 12,14-dideoxy-cyclothialidine undoubtedly pointed to the important role of the phenolic hydroxy groups. Activity for a synthetic analogue of cyclothialidine was indeed observed for the first time for a bicyclic lactone (7, Ro 42-9416) bearing the aromatic substitution pattern of the lead structure. In order to probe the impact of the various substituents of the bicyclic core structure, the individual substituents of Ro 42-9416 were systematically disconnected to afford five tetra-substituted derivatives (Table 1). It turned out that with the exception of the phenol group in the 14-position (R^3), removal of all the other substituents did not impair the activity of the compounds. The partial structure common to all compounds found active is the bicyclic 14-hydroxylated lactone X.
The observation that either of the two substituents at the lactone ring could be removed without signifi-cant loss of activity raised the question if the entire lactone ring was recognised at all by the enzyme . The answer was provided by the finding that the enantiomer of Ro 42-9416 showed only borderline activity, emphasising the importance of the 3-dimensional shape of the lactone (9). From the comparison of the activity of Ro 42-9416 with that of its three diastereomers we learned that it is in particular the R-confi-guration at C4 which ensures the enzyme inhibitory activity of the ring system, presumably by controlling the lactone conformation. At C7, R- and S-configuration is tolerated. The simultaneous removal of both C4- and C7-substituents virtually abolishes the biological activity (9).

Table 1. Structural Requirements for DNA Gyrase Inhibition: The Importance of the Side Chains

Ro #	R^1	R^2	R^3	R^4	R^5	MNEC[1] (µg/ml)	"Minimal Structural Requirement" [2]
42-9416	Me	OH	OH	COOMe	NHBoc	0.5	
43-2634	(H)	OH	OH	COOMe	NHBoc	0.5	
43-1683	Me	(H)	OH	COOMe	NHBoc	1	
43-1173	Me	OH	[H]	COOMe	NHBoc	>100	
43-0905	Me	OH	OH	(H)	NHBoc	1	
43-4109	Me	OH	OH	COOMe	(H)	0.2	

1) Supercoiling assay. MNEC = Maximum non effective concentration. 2) Partial structure common to compounds active in the supercoiling assay.

On the basis of the bicyclic lactone **X**, we have prepared many potent DNA gyrase inhibitors with activities similar to that of the lead compound. However, with regard to the antibacterial activity, progress was rather modest (Table 2). Cyclothialidine itself showed minimal antibacterial activity in our test system and, with the exception of *Streptococcus pyogenes* β15, only very sensitive strains were susceptible. The first synthetic compound (7, Ro 42-9416) found active in the enzyme assay was only slightly better. Significant progress was achieved by the methylation of one phenol group. Compounds of the 12-methoxy series (R^2 = OMe) exhibited moderate anti-staphylococcal activity. Systematic variations of the lactone substituents eventually led to compounds of improved antibacterial potency, among the best compounds being the 1,2,4-oxadiazole derivative Ro 47-3904 and the thioacetyl derivative 46-9288. These congeners of cyclothialidine display a moderate but broad-spectrum activity against Gram-positive bacteria. In general, Gram-negative bacteria such as *Klebsiella oxytoca, Enterobacter cloacae, Pseudomonas aeruginosa* and wild type strains of *Escherichia coli* were not susceptible, although activity was found against selected strains of *Neisseria meningitidis* and *Moraxella catarrhalis* .

Ro 46-9288 was tested in comparison with vancomycin and novobiocin against a larger number of Gram-positive isolates. Although less active than vancomycin and novobiocin, Ro 46-9288 showed consistent activity with MIC values in a narrow range. Its activity against methicillin sensitive and methicillin resistant strains of *Staphylococcus aureus* was found to be 2-4 times, against strains of *Staphylococcus epidermidis* 1-2 times, and against strains of *Streptococcus pyogenes, Streptococcus agalactiae,* the viridans group *streptococci, Enterococcus faecalis,* and *Enterococcus faecium* about 8-32 times lower than that of vancomycin. Growth inhibition kinetics of *Staphylococcus aureus* ATCC 25923 revealed that at a concentration 4 times higher than its MIC, Ro 46-9288 showed slow killing of the bacterial cells similar to the effect of novobiocin. In comparison, vancomycin was more rapidly bactericidal under the test conditions used (9).

TABLE 2. *In Vitro* Antibacterial Activity of Cyclothialidine Analogues

Ro #	R^1	R^2	R^3	Gyrase Inhibitory Activity[1] MNEC (µg/ml)	*E. coli* DC2	*N. meningitidis* 69480	*M. catarrhalis* RA 117	*S. aureus* 887	*S. pyogenes* β15	*M. luteus* ATCC8340
Cyclothialidine	OH	NH-Ala-OH	NH-3Hyp-Ser	0.05	>	128	>	>	32	128
43-4106	OH	COOMe	NH-3Hyp-Boc	0.05	>	64	>	>	32	8
42-9416	OH	COOMe	NHBoc	0.5	>	16	>	>	32	16
43-4109	OH	COOMe	H	0.2	>	32	>	>	128	16
43-9052	OMe	COOMe	NHBoc	0.2	>	16	64	32	64	16
46-2252	OMe	COOMe	NHCHO	0.1	32	2	32	32	4	8
46-5734	OMe	COOMe	NH$_2$	1	>	128	>	>	128	64
47-0268	OMe	COOMe	OH	0.1	64	8	32	16	32	16
44-4728	OMe	COOMe	H	0.5	64	16	32	32	32	32
44-4729	OMe	CH$_2$OH	H	>5	>	128	>	>	64	128
44-4730	OMe	CH$_2$OAc	H	2.5	128	16	64	64	64	64
47-3916	OMe	CONH	H	0.1	64	4	64	64	16	16
47-4491	OMe		H	0.2	>	16	64	64	16	16
47-3899	OMe		H	0.05	16	2	8	4	8	4
46-9288	OMe	COOEt	NHCSMe	0.2	64	2	8	8	8	4
47-3904	OMe		OH	0.05	8	1	4	4	4	2

1) Supercoiling assay; MNEC= Maximal non effective concentration. 2) Agar dilution (BB2 medium supplemented with 1% IsoVitalex and 5% sheep blood). Inoculum 10^4 CFU/spot. MIC = Minimal inhibitory concentration; ">" = >128 µg/ml.

2. Structural modifications of the bicyclic core structure of cyclothialidine

The fact that the oxidation of the ring sulfur to a sulfoxide or a sulfone group markedly reduced the biological activity of cyclothialidine analogues raised the question whether this potential metabolical target could be replaced by a more stable group. Therefore, we synthesized two isosteric analogues of Ro 44-8000, i.e. Ro 46-5222 and Ro 47-0146, in which the sulfur atom was replaced by methylene or oxygen, respectively (Table 3). It was found that the DNA gyrase inhibitory activity for the methylene analogue was diminished by a factor 50, and that of the oxygen analog was abolished completely. Conformational studies (NMR, CD) suggested the weakly active methylene analogue to adopt a similar conformation as the parent compound, whereas the oxygen analogue seems to prefer a different conformation in solution (vid. $J_{3,4'}$). These data, however, do not provide a satisfying explanation for the observed biological results.

In order to probe the importance of the bicyclic core structure **X**, we synthesized a homologous series of 11- to 18-membered bicyclic lactones (Table 4). As we had observed earlier that the biological activity of cyclothialidine analogues was improved by converting the lactam function into a thiolactam, we also prepared the analogous thiolactam series. DNA gyrase inhibitory activity was found over a rather large structural scope, i.e. 11- to 16-membered lactones, but showed an optimum for 14-membered lactones. Considering this high degree of structural variability tolerated by the enzyme, it was not unexpected that we eventually identified "*seco-compounds*" - analogues lacking a lactone entity - that also inhibited DNA gyrase. The partial structure **XI** features the common element present in all active cyclothialidine analogues. As shown in Table 4, the antibacterial activities found for these homologous series reflect in a qualitative fashion the enzyme inhibitory activity. The *in vitro* antibacterial activity against repesentative strains of two 14-membered lactones, Ro 47-5990 and Ro 48-2865, are given in Table 5. These congeners of cyclothialidine still do not exhibit growth inhibitory activity against Gram-negative strains - *Xanthomonas maltophilia* being an exception. They do, however, cover broadly the spectrum of Gram-positive bacteria, and the 1,2,4-oxadiazole Ro 48-2865 exhibits considerable antibacterial potency.

TABLE 3. Replacement of the Ring Sulfur by Methylene or Oxygen

	X = S (Ro 44-8000)	X = CH2 (Ro 46-5222)	X = O (Ro 47-0146)
DNA gyrase inhibition: MNEC (µg/ml)[1]	0.2	10	>100
Antibacterial activity *in vitro*: MIC (µg/ml)[2]			
Staphylococcus epidermidis 16-2	8	>64	>64
Streptococcus pyogenes β15	16	>64	>64
^{1}H-NMR (d4-MeOH) coupling constants (Hz):			
$J_{3,4}$ / $J_{3',4}$	4.4 / 10.4	5.2 / 12.4	4.6 / 4.6
$J_{7,8}$ / $J_{7,8'}$	2.9 / 2.5	3.5 / 1.5	3.9 / 2.5

1) DNA gyrase inhibition. Supercoiling assay. MNEC = Maximum non effective concentration. 2) Antibacterial activity *in vitro*. Agar dilution (Mueller-Hinton medium). Inoculum 10^4 CFU/spot. MIC = Minimum inhibitory concentration.

Table 4. Inhibitor Subclasses Structurally Related to Cyclothialidine

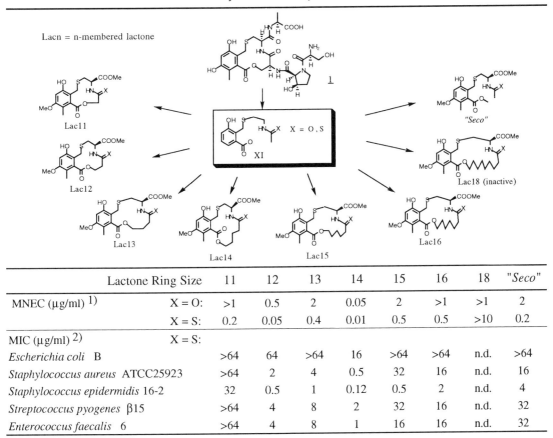

Lactone Ring Size		11	12	13	14	15	16	18	"Seco"
MNEC (µg/ml) [1]	X = O:	>1	0.5	2	0.05	2	>1	>1	2
	X = S:	0.2	0.05	0.4	0.01	0.5	0.5	>10	0.2
MIC (µg/ml) [2]	X = S:								
Escherichia coli B		>64	64	>64	16	>64	>64	n.d.	>64
Staphylococcus aureus ATCC25923		>64	2	4	0.5	32	16	n.d.	16
Staphylococcus epidermidis 16-2		32	0.5	1	0.12	0.5	2	n.d.	4
Streptococcus pyogenes β15		>64	4	8	2	32	16	n.d.	32
Enterococcus faecalis 6		>64	4	8	1	16	16	n.d.	32

1), 2) see below

TABLE 5. *In Vitro* Antibacterial Activity of Ro 47-5990, Ro 48-2865, Novobiocin and Vancomycin

	Ro 47-5990	Ro 48-2865	Novo-biocin	Vanco-mycin
DNA gyrase inhibition: MNEC (µg/ml) [1]	0.01	0.002	0.1	-
Antibacterial activity *in vitro*: MIC (µg/ml) [2]				
Escherichia coli ATCC 25922	>16	>16	>16	>16
Klebsiella pneumonia NCTC 418	>16	>16	>16	>16
Enterobacter cloacae P99	>16	>16	>16	>16
Xanthomonas maltophilia IAC 739	4	1	>16	>16
Staphylococcus aureus ATCC 25923	0.25	0.06	0.25	2
Staphylococcus epidermidis ATCC 14990	0.12	0.016	0.12	2
Enterococcus faecalis 6	1	0.12	8	2
Enterococcus faecium DUMC 73-92	4	0.5	2	>16
Streptococcus pneumoniae 907	1	0.25	2	0.5
Streptococcus pyogenes β15	1	0.25	2	0.5

1) DNA gyrase inhibition. Supercoiling assay. MNEC = Maximum non effective concentration. 2) Antibacterial activity *in vitro*. Agar dilution (Mueller-Hinton medium). Inoculum 10^4 CFU/spot. MIC = Minimum inhibitory concentration. n.d. = not determined.

DISCUSSION

The screening technique applied in the discovery of the DNA gyrase inhibitor cyclothialidine has provided a lead compound for which its outstanding inhibitory activity against the isolated bacterial enzyme was not translated into an adequate growth inhibitory activity against intact bacteria. To the medicinal chemist, the task of removing this discrepancy provided a worthwile challenge, all the more since the unique structure of this natural product seemed to be synthetically easily accessible. An extensive modification programme has indeed revealed cyclothialidine to be a progenitor of a number of subclasses of DNA gyrase inhibitors including bicyclic lactones of various ring sizes and also comprising "*seco*-compounds" lacking the lactone moiety. A benzyl sulfide entity XI bearing a single phenolic hydroxy function was identified to be the structural element common to all congeners of this inhibitor family. It can be concluded that in all compounds found active, this "pharmacophore" is able to adopt a similar conformation when binding to DNA gyrase. The detrimental effect on the biological activity caused by the replacement of the sulfur atom of this entity remains unexplained and might be attributed to either impairing of the required binding conformation or a generally weaker interaction of the sulfur substitutes with the enzyme.

As a result of the extensive structural modifications, congeners of cyclothialidine were found exhibiting potent and broad-spectrum *in vitro* antibacterial activity against Gram-positive species including staphylococci, streptococci, pneumococci and enterococci. Therefore, the DNA gyrase inhibitory principle contained in cyclothialidine can be considered as a basis for a new class of antibacterial agents. It remains to be shown whether this class of antibacterials can provide clinically useful drugs.

Acknowledgements: The authors thank their co-workers for the excellent technical assistance: P. Schneider, R. Villard, G. Wassner, and D. Wechsler for the chemical syntheses. R. Blum, K. Kuratli, J. Nell, K. Outten, I. Pfister, P. Schmitz, and U. Weis for the biological testing .

REFERENCES

1. D. C. Hooper and J. S. Wolfson, Eur. J. clin. Microbiol. Infect. Dis., **10**, 223 (1991).
2. K. Mizuuchi, M. H. O'Dea, and M. Gellert, Proc. Natl. Acad. Sci. USA **75**, 5960 (1978).
3. J. Watanabe, N. Nakada, S. Sawairi, H. Shimada, S. Ohshima, T. Kamiyama, and M. Arisawa, J. Antibiot. **47**, 32 (1994).
4. T. Kamiyama, N. Shimma, T. Ohtsuka, N. Nakayama, Y. Itezono, N. Nakada, J. Watanabe, and K. Yokose, J. Antibiot. **47**, 37 (1994).
5. N. Nakada, H. Shimada, T. Hirata, Y. Aoki, T. Kamiyama, J. Watanabe, and M. Arisawa, Antimicrob. Agents Chemother. **37**, 2656 (1993).
6. N. Nakada, H. Gmuender, T. Hirata, and M. Arisawa, Antimicrob. Agents Chemother. **38**, 1966 (1994).
7. M. Arisawa, E. Goetschi, P. Hebeisen, T. Kamiyama, H. Link, R. Masciadri, H. Shimada, and J. Watanabe, Patent Application No. EP92/00809 (1992)
8. R. Otter, and N. R. Cozzarelli, Methods Enzymol. **100** , 171 (1983).
9. E. Goetschi, P. Angehrn, H. Gmuender, P. Hebeisen, H. Link, R. Masciadri, and J. Nielsen, Pharmacol. Ther. **60**, 367 (1993).

Novel AMPA/kainate receptor antagonists

Shuichi Sakamoto,* Junya Ohmori, Jun-ichi Shishikura, Masao Shimizu-Sasamata and Masamichi Okada

Institute for Drug Discovery Research, Yamanouchi Pharmaceutical Co.Ltd., 21 Miyukigaoka, Tsukuba, Ibaraki, 305, JAPAN

Abstract

A novel series of quinoxalinediones substituted with imidazole or other heteroaromatics was synthesized and evaluated for their activity to inhibit [^{3}H]AMPA binding. A critical early study proved that 1H-imidazol-1-yl and 1H-1,2,4-triazol-1-yl moieties can function as bioisosters for the cyano and nitro groups substituted at the 6-position on the quinoxalinedione. This investigation led to the discovery of YM90K [6-(1H-imidazol-1-yl)-7-nitro-2,3(1H,4H)-quinoxalinedione], being now in the clinical trials, as a potent AMPA receptor antagonist (Ki = 0.084 μM). The further structure-activity relationships conducted to more potent derivatives, represented by 4-hydroxy derivatives 27; (0.021 μM), 29; (0.042 μM), and fused tricyclic compounds 33; (0.048 μM), 36; (0.020 μM). Compounds 27 and 36 demonstrated the most potent inhibition reported to date. YM90K also showed potent activities *in vivo* models including sound-induced seizure in DBA/2 mice and delayed neuronal death in gerbil global ischemia.

Introduction

Although L-glutamic acid is a major neurotransmitter mediating synaptic excitation in the mammalian central nervous system, it can become toxic to the neurons and lead to their degeneration (1). Consequently, overstimulation of the glutamate receptors is thought to link to a

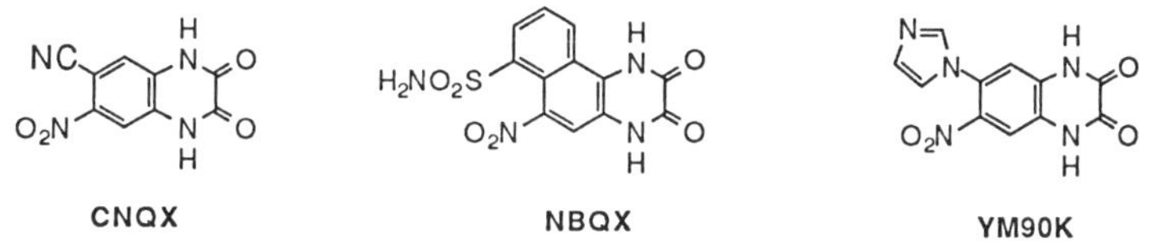

Fig. 1 AMPA antagonists

number of neurodegenerative disorders such as stroke and Alzheimer's disease, and the glutamate receptor antagonists are anticipated to have high therapeutic potential (2). Excitatory amino acid receptors have been divided into four subtypes (3), namely N-methyl-D-aspartic acid (NMDA), α-amino-3-hydroxy-5-methylisoxazolepropionic acid (AMPA), kainic acid (KA) and metabotropic glutamate receptors. Although AMPA is the most potent and selective agonist for the AMPA receptor, kainate produces the largest currents. Thus AMPA receptor is also referred to as AMPA / kainate receptor. Among these four subtypes, the NMDA receptor subtype has been well studied because of the early discovery of selective and potent antagonists, represented by CGS19755 as a competitive receptor antagonist and MK801 as a noncompetitive channel blocker. Although many NMDA antagonists have been reported to have cerebroprotective effects in focal ischemic models (4), their effects were inconclusive in global ischemia models (5). Therefore, we have focused on the study of AMPA receptor subtype as following reasons, i) AMPA receptor also mediates permeability of calcium ion that may trigger neurodegenerative processes (6), ii) disclosure of AMPA antagonists such as 6-cyano-7-nitroquinoxaline-2,3-dione (CNQX) (7) and 2,3-dihydro-6-nitro-7-sulfamoylbenzo(f)quinoxaline (NBQX) (8), iii) NMDA receptor antagonists are reported to produce some side effects such as a psychotomimetic action (9), an impairment of learning behavior (10) and ultra structural changes in cortical neurons (11). These reports have rendered us an intensive research for exploring more potent AMPA antagonists to approach their therapeutic potential free from serious side effects. This paper will describe the synthetic strategy, synthesis, structure-activity relationships and some in vivo studies of YM90K and reference compounds (12).

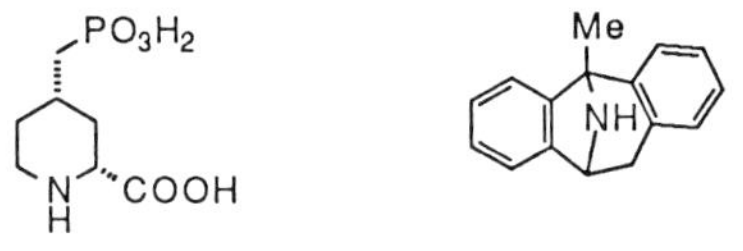

CGS19755   MK801

Fig. 2 NMDA antagonists

Synthetic strategy

It was reported that quinoxalinedione **1** had a weak affinity for AMPA receptor and that introduction of cyano or nitro group at the 6-position resulted in enhancement of it's affinity (7).

Scheme 1 Synthetic strategy

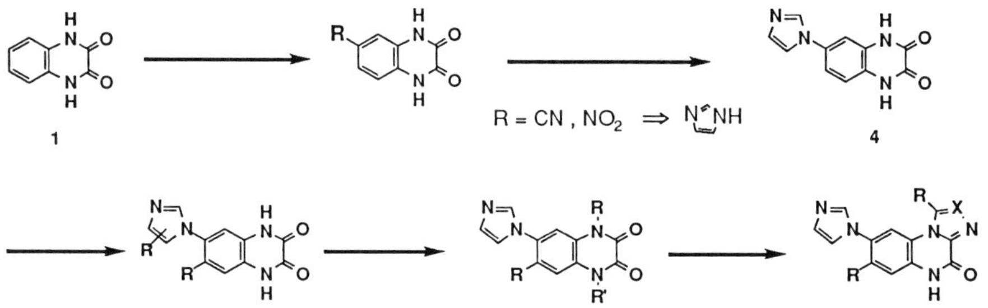

Then, our initial efforts were invested in development of new bioisosters for the cyano and nitro groups on the quinoxalinedione based on the following consideration. These groups have i) π-bond system, ii) planar structure, and iii) adequate hydrophilicity ($\pi = -0.57$ and -0.28 respectively). Our hypothesis that imidazole has all these properties and coefficient π (-0.65) (13) was proved by evaluation of the 6-(1H-imidazol-1-yl)-quinoxalinedione **4**. This breakthrough compound **4** became a new template for further design, such as introduction of other heteroaromatics at the 6-position together with altering functional groups at the 7-position, consequently, this contrivance led us to investigate YM90K. Then subsequent studies were directed to modify the diketopiperazine ring, such as incorporation of substituents at N1 or N4 position and construction of the fused ring system (Scheme 1).

Chemistry

Representative compounds were prepared by method 1) ~ 5) as shown in Scheme 2. 1) Derivatives with heteroaromatic ring at the 6-position were prepared as follows. Quinoxalinedione ring was formed by treatment with an appropriate diamine and oxalic acid in refluxing 4N-HCl solution. After nitration of the 6-fluoro-quinoxalinedione, epso-substitution with appropriate heteroaromatics afforded the desired compounds. 2) The synthesis of YM90K was started from the substitution reaction of chloro-nitroaniline by imidazole. Hydrogenation of the nitro group followed by cyclization of the resulting diamine gave the imidazolyl-quinoxalinedione, which nitration was mainly occurred at the 7-position to yield YM90K after recrystallization from dil. HCl. 3)

Hydrogenation of trifluoromethyl-nitrobenzene, acetylation of the resulting aniline followed by nitration gave the 1,2,4,5-substituted derivative. Substitution with imidazole followed by treatment with usual methods provided the 7-trifluoromethyl derivative. 4) 4-Hydroxy derivatives were prepared from intramolecular cyclization between the oxalylamide, which was obtained from nitroaniline by acylation with ethyl oxalyl chloride, and the hydroxylamine, which was prepared by

Scheme 2 Chemistry

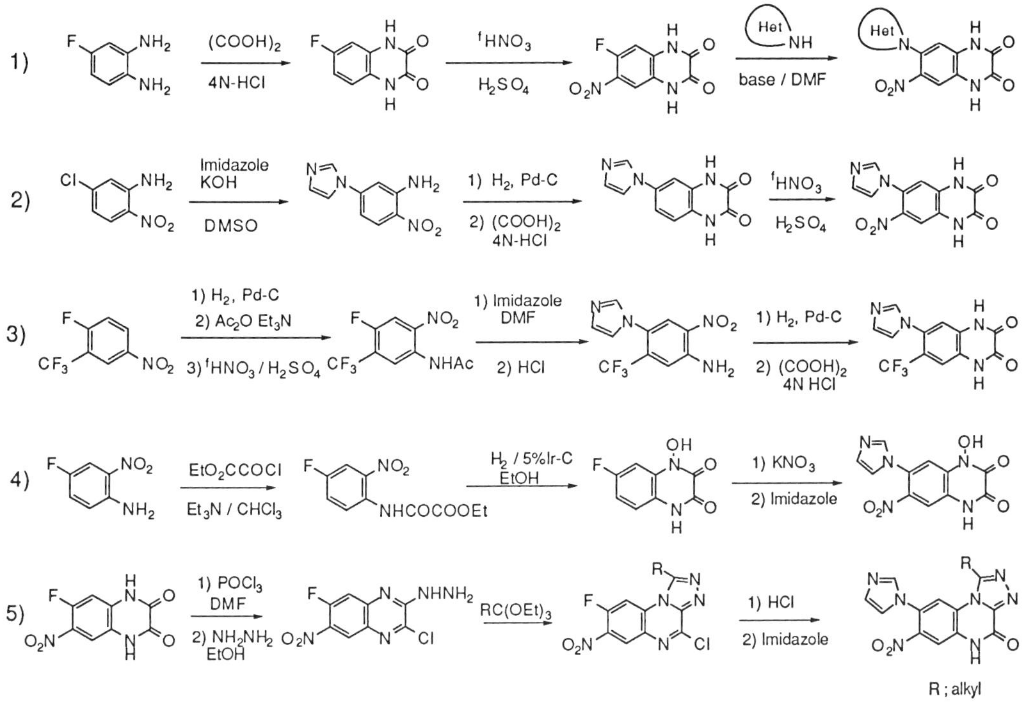

partial hydrogenation of the nitro group using Iridium-charcoal as a catalyst. 6-Fluoro-4-hydroxy-quinoxalinedione was nitrated and treated with imidazole to transform the target compound. 5) A new fused ring system at the piperazine portion was successively constructed as follows. Selective hydrazination of the 2,3-dichloroquinoxaline which was obtained by chlorination of 6-fluoro-7-nitro-quinoxalinedione, subsequent cyclization using appropriate orthoesters afforded the triazo[4,3a]quinoxaline system. Acid hydrolysis of the remaining chloro group followed by substitution with imidazole yielded the desired compounds.

Structure–activity Relationships

The compounds evaluated their affinity for AMPA receptor are summarized in Tables 1-6. Quinoxalinedione **1** itself had a weak affinity, however, introduction of cyano **2** or nitro **3** group at the 6-position enhanced its potency (Table 1). Replacement of the cyano or nitro group by imidazole resulted in a compound **4** ($Ki = 1.6$ μM) nearly equal to or more potent than those of **2** and **3**. Introduction of pyrollidine group **5** led to the decreased affinity. These binding data indicate that $1H$-imidazol-1-yl ring can function as a new bioisoster for the cyano and nitro groups for AMPA receptor. Since the presence of the nitro group at the 7-position seemed to be essential for the enhancement of their inhibitory activity from the data of CNQX and NBQX, several compounds possessing imidazole or other heteroaromatics at the 6-position with keeping the nitro group at the

7-position were prepared (Table 2).   The 6-(1*H*-imidazol-1-yl) compound, YM90K, exhibited markedly potent activity (0.084 μM) indicating over 1000-fold improved affinity related to quinoxalindione **1**, 3 times higher affinity than CNQX.   1*H*-1,2,4-Triazol-1-yl derivative **6** also showed good affinity almost same as CNQX, suggesting that this heteroaromatic also become a

Table 1. 6-Substituted Quinoxalinedione Derivatives

	R	AMPA receptor affinity Ki (μM) / % inhibition *
1	H	24% (100) *
2	CN	5.0
3	NO₂	2.0
4	(imidazol-1-yl)	1.6
5	(pyrrolidin-1-yl)	33% (100) *

*These compounds were tested at the concentration (μM) indicated with the percent inhibition of binding shown.

Table 2. 6-Substituted-7-nitro Derivatives

	R	AMPA receptor affinity Ki (μM)
CNQX	CN	0.27
YM90K	(imidazol-1-yl)	0.084
6	(1,2,4-triazol-1-yl)	0.23
7	(pyrazol-1-yl)	26 % (1.0) *
8	(imidazol-4-yl)	0.60
9	(pyrid-3-yl)	0.60

Table 3. 7-Nitro-Quinoxalinedione Derivatives

	R	AMPA receptor affinity Ki (μM)
10	(O₂N-triazolyl)	0.10
11	(Me-triazolyl)	0.22
12	(Ph-triazolyl)	0.23
13	(Me-triazolyl)	0.47
14	(Et-triazolyl)	0.40
15	(benzotriazolyl)	24 % (1.0)

Table 4. 6-(1*H*-Imidazol-1-yl) Quinoxalinedione Derivatives

	R	AMPA receptor affinity Ki (μM)
16	(imidazolyl)	0.82
17	CN	0.19
18	CF₃	0.20
19	MeSO₂	2.3
20	F	3.0
21	Cl	1.2
22	Br	0.73
23	I	0.29
24	MeO	0.4

bioisoster of those groups.   On the other hand, the pyrazole derivative **7** lost it's activity and imidazol-4-yl **8** or pyrid-3-yl **9**, regardless of having beta-nitrogen like imidazol-1-yl group

(YM90K), exhibited only poor activities. It suggests the precise positioning of the nitrogen atoms in the heteroaromatic ring and the specific electronic topography of the ring are both important for significant binding to the AMPA receptor. Next assessment was performed for the derivatives with substituted imidazoles at the 6-position (Table 3). At first the effects of the 4-position of imidazole resulted that the nitro **10**, methyl **11** and even phenyl **12** derivatives tolerated their potency, so it seemed that there was a steric pocket at this site of the receptor. In contrast, 2-substituents (Me; **13**, Et; **14**) led to 5-fold decreases related to that of YM90K, while benzimidazolyl derivative **15** showed marked decrease on potency. We next turned our attention to optimize the substituents at the 7-position with maintaining the imidazol-1-yl group at the 6-position (Table 4).

Table 5. 6-(1*H*-Imidazol-1-yl) Quinoxalinedione Derivatives

	R	AMPA receptor affinity Ki (μM)
25	Pr	0.14
26	(cyclohexyl)	0.11
27	OH	0.021
28	OMe	0.38
29	(structure)	0.042
30	(structure)	0.43

Table 6. 6-(1*H*-Imidazol-1-yl) Fused Derivatives

	R	AMPA receptor affinity Ki (μM)
31	H	0.19
32	Me	0.096
33	Et	0.048
34	Pr	0.099
35	H	0.057
36	Et	0.020

The second imidazole moiety (see **16**) could not play a role as a bioisoster of the nitro group. Since nitro group is one of the strongest electron-withdrawing group, the effects of other electron-withdrawing group were examined. Introduction of cyano or trifluoromethyl group afforded compounds (**17** and **18**) exhibiting good activities similar to that of CNQX, however, methylsulfone **19** which has also strong electron-withdrawing group lost it's activity. Since it was not clear of the relationship between the AMPA activity and electron-withdrawing ability, we evaluated the effects of halide group. Fluorine (**20**) having the strongest electron-negativity displayed only a weak binding activity and, surprisingly, the iodide **23** showed most potent activity in halide series close to that of CNQX. These data indicated that the activity was in inverse proportion to the rank order of the electron-negativity and suggested that the activity might parallel with the volume size of the 7-substituent. Therefore, we prepared derivatives possessing electron-donating group with moderate size at the 7-position. Methyl, hydroxy and amino derivatives showed little activity (data not shown), however methoxy **24**, methythio, and methylamino derivatives were found to display moderate activity after evaluation of several derivatives. The structure-activity relationships shown in Table 1-4 suggest that the AMPA activity might correlate with not only the electron-withdrawing ability but also the volume size of substituents at the 7-position which would influence the angle between imidazole ring and quinoxaline ring (**14**). Next, we examined the effects of substituent at

N1 or N4 position (Table 5). Introduction of alkyl group such as propyl **25** and cyclohexyl **26** group at N4 position allowed to tolerate their activity, which indicates the existence of a large pocket around N4 position. One of the most significant improvement on potency involved that introduction of hydroxy group onto the N4 position yielded a hydroxamic derivative **27** (0.021 μM) which is more potent than NBQX (0.060 μM). This change effected 4-fold increase related to the

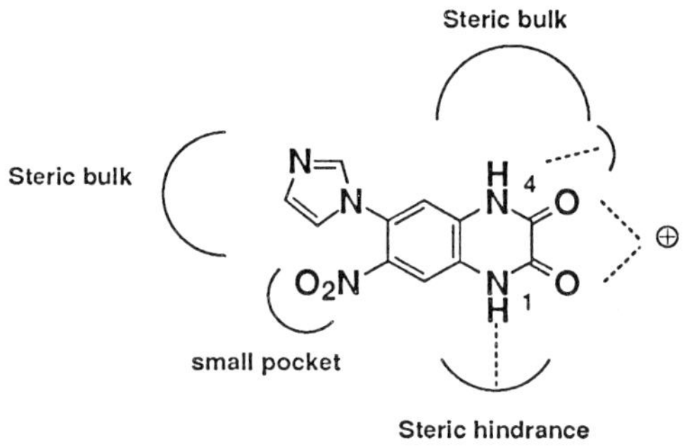

Fig. 3 Proposed pharmacophore for the binding of imidazolyl-quinoxalinediones to the AMPA receptor

affinity of YM90K. Introduction of methoxy **28** at N-4 position or hydroxy **30** group at N-1 position resulted in decrease of the potency. This investigation was applied to the compounds showing potent activity in the series of imidazolyl-quinoxalinediones, and 4-hydroxy-6-trifluoromethyl derivative **29** was also confirmed to show the potent activity (0.042 μM). Since forming of a hydrogen bonding between the 4-hydroxy group and the ketone at the 3-position seemed to construct a pseudo 5-membered ring (Table 6), next modification was focused on the fused compounds such as triazoquinoxalines and imidazoquinoxalines. This transformation provided one of the most potent compound **36** (0.020 μM) as well as **27** identified to date. A crystalline structure determination (14) and the structure activity-relationships we have obtained so far led to propose a hypothetical pharmacophore model of the imidazolyl-quinoxalinedione for the AMPA receptor as shown in Fig.3. Table 7 shows the inhibitory activity of YM90K, CNQX and NBQX for some excitatory amino acid receptor subtypes (KA, NMDA-sensitive glutamate and strychnine-insensitive glycine). YM90K exhibited no inhibitory activity for NMDA receptor subtype and high selectivity for the AMPA receptor.

Table 7. The affinities of **YM90K, CNQX** and **NBQX** to various glutamate receptor subtypes

Glutamate Receptor Subtypes	AMPA receptor	Kainate receptor (High affinity site)	NMDA receptor-ion channel complex	
			NMDA binding site	Strychnine-insensitive glycine site
Ligand	[³H]-AMPA Ki(μM)	[³H]-Kainate Ki(μM)	[³H]-Glutamate Ki(μM)	[³H]-Glycine Ki(μM)
YM90K	0.084	2.2	>100	37
CNQX	0.27	1.8	25	5.6
NBQX	0.060	4.1	>100	>100

In vivo studies

YM90K was characterized its antagonistic ability in vivo by prevention of a decrease in the activity of choline acetyltransferase (ChAT) produced by AMPA in the rat striatum. AMPA (25 nmol) caused a 40-55 % decline in ChAT activity at 7 days after intrastriatal injection. Coinjection of

YM90K protected rat striatum from this damage in a dose-dependent manner. This result indicates that YM90K has an antagonistic action against AMPA at its receptor in the striatum (Fig.4).

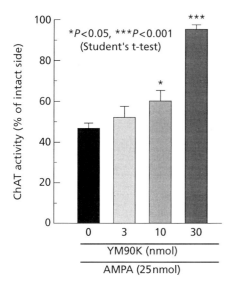

Fig. 4 Dose-dependent effects of YM90K on AMPA-induced striatal neurodegeneration in rats.

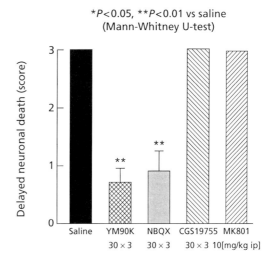

Fig. 6 Effects of AMPA/KA antagonists, YM90K and NBQX, and NMDA antagonists, CGS19755 and MK-801 on the delayed neuronal death in gerbils after 5 min bilateral carotid artery occlusion.

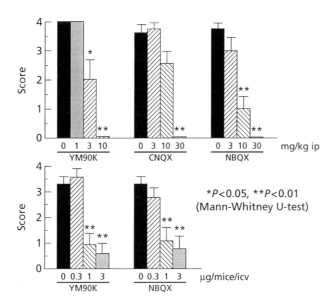

Fig 5. Anticonvulsant effects of YM90K, CNQX and NBQX on sound-induced seizure in DBA/2 mice.

The primary in vivo screen is blockade of sound-induced seizure in the DBA/2 mice. YM90K at a dose of 3 mg/kg ip showed almost comparable activity to NBQX at a dose of 10 mg/kg when administered at 15 min prior to sound exposure, however, CNQX showed only weak activity. On the other hand, activity by icv dosing of YM90K and NBQX showed a good correlation between affinity (K_i values) and icv anti convulsant potency (Fig.5). On a global ischemic model (delayed neuronal death) in gerbils, treatment with YM90K administered three time at 60, 70 and 85 min after 5 min occlusion significantly reduced CA1 hippocampal damage and the minimum effective dose was 15 mg/kg (ip) x 3 (NBQX; 30 mg/kg x 3) (data not shown). On the other hand in this model, NMDA antagonists, CGS19755 and MK801, did not exhibit any significant prevention (Fig.6) (5b). Furthermore YM90K is effective against this neuronal damage even after 6 hour administration with the same intervals mentioned before (data not shown).

Conclusion

We have investigated that 1*H*-imidazol-1-yl and 1*H*-1,2,4-triazol-1-yl are efficient bioisosters for the cyano and nitro groups on the quinoxalinedione to the AMPA receptor. This investigation led to the discovery of YM90K [6-(1*H*-imidazol-1-yl)-7-nitro-2,3(1*H*,4*H*)-quinoxalinedione] as a potent and selective AMPA receptor antagonist. YM90K showed potent in vivo activities, that is, it blocked the sound-induced seizure in DBA/2 mice at the dose of 3 mg/kg ip and protected the delayed neuronal death in gerbil global ischemia at 15 mg/kg(ip) x 3. YM90K is effective against the neuronal damage as far as 6 hour after ischemia. The further structure-activity relationship studies conducted to more potent derivatives, represented by 4-hydroxy derivatives **27, 29** and fused type compounds **33, 36**.

References

1. a) Curtis, D. R.; Watkins, J. C. *J. Neurochem.* **1960**, *6*, 117-141. b) Krnjevic, K.; Phillis, J. W. *J. Physiol.* **1963**, *165*, 274-304. c) Takeuchi, A; Takeuchi, N. *J. Physiol.* **1964**, *170*, 296-317.
2. a) Choi, D. W. *Neuron* **1988**, *1*, 623-634. b) Olney, J. W *Annu. Rev. Pharmacol. Toxicol.* **1990**, *30*, 47-71.
3. Monghan, D. T.; Bridges, R.J.; Cotman, C. W. *Annu. Rev. Pharmacol. Toxicol.* **1989**, *29*, 365-402.
4. Park, C. K.; Nehls, D. G.; Graham, D. I.; Teasdale, G. M. *Ann. Neurol.* **1988**, *24*. 543-551.
5. a) Buchan, A. M.; Li, H.; Cho, S.; Pulsinelli, W. A. *Neurosci. Lett.* **1991**, *132*. 255-258. b) Yatsugi, S.; Kawasaki, S.; Katoh, M.; Takahashi, M.; Koshiya, K.; Shimizu-Sasamata, M *J. Cereb. Blood Flow Metab.* **1993**, *13*, Suppl.1, S635.
6.a) Hollmann, M.; Hartley, M.; Heinemann, S. *Science* **1991**, *252*, 851-853. b) Gilbertson, A.T.; Scobey, R.; Wilson, M. *Science* **1991**, *252*, 1613-1615.
7. Honore, T.; Davies, S. N.; Drejer, J.; Fletcher, E. J.; Jacobsen, P.; Lodge, D.; Nielsen, F.E. *Science* **1988**, *241*, 701-703.
8. a) Sheardown, M. J.; Nielsen, E. Φ.; Hansen, A. J.; Jacobsen, P.; Honore, T. *Science* **1990**, *247*, 571-574. b) Kaku, D. A.; Goldberg, M. P.; Choi, D.W. *Brain Res.* **1991**, *554*, 344-347. c) Judge, M. E.; Sheardown, M.J.; Jacobsen, P.; Honore, T*Neurosci. Lett.* **1991**, *133*, 291-294.
9. Koek, W.; Woods, J. H.; Winger, G.D.*J. Pharmacol. Exp. Ther.* **1988**, *245*, 969-974.
10. Morris, R. G. M.; Anderson, E.; Lynch, G.S.; Baudry, M.*Nature* **1986**, *319*, 774-776.
11. Olney, J. W.; Labruyere, J.; Price M.T. *Science*,**1989**, *244*, 1360-1362.
12. a) Sakamoto, S.; Ohmori, J.; Shimizu-Sasamata, M.; Okada, M.; Kawasaki, S.; Yatsugi, S.; Hidaka, K.; Togami, J.; Tada, S.; Usuda, S.; Murase, K. *XIIth International Symposium on Medicinal Chemistry Basel,* Switzerland Sep. **1992**. Abst 28. b) Shimizu-Sasamata, M.; Kawasaki, S.; Yatsugi, S.; Ohmori, J.; Sakamoto, S.; Koshiya, K.; Usuda, S.; Murase, K. *22nd Annual Meeting Society for Neuroscience* Anaheim USA Oct. **1992**. Abst 44.14. c) Okada, M.; Hidaka, K.; Togami, J.; Ohno, K. Tada, S.; Ohmori, J.; Sakamoto, S.; Yamaguchi, T. *22nd Annual Meeting Society for Neuroscience* Anaheim USA Oct. **1992**. Abst 44.15. d) Shimizu-Sasamata, M.; Kawasaki, S.; Yatsugi, S.; Ohmori, J.; Sakamoto, S.; Koshiya, K.; Usuda, S.; Murase, K. *J. Cereb. Blood Flow Metab.* **1993**, *13*, Suppl.1, S664. e) Ohmori, J.; Sakamoto,S.; Kubota, H.; Shimizu-Sasamata,M.; Okada, M.;Kawasaki, S.; Hidaka,K.; Togami, J.; Furuya, T.; Murase, K. *J. Med. Chem.* **1994**, *37*, 467-475.
13.a) Hansch, C.; Lien, E. *J. Med. Chem.* **1971**, *14*, 635-670. b) James,M.; Sloan, K.B.;*J. Med. Chem.* **1985**, *28*, 1120-1124.
14. Ohmori, J.; Kubota, H.; Shimizu-Sasamata, M.; Okada, M.; Sakamoto, S. *J.Med.Chem.* in press.